MASS-LOSING PULSATING STARS AND THEIR CIRCUMSTELLAR MATTER

ASTROPHYSICS AND SPACE SCIENCE LIBRARY

VOLUME 283

MASS-LOSING PULSATING STARS AND THEIR CIRCUMSTELLAR MATTER

Observations and Theory

Edited by

Y. NAKADA

Kiso Observatory,
University of Tokyo, Agematsu, Japan

M. HONMA

National Astronomical Observatory,
Mitaka, Japan

and

M. SEKI

Astronomical Institute,
Tohoku University, Sendai, Japan

KLUWER ACADEMIC PUBLISHERS

DORDRECHT / BOSTON / LONDON

A C.I.P. Catalogue record for this book is available from the Library of Congress.

ISBN 1-4020-1162-8

Published by Kluwer Academic Publishers,
P.O. Box 17, 3300 AA Dordrecht, The Netherlands.

Sold and distributed in North, Central and South America
by Kluwer Academic Publishers,
101 Philip Drive, Norwell, MA 02061, U.S.A.

In all other countries, sold and distributed
by Kluwer Academic Publishers,
P.O. Box 322, 3300 AH Dordrecht, The Netherlands.

Printed on acid-free paper

Printed in the Netherlands.

Contents

Preface

Editing the proceedings of a scientific meeting is not an easy task. Sometimes people who give an excellent talk do not send the manuscript by the deadline. However, this time, thanks to the punctuality of all the participants, we have this excellent volume for the workshop on mass-losing pulsating stars and their circumstellar matter prepared in time. Almost all of the oral presentations including the summary are collected in this volume. We regret that we cannot put in this volume a few posters that we failed to receive before the editorial work.

The workshop was planned as a small meeting with less than fifty attendants because the city of Sendai was far from the most of the active institutions. However, the number of submitted papers exceeded the SOC's expectation; many interesting contributions had to be scheduled in the poster session. Still, the oral sessions were so tight that many participants might have felt frustrated for the shortage of discussions. The organizers of the workshop have to apologize to the attendants for the inconvenience caused from such a happy underestimate about the size of the workshop.

Mass-losing pulsating stars are characterized by the strong stellar wind or the enormous circumstellar matter. The collaboration of the astronomers in the fields of optical, infrared and radio astronomy is essential to resolve the nature of these objects. The workshop clarified the importance of such collaboration with so many newest observational and theoretical results. The importance of the high precision interferometry to study the asymptotic giant branch stars was stressed in this workshop. Peter Wood expressed his hopes on the astrometry such as VERA in the radio, GAIA in the optical and JASMINE in the near-infrared in his summary.

The Scientific Organizing Committee consisted of:

S. H. Cho (Korea), S. Deguchi (Japan)
K. Y. Lo (China R.), Y. Muraki (Japan)
Y. Nakada (Japan), M. J. Reid (USA)
T. Sasao (Japan), M. Takeuti (chair, Japan)
H. J. van Langevelde (The Netherlands), C. Waelkens (Belgium)
P. A. Whitelock (South Africa), A. Winnberg (Sweden)
P. R. Wood (Australia).

The members of the SOC enjoyed the exchange of ideas for preparing the workshop. I wish to express hearty thanks to them for their collaboration which was essential for the success of the workshop.

I believe the participants enjoyed the splendid May weather of Sendai and the excursion to Mizusawa Observatory. I wish to convey our gratitude and our appreciation to Astronomical Institute of the Tohoku University and the Earth Rotation Division and Mizusawa Astrogeodynamics Observatory of the National Astronomical Observatory of Japan for the hospitality and assistance rendered.

The Local Organizing Committee consisted of:

T. Aikawa (Tohoku Gakuin U.), O. Kameya (NAO)
U. Lee (Tohoku U.), M. Miyoshi (NAO)
T. Murayama (Tohoku U.), Y. Nakada (U. of Tokyo)
Y. Nakamura (Fukushima U.), H. Saio (Tohoku U.)
T. Sasao (NAO), M. Seki (chair, Tohoku U.)
S. Tamura (Tohoku U.)

They worked very efficiently for the success of the workshop. On behalf of the SOC, I wish to express our thanks to them for their hard work.

I also express our thanks to Drs Y. Nakada (University of Tokyo), M. Honma (National Astronomical Observatory of Japan), and M.Seki (Tohoku University) for editing this volume.

The SOC and LOC would like to express their gratitude to the Commemorative Association for the Japan World Exposition (1970), the Saito Gratitude Foundation, Sendai Tourism and Convention Bureau, and Ministry of Education, Culture, Sports, Science and Technology (MEXT) for their kind financial support. The workshop could not have been held without the support of these organizations.

At the end of this short preface, I wish to note my hope that this volume will be used for further development of the study of variable stars, especially by young astronomers, beyond the classical point of views.

M. TAKEUTI

Sendai, Japan. October 2002

Mass-losing Pulsating Stars
and
their Circumstellar Matter
- Newest Results of Observations and Theory -
(質量放出脈動星と星周物質)
May 13-16, 2002
SAITO Ho-on Kaikan, Sendai, Japan
Supported by:
The Commemorative Association for the Japan World Exposition (1970)
Sendai Tourism & Convention Bureau
Saito Ho-on Kai
National Astronomical Observatory, Japan
Astronomical Institute, Tohoku University

I

RECENT RESULTS OF MASSIVE PHOTOMETRY

THE SEQUENCE OF MASS-LOSING AGB STARS IN THE MAGELLANIC CLOUDS

From small-amplitude pulsators to post-AGB stars

P.R. Wood
Research School of Astronomy and Astrophysics, Australian National University
wood@mso.anu.edu.au

Abstract Surveys of variable stars in the Magellanic Clouds, such as those by MACHO, EROS, OGLE and MOA, have revealed red variable stars with amplitudes from a few hundredths of a magnitude (SR variables) up to roughly six magnitudes (Mira variables). The MSX survey found large numbers of mass-losing, dust-enshrouded AGB stars in the Magellanic Clouds and the MACHO data shows them to be pulsating stars in a more advanced evolutionary state than the Mira variables. In this paper, these results are described and compared with theories of the pulsation and evolution of AGB stars. One particularly interesting group of variables in the LMC are the SR variables with long secondary periods (sequence D stars). It is argued that these are red giants with brown dwarf or planetary companions in elliptic orbits. A cloud of dust and gas loosely associated with the orbiting companion is assumed to cause the photometric variability at the long secondary period.

Keywords: Stars: AGB, stars: variable, stars: binary

Introduction

Long-term monitoring of stars in the Magellanic Clouds by the various microlensing surveys has revolutionized our understanding of variability in red giant stars. In particular, for the first time, multiple, parallel period luminosity sequences have been found (although Wood and Sebo 1996 had tentatively detected some of these sequences before the MACHO results appeared). All the variable red giants appear to be AGB stars except for a very small group of close binaries on the first giant branch.

Once heavy mass loss starts in the pulsating AGB stars, their remaining lifetime as AGB stars becomes very short (hence they are rare).

Y. Nakada et al. (eds.), Mass-Losing Pulsating Stars and their Circumstellar Matter, 3–10.

They become enshrouded in dusty circumstellar shells and they become faint optically but bright in the mid-infrared. Such stars are best found by mid-infrared surveys such as the MSX survey (Egan et al. 2001). In many cases, the MACHO experiment was able to detect a variable star near these mid-infrared sources in the LMC.

In this paper, the various variable star groups from the small amplitude variables to the dust-enshrouded MSX sources are discussed. Finally, some new radial velocity observations of the mysterious semi-regular variables with long-secondary periods are discussed.

Period-luminosity laws

Examples of the period-luminosity sequences exhibited by variable red giants in the LMC can be found in Wood et al. (1999) and Wood (2000a) (using MACHO data), Cioni et al. (2001) (using EROS data) and Noda et al. (2002) (using MOA data). Fig. 1 shows the (K, $\log P$) sequences for MACHO variables in a 0.5×0.5 degree area of the LMC bar from Wood (2000a) along with the variable MSX sources detected by the MACHO experiment in a 3×3.5 degree area centred on the LMC bar.

Looking at the MACHO-selected variables first (the circles in Fig. 1), it can be seen that these fall on 4 parallel sequences A–D with luminosities above the minimum luminosity on the thermally-pulsing AGB stars (sequence E variables are not discussed here: they appear to be close binaries on the first giant branch - see Wood et al. 1999). Sequence C is the sequence on which the Mira variables lie: however, most of the variables on this sequence have V amplitudes less than 2.5 mag. so they would be classified as SR variables in the General Catalog of Variable Stars.

Wood et al. (1999) showed that the variables on sequences A–C can be explained as stars pulsating in low-order radial modes, with the sequence C variables being fundamental mode pulsators and the sequence B and C pulsators being 1st, 2nd or 3rd overtone pulsators. The non-appearance of higher overtones is because their frequencies exceed the acoustic cut-off frequencies in the stars (e.g. Fox and Wood 1982). Note that if the sequence C variables are fundamental mode pulsators, then it is impossible for the sequence D variables to be radial pulsators since the longest period of radial pulsation is the fundamental mode. The sequence D variables will be discussed below.

About 50% of the AGB stars detected by the MSX experiment were seen by MACHO. Given the large ratio (~42) of the areas from which the MSX sources and the MACHO-selected variables, respectively, in

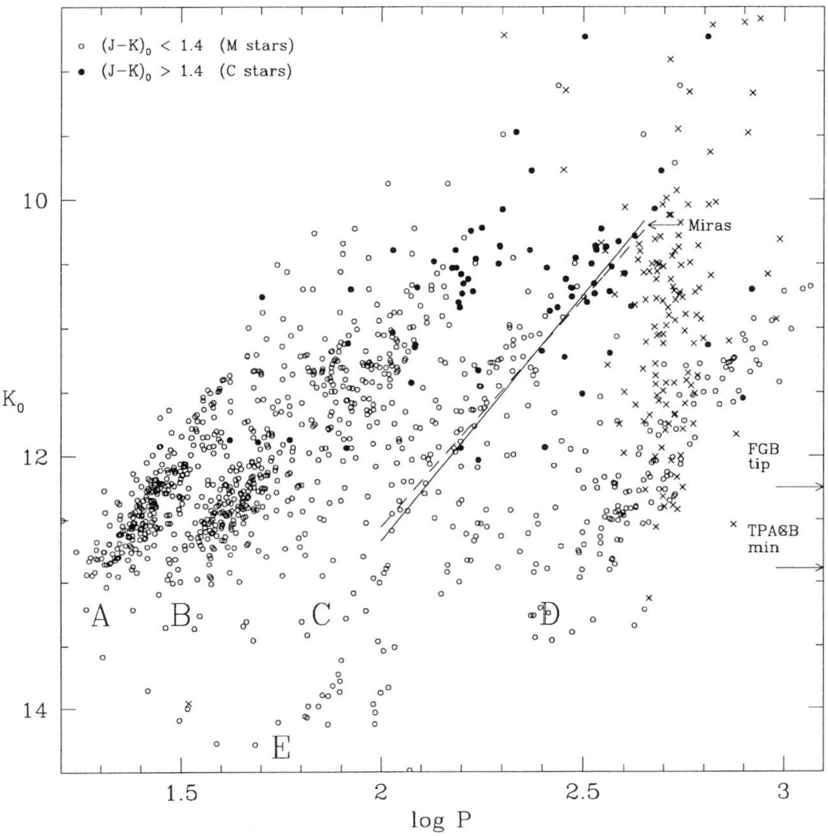

Figure 1. Variable red giant stars in the LMC in the K,log P plane. The periods are derived from MACHO data and the K magnitudes are from observations obtained with the ANU 2.3m telescope. The circles are low mass loss rate, optically-visible objects from Wood (2000a) while the crosses are high mass loss rate objects discovered by the MSX satellite, and also seen by MACHO. The lines are the Mira K,log P relations from Hughes & Wood (1990) and Feast et al. (1989). The sequence labels A–E are those given by Wood et al. 1999).

Fig. 1 are drawn, it is clear that the MSX sources are relatively very rare, and that they must have very short lifetimes. This is as expected given the large mid-infrared fluxes of the MSX sources which indicate thick circumstellar shells and large mass loss rates. It is clear from Fig. 1 that the majority of the MSX sources form a sequence from the long-period, high-luminosity tip of the optically-selected Mira sequence down to much fainter K magnitudes with little or no increase in period. Examination of the colours of the MSX sources shows that the bluer objects fall at the tip of the Mira sequence and that the very red objects

(J-$K \sim 6$) are the faintest ones (the K,J-K diagram for these stars can be found in Fig. 1 of Wood and Cohen 2001).

The MACHO- and MSX-selected variables cover the complete evolution of mass-losing, variable AGB stars. Low on the AGB, stars pulsate with low amplitude in a low-order overtone (2nd or 3rd), or perhaps both (there are many multimode pulsators on sequences A–C). As luminosity increases, the dominant mode of pulsation moves to a lower order until the fundamental mode dominates and Mira-like pulsation of large amplitude occurs. At this time, the mass loss rate increases dramatically, the star becomes enveloped in a dusty envelope whose apparent effective temperature is much less than that of the photosphere, the flux peak in the spectrum shifts from $\sim$1 μm to $\sim$5 μm, and the K magnitude declines. The star continues to pulsate in the fundamental mode with a period of $\sim$450-600 days until the AGB envelope is dissipated and the star becomes a post-AGB star (Wood and Cohen 2001). Full details of the MSX star behaviour will be published in Wood and Cohen (2002).

The sequence D variables

Since the sequence D variables can not be pulsating in radial modes, Wood et al. (1999) postulated several alternative explanations:

- they are some sort of binary system where the companion is associated with a loosely-bound cloud of dust and gas that periodically obscured the red giant
- they undergo episodic dust ejection events as predicted in luminous carbon star models of Winters et al. (1994) and Höfner et al. (1995), and perhaps as seen in RCB stars
- they are pulsating in some "strange" mode, an example of which is given by Wood (2000b)
- they are pulsating in a non-radial mode (although this was regarded as unlikely).

If these stars are in binary systems, or are pulsating in a strange mode, they will exhibit radial velocity variations with a period equal to that of the long secondary period. There are some analogs of the sequence D variables in the solar vicinity and some of these stars have recently been tested for small radial velocity variations by Hinkle et al. (2002) and Olivier and Wood (2002b). The observations show that all these stars have radial velocity variations with full amplitudes of $\sim$4 km s^{-1}: the example of Z Eri, from Olivier and Wood (2002b), is shown in Fig. 2.

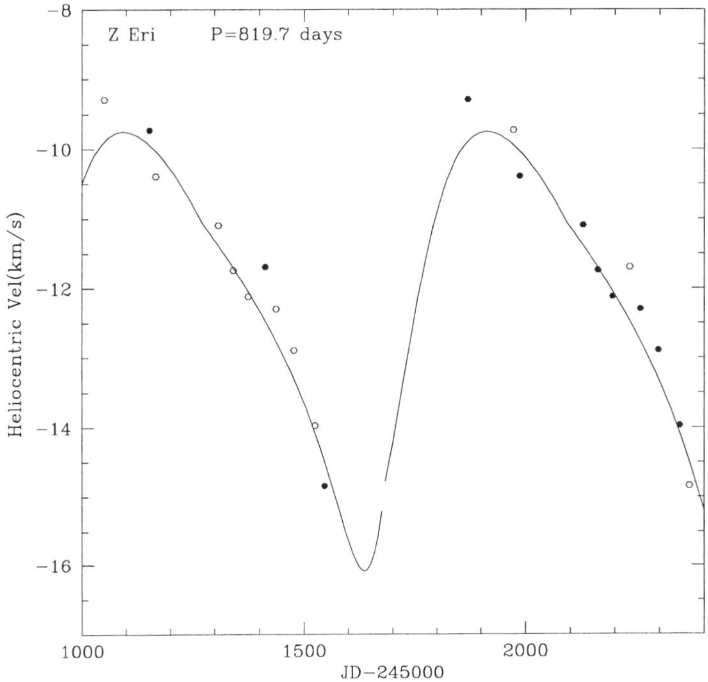

Figure 2. The radial velocity curve for the local SR variable Z Eri which has a short period of ~80 days and a long secondary period of ~800 days (Kholopov et al. 1988; Sacharow 1953). Solid points are velocities plotted at the dates measured while open points are velocities shifted by ±1 period. The curve is a fit of a binary star radial velocity curve to the data.

It is very notable that all the radial velocity curves obtained so far are very asymmetric and of small amplitude. In the binary model, this indicates a companion with a brown-dwarf-like mass in a highly eccentric orbit ($\epsilon \sim 0.4$). The similarity of the velocity curves of different stars led Hinkle et al. (2002) to suggest that a binary explanation was unlikely and they favoured an explanation involving some sort of pulsation. If this is the case, the change in radius over one pulsation cycle (obtained by integrating the velocity curve) is a large fraction of the mean stellar radius. This would mean a significant change in T_{eff} over the long pulsation cycle and hence a significant change in spectral type. However, in the case of Z Eri, the spectral types derived throughout the long secondary period are remarkably constant and indicate almost no change in stellar radius. In addition, a large change in radius would mean a large change in the length of the short primary pulsation period: however, this is not observed in the MACHO light curves (Wood et al. 1999). These points seems to cast doubt on the radial pulsation hypothesis. The nonradial pulsation hypothesis suffers from similar problems, as well as problems of existence and visibility at the stellar surface (Wood et al. 1999).

The binary model also has many problems. In the case of Z Eri, the two stars are so close at periastron that the inner Lagrangian point is within the red giant envelope (at least for orbital inclinations $\sin i \sim 1$)! In addition, the tidal circularization time for the orbit is only $\sim 10^5$ yr and the tidal synchronization time is even smaller at $\sim 10^4$ yr. More details are given in Olivier and Wood (2002b).

If neither binary nor radial pulsation models apply, then another possibility is the episodic dust ejection model. However, since many of the LMC stars on sequence D are of low luminosity (~2000 $L_{\odot}$), it is hard to see how this mechanism could work. In addition, dust formation is a circumstellar phenomenon and it is not obvious why photospheric velocity variations of several km s^{-1} should be seen.

Another possibility is some kind of rotating, spotted star model. The "spots" would need to be very large indeed in this case as the stellar light varies by a factor of two in some of the sequence D stars. Furthermore, it is not obvious why rotation would give a characteristic, asymmetric velocity curve as shown in Fig. 2. Finally, Olivier and Wood (2002a) estimate that the rotation velocity of the local analogs of the stars on sequence D have rotation velocities of <3 km s^{-1} whereas the equatorial velocity required for a star of radius ~200 $R_{\odot}$ to rotate once in ~800 days (these values are appropriate for Z Eri) is ~12.6 km s^{-1}. Clearly, the origin of the light variations in the sequence D variables remains a mystery. However, overall, the radial velocity variations tend to favour a binary hypothesis.

References

Cioni, M.-R.L., Marquette, J.-B., Loup, C., Azzopardi, M., Habing, H.J., Lasserre, T., & Lesquoy, E. 2001, A&A, 377, 945

Egan, M.P., Van Dyk, S.D., & Price, S.D. 2001, AJ, 122, 1844

Feast, M.W., Glass, I.S., Whitelock, P.A., & Catchpole, R.M. 1989, MNRAS, 241, 375

Fox, M.W., & Wood, P.R. 1982, ApJ, 259, 198

Hinkle, K.H., Lebzelter, T., Joyce, R., & Fekel, F.C., 2002, AJ, 123, 1002

Höfner, S., Feuchtinger, M.U., & Dorfi, E.A. 1995, A&A, 297, 815

Houk N. 1963, AJ, 68, 253

Hughes, S.M.G., & Wood, P.R. 1990, AJ, 99, 784

Kholopov, et al. 1988, General Catalog of Variable Stars, 4th Edition

Olivier, R., & Wood, P.R. 2002a, submitted

Olivier, R., & Wood, P.R. 2002b, in preparation

Sacharow, G.P. 1953, Variable Stars 9, 175
Noda, S. & the MOA Consortium, 2002, MNRAS, 330, 137
Winters, J.M., Fleischer, A.J., Gauger, A., & Sedlmayr, E. 1994, A&A, 290, 623
Wood, P.R. 2000, PASA, 17, 18
Wood, P.R. 2000, in ASP Conf. Ser. 203, The Impact of Large-Scale Surveys on Pulsating Star Research, eds. Szabados, L. & Kurtz, D. (San Francisco: ASP), 379
Wood, P.R. & the MACHO Collaboration, 1999, in IAU Symp. 191, Asymptotic Giant branch Stars, eds. T. Le Bertre, A. L'ebre, C. Waelkens (ASP: San Francisco), 151
Wood, P.R., & Cohen, M. 2001, in ApSSL 265, Post-AGB Objects as a Phase of Stellar Evolution, eds. R. Szczerba, R. & G'orny, S.K. (Dordrecht: Kluwer), 71
Wood, P.R., & Cohen, M. 2002, in preparation
Wood, P.R., & Sebo, K.M. 1996, MNRAS, 282, 958

Discussion

Dorfi: Could you explain the shaping of PNe by these orbiting brown dwarfs?

Wood: One of the favourite explanations for producing bipolar PNe is the influence of a binary companion on the wind flow during the ejection of the matter which will form the planetary nebula. It is interesting that about 25% of AGB stars in the LMC show the long secondary periods (which we suggest indicate a close companion) and 25% is roughly the fraction of planetary nebulae that show bipolarity.

Glass: In the double-period variables, are the long periods found to be stable?

Wood: The light curves certainly do not repeat accurately from period to period. Sometimes the long secondary period almost disappears, while it is strong at other times. Even the phase seems to vary from cycle-to-cycle although the long-term phase may be stable. Such a situation requires the source of obscuration (an orbiting cloud) in the binary model to be very loosely bound to the orbiting companion, or material ejected at periastron to come between us and the AGB star in a manner which does not repeat reliably from cycle-to-cycle.

Jura: Have you computed the effects of accretion on the total mass of the putative secondary?

Wood: No.

van Langevelde: (1) In the model for orbiting planets and brown dwarfs, how would you get a P-L relation for the stars displaying long secondary periods? (2) Where would the very long period OH/IR stars fit in: are they the variables with long secondary periods?

Wood: (1) The period of a low-mass orbiting companion of an AGB star $P \propto a^{\frac{3}{2}} M^{-\frac{1}{2}}$, where a is the semi-major axis, while the period of a pulsating star is $P \propto R^{\frac{3}{2}} M^{-\frac{1}{2}}$. Provided $a \propto R$, both binaries and

pulsating stars will follow parallel *PL* laws. In the model proposed, a is indeed proportional to R since we are demanding that the companion is very close to the stellar surface, at least at periastron.
(2) The sequence D variables (those with long secondary periods) are not related to the OH/IR stars - most sequence D variables are of low luminosity ($<3000L_\odot$) and are not very red so they do not have dense dust shells like the OH/IR stars. There are occasional OH/IR stars (or their C star analogs) on sequences D as shown by the crosses in Fig. 1, but such stars are very rare.
Habing: In the HR-diagram for MSX sources in the LMC, what are the relatively warm stars ($J - K \sim 1.0$) at high luminosity ($K < 9$)?
Wood: Red supergiants (masses > 8 $M_\odot$) - see Wood and Cohen (2001).
Tanabé: We see a gap between normal AGB stars and MSX stars in Fig. 1 of Wood and Cohen (2001). Is this a real one or just a selection effect?
Wood: This is a selection effect. The common AGB stars detected in the near infrared at J and K are mostly confined to J-K < 2.5. On the other hand, MSX only saw AGB stars with relatively high mass loss rates and J-K > 2.5.

THREE ASPECTS OF RED GIANT STUDIES IN THE MAGELLANIC CLOUDS

Maria-Rosa L. Cioni

European Southern Observatory

Karl-Schwarzschild-str. 2, 85748 Garching bei Muenchen (Germany)

mcioni@eso.org

Abstract There are three important aspects concerning the study of the red giant and in particular of the asymptotic giant branch (AGB) stars in the Magellanic Clouds. These are: the surface distribution, the luminosity function and the variability. The spatial distribution of AGB stars is an efficient tool to study the structure of the galaxies and their metalicity by analysing the ratio between carbon– and oxygen–rich AGB stars. The shape of the luminosity function carries informations about the star formation rate in the Clouds and it can be mathematically related to their history. Most AGB stars vary their magnitude in a few to several hundred years time; the one epoch DENIS magnitudes for both Large and Small Magellanic Cloud AGB stars outline the same relations as a function of period.

Keywords: Magellanic Clouds, red variables, massive photometry

Introduction

Studying the stellar content of the Magellanic Clouds has some advantages: these are nearby galaxies that can be fully resolved in stars, they are relatively un–obscured and all their stars are at about the same distance. Before the publication of the catalogues and follow–up studies from large scale surveys what was known about red giants in the Clouds were spatially and magnitude limited informations, often focalised on particular objects (i.e. Mira variables). The past few years have seen the release of the DENIS catalogue towards the Magellanic Clouds (DCMC – Cioni et al. 2000a) which provides photometry in the broad I, J and K_S bands. These filters are particularly suitable for the study of the red giant population (Cioni et al. 2000b, 2000c, 2001). The release of

Y. Nakada et al. (eds.), Mass-Losing Pulsating Stars and their Circumstellar Matter, 11–18.

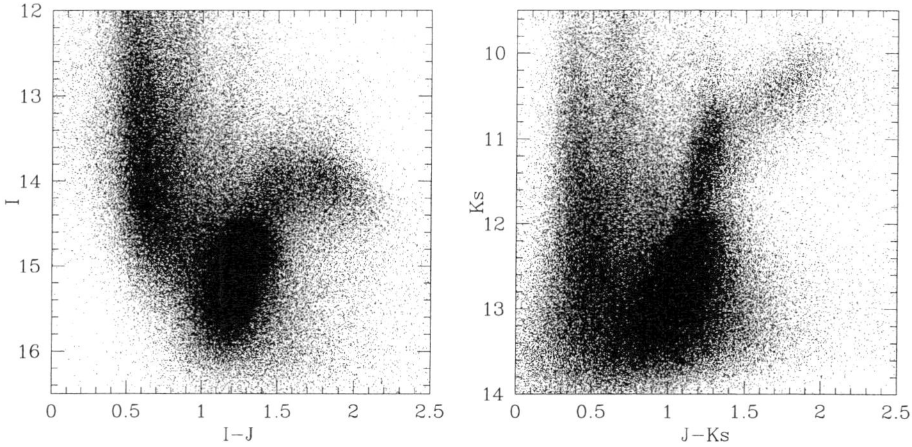

Figure 1. Colour–magnitude diagrams of DCMC sources detected simultaneously in three DENIS wave bands. Left: $(I - J, I)$. The TRGB is at $I = 14.54$. Right: $(J - K_S,\ K_S)$, the TRGB is at $K_S = 11.98$, O–rich AGB stars have $(J - K_S) = 1.2$ and C–rich AGB stars have $(J - K_S) > 1.4$.

the 2MASS catalogue (Nikolaev & Weinberg 2000) provides photometry in the J, H and K_S broad bands and reaches slightly fainter magnitudes than the DENIS measurements. The three micro-lensing surveys OGLE, MACHO & EROS provide light–curves over several years in two very broad blue and red filters that allow to characterise the complex variability of the red giants, as discussed also in the contribution by Wood (this proceeding). The OGLE and MACHO datasets are at present publically available. The combination of the near–infrared observations and light–curves is a key to understand the evolution and properties of long–period variables. In addition Zaritsky (2002) published his $UBVI$ survey of the SMC providing also an extinction map of the Cloud. Finally, Massey (2002) presented less sensitive $UBVR$ measurements in both Clouds to characterise the more massive and brighter stellar populations.

In the following sections the discussion is concentrated on the DENIS measurements described in my Ph.D. thesis (2001), I am also using EROS and MACHO light–curves, and ISO (LW2, LW10) measurements covering small fields in the Clouds (Loup et al. *in preparation*).

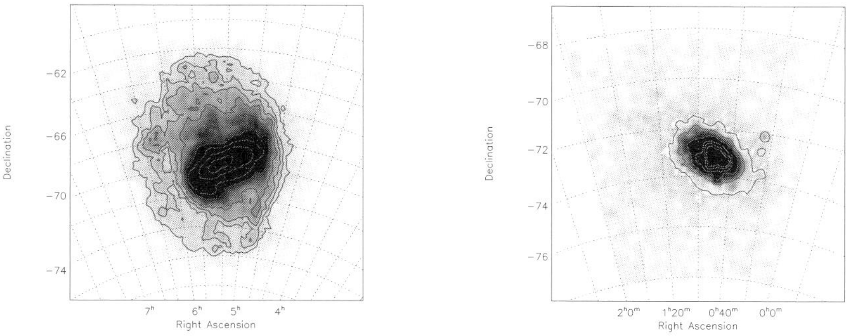

Figure 2. Star counts of LMC (left) and SMC (right) AGB stars in bins of 0.04 square degrees.

1. The spatial distribution of AGB stars

The colour–magnitude diagrams (CMDs – Fig. 1) made using the DENIS bands allow us to distinguish different groups of stars. In the $(I-J, I)$ diagram the densest region is the populous red giant branch of the Large Magellanic Cloud (LMC). It contains stars less massive then about 2 solar masses and older than about $1.5-2$ Gyr. These stars reach their maximum luminosity at the tip of the red giant branch (TRGB), a well defined feature that depends theoretically only on the mass of the stellar core. The plume of objects brighter than the TRGB are AGB stars, which are statistically separated into carbon (C–) and oxygen rich (O–rich) in the diagram $(J-K_S, K_S)$ because of the absorption effect of the different molecular bands in their atmosphere (see figure caption for the distinction). Cepheids and other younger stars in the Clouds overlap the vertical sequences of our own Galaxy. These CMDs made of sources detected in the direction of the LMC by DENIS simultaneously in all three wave bands show how easily the different stellar groups can be characterised from broad band measurements.

If we select AGB stars from the $(I-J, I)$ CMD, where the galactic foreground contamination is negligible, and plot them in the sky we obtain an extended smooth structure (Fig. 2). Note that the DCMC is complete for optically thin AGB stars; only the most obscured AGBs are not detected within the DENIS limits. In the LMC we clearly distinguish the bar and an almost circular disk structure of about 10° in diameter. A clearly elliptical structure is seen in the SMC with a semi–major axis of about 3° and a semi–minor axis of about 2°. Bins are of 0.04 square degrees. See Cioni et al. (2000c) for more details.

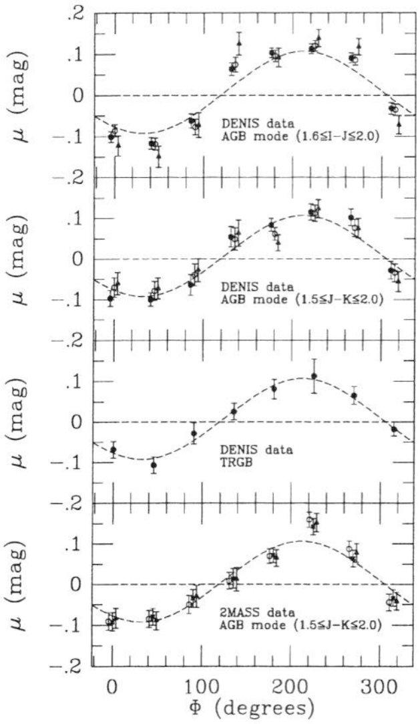

Figure 3. Variation of magnitude as a function of position angle in 8 disjunctive sectors between 2.5 and 6.7 degrees from the center of the LMC.

If we consider the shape of the LMC and imagine to divide the outermost part (ring that excludes the bar) in four sectors, then the modal magnitude of a selected sample of stars which span a small interval in magnitude will outline the projected morphology. In figure 3 as a function of position angle it is shown the variation of the magnitude of, from the top: AGB stars selected from the $(I - J, I)$ diagram having $1.6 < (I - J) < 2.0$, C–rich AGB stars selected from the $(J - K_S, K_S)$ diagram having $1.5 < (J - K_S) < 2.0$, the TRGB obtained from the magnitude distribution in the I band, and the C–rich AGB stars selected form the $(J - K_S, K_S)$ diagram within the same colour range as before but using this time 2MASS data. There is one point per sector and per band. Results are in very good agreement. The dashed curve represents the best fit model for an inclination angle of 34.7° and a position angle of 122.5°. The sinusoidal behaviour shows the effect of differences in distance coupled with the inclination of the Cloud. The conclusion of this work, described in detail in Van der Marel & Cioni (2001) is that the intrinsic shape of the LMC disk is elliptical and it shows that AGB stars are useful indicator of the extended structure of a galaxy.

The fact that there are AGB stars both C– and O–rich (M–type) is believed to be primarily a consequence of the third dredge–up that enriches the stellar atmosphere with carbon products. This dredge up seems to be more efficient in metal–poor environments and that is why we find more C–rich AGB stars in the SMC than in the LMC and in the Galaxy. The C/M ratio can thus be used as a metalicity indicator. Figure 4 shows the distribution of the C/M ratio in both Clouds. The

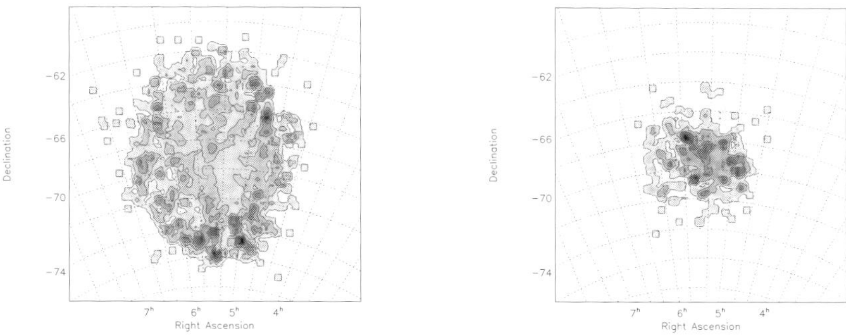

Figure 4. Distribution of the ratio between C–rich and O–rich AGB stars in the LMC (left) and in the SMC (right). Highest contours correspond to a C/M of 0.55, the lowest to 0.1

separation between C– and O–rich AGB stars has been done using the $(J - K_S)$ colour. The darkest regions correspond to a higher C/M ratio. In the LMC there are more C–rich AGB stars in the external parts, thus indicating a decreasing metalicity with increasing radius. In the SMC the situation is more complicated and irregular (Cioni & Habing, *in preparation*).

Using the VLT telescope when the FLAMES instrument will be available, the metalicity of red giants could be measured more precisely in order to study the chemical history enrichment of the Magellanic Clouds and the presence of a radial gradient. FLAMES is a multi–fibre spectrograph which will allow to observe 130 optical spectra with a resolution of $\Delta\lambda/\lambda = 28000/17000$ and up to 45000 for the eight fibres connected to the existing UVES instrument.

2. The luminosity function (K_S)

Fig. 5 shows the histogram of the magnitude of AGB stars. The continuous line indicates the whole distribution, the dashed line indicates only the O–rich AGB stars and the dotted line only the C–rich AGB stars selected on the basis of their $(J - K_S)$ colours. In the case of the SMC there is another thicker continuous line that indicates the C–rich AGB stars spectroscopically confirmed by Rebeirot (1983); this shows that the selection based on the colour is good enough for statistical studies. At the faintest mag the sharp increase of star counts marks the position of the TRGB. The global shape of the AGB histogram differs between the Clouds in the sense that it is rather monotonically decreasing for increasing mag in the SMC and it resembles a bump, thus with a flatter

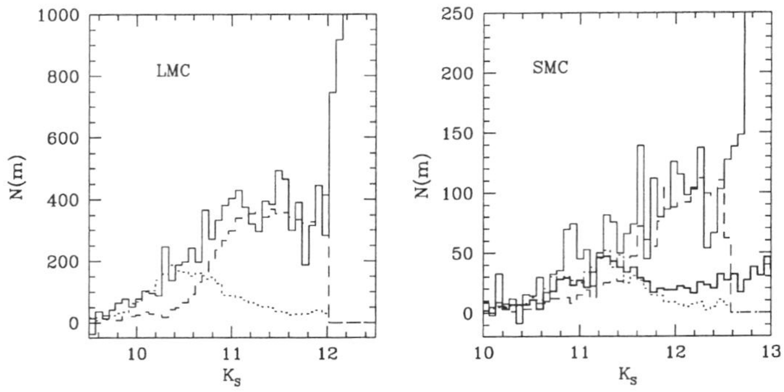

Figure 5. Histograms of the AGB stars in the LMC (left) and in the SMC (right). See text for details.

top, in the LMC. This difference in shape can be attributed to a different star formation history between the Clouds. Habing derives the following relation:

$$\frac{dL}{dM_{bol}} \propto SFR(t_{NOW} - t_{TO}) \times \frac{dM_0}{dM_{bol}} \tag{1}$$

where the slope of the luminosity function is proportional to the star formation rate at a time t_{TO} ago. By considering two different star formation rates, assuming the Salpeter initial mass function and that the stars decrease in M_{bol} at the same rate, we can reproduce the shape of the luminosity functions in the Clouds. We conclude that for the first 11 Gyr of our Universe there has been little formation of stars in the LMC with mass higher than about 1.0 solar mass. However, the SMC has older AGB stars. This is in agreement with the prediction by Marigo et al. (1999) based on the study of the carbon star luminosity functions. Our undergoing study aims at deriving the SFH as a function of position in the Clouds (Habing & Cioni, *in preparation*).

3. AGB period–luminosity relation

Figure 6 shows the period–magnitude diagram for sources in the LMC and in the SMC. The LMC sources are in the bar close to the optical center and are part of a study of cross–identification between DCMC sources and EROS2 light–curves (Cioni et al. 2001). The sources in the SMC were selected from the ISO mini–survey catalogue of Loup et al. (2002, soon to be released) to have a confident cross–identification with DCMC or 2MASS. Their light–curves were obtained from the MACHO database. The first conclusion we can deduce from this figure is that there seem to be no clear differences between the two Clouds. Both sources occupy the four slanted relations, of which relation C is where

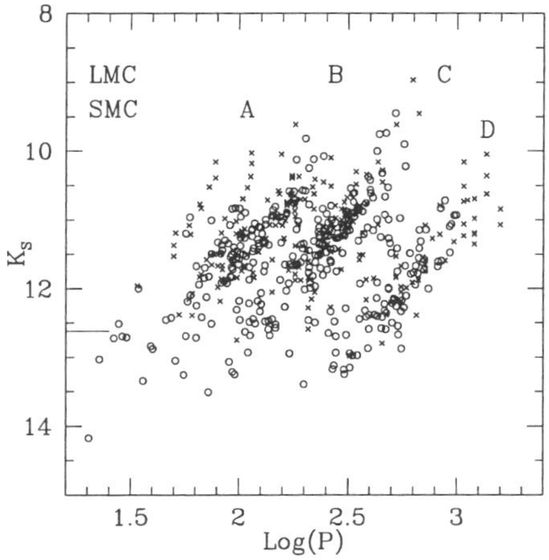

Figure 6. Period–magnitude diagram of long period variables in the Magellanic Clouds. Circles are for LMC stars and crosses for SMC stars.

Mira variables are also located. Note that all the sources have been placed at the distance of the SMC and that the K_S magnitude is from DCMC. The small horizontal line indicates the position of the TRGB. Sources occupying relation D are multi-periodic and the second period, also plotted, usually belong to relation A or B. In the SMC there are sources with longer periods, this is because the MACHO light–curves cover a longer time range with respect to the EROS2 data. In addition there is a lack of SMC sources around the TRGB, this is probably a selection effect because these sources do have an emission at 7 and 12 micron. ISO data exist also for sources in the LMC and they are going to be used also to check for variables at the TRGB magnitude.

4. Summary

We have seen that different groups of red giants can be distinguished in the near–infrared colour–magnitude–diagrams, we have also seen that the AGB stars are very good indicators of the extended structure of a galaxy and that from their luminosity function we can obtain indications on the stars formation history and the metalicity, the latter from the ratio of C– over O–rich AGB stars. Finally, these long period variables follow the same period–luminosity relations in both Clouds.

5. Asked questions

P. Wood. Do you take into account the variation of mass loss rate between the SMC and the LMC when you estimate the SFR from the AGB luminosity function?

Cioni No, it is not taken into account. On the other hand these AGB stars are optically thin and do not loose mass at a very high rate, thus we do not expect a strong effect due to varying mass loss rate.

M. Jura. For the LMC, is there any relationship in terms of ellipticity and/or positional angle of the "elliptical" disk and the inner bar?
Cioni The inner bar requires a separate study which van der Marel is currently undertaking, thus I cannot at present answer your question which is indeed very interesting.

S. Deguchi. How is the P–K relation of MSX sources?
Cioni Among the sources discussed here there are very few MSX sources. The contribution by Wood discusses them in more detail.

Acknowledgments

I would like to kindly thank Cecile Loup for making available the ISO data prior publication and Harm Habing for a fruitful discussion and collaboration on the newest results. I am also grateful to M. Zoccali for comments on the text.

References

Cioni, M.-R.L., Loup, C., Habing, H.J., et al. 2000a, A&AS, 144, 235
Cioni, M.-R.L., van der Marel, R.P., Loup, C., & Habing, H.J. 2000b, A&A, 359, 601
Cioni, M.-R.L., Habing, H.J., & Israel, f.P. 2000c, A&A, 358, L9
Cioni, M.-R.L., Marquette, J.-B., Loup, C., et al. 2001, A&A, 377, 945
Cioni, M.-R.L. 2001, Ph.D. Leiden Observatory
Marigo, P., Girardi, L., & Bressan, A. 1999, A&A, 344, 123
Massey, P. 2002, ApJS, 141, 81
Nikolaev, S., & Weinberg, M.D. 2000, ApJ, 542, 804
van der Marel, R.P., & Cioni, M.-R.L. 2001, AJ, 122, 1807
Rebeirot, E., Martin, N., Mianes, P., et al. 1983, A&AS, 51, 277
Zaritsky, D., Harris, J., Thompson, I.B., et al. 2002, AJ, 123, 855

LUMINOSITIES OF AGB VARIABLES

Patricia Whitelock
SAAO, PO Box 9, Observatory, 7935, South Africa.
paw@saao.ac.za

Abstract The prevailing evidence suggests that most large-amplitude AGB variables follow the period luminosity (PL) relation that has been established for Miras in the LMC and galactic globular clusters. Hipparcos observations indicate that most Miras in the solar neighbourhood are consistent with such a relation. There are two groups of stars with luminosities that are apparently greater than the PL relation would predict: (1) in the LMC and SMC there are large amplitude variables, with long periods, $P > 420$ days, which are probably undergoing hot bottom burning, but which are very clearly more luminous than the PL relation (these are visually bright and are likely to be among the first stars discovered in more distant intermediate age populations); (2) in the solar neighbourhood there are short period, $P < 235$ days, red stars which are probably more luminous than the PL relation. Similar short-period red stars, with high luminosities, have not been identified in the Magellanic Clouds.

Keywords: Mira variables, variable stars, carbon stars, LMC, Sagittarius Dwarf Spheroidal, Local Group Galaxies, Solar Neighbourhood, Hipparcos, luminosities, PL relation

Introduction

This review concentrates on the luminosities of large amplitude Mira-like variables. These are of particular interest because it is during this large-amplitude phase that most of the mass loss occurs. Furthermore, the kinematics of these stars depend upon their pulsation period; thus, if we can measure the period, we can tell a great deal about the star and its parent population (Feast & Whitelock 2000a). I discuss both O- and C-rich variables, but concentrate on the O-rich ones about which we know most. Because of its importance to luminosities, some emphasis is put on the period luminosity (PL) relation, drawing on observations of globular clusters, the Magellanic Clouds, other local group galaxies and last but not least, the solar neighbourhood.

Y. Nakada et al. (eds.), Mass-Losing Pulsating Stars and their Circumstellar Matter, 19–26.

Before examining what we know about luminosities it is worth emphasizing the importance of AGB variables in the understanding of extragalactic populations. The most luminous stars present in old or intermediate age populations are the large-amplitude AGB variables. Thus, as we become able to resolve individual stars in ever more distant stellar populations, those we see first and best will be this type of AGB variable. So, if we are to use such stars as probes of their parent population, it is crucial that we understand how their properties depend on age, metallicity etc.

1. Globular Cluster Miras

The Miras in globular clusters have always been key to calibrating the luminosity of the tip of the AGB. Unfortunately, because of their short lifetimes there are rather few Miras in globular clusters and fewer still that have been well studied. Let me remind you that the Miras are the most luminous stars found in the clusters; in fact they are the only stars with luminosities above the tip of the red giant branch (RGB). They are only found in metal-rich clusters ([Fe/H] > -1), and we presume that the AGB in metal-deficient systems terminates below the tip of the RGB. The pulsation period of a Mira is a function of the metallicity of its parent cluster (e.g. Feast & Whitelock 2000b). In fact it is only for the Miras in clusters, and a very few in binary systems, that we can determine metallicities. The Miras in galactic globular clusters are all O-rich and there is no particular evidence to suggest that they have reached the thermally pulsing part of the AGB.

Feast et al. (2002) recently reexamined the luminosities of the Miras in globular clusters using a new distance calibration based on Hipparcos parallaxes of sub-dwarfs and published photometry for 6 galactic globular clusters, together with new observations of NGC 121 v1, a short period low metallicity Mira in the SMC. They demonstrated that these cluster Miras fit the same PL relation as do the LMC Miras, and derived a zero-point for the PL relation. There are many globular clusters, particularly near the galactic centre, which have not yet been properly surveyed for Miras. We are in the process of rectifying this situation using the Infrared Survey Facility in South Africa in collaboration with astronomers from the University of Tokyo. Any new cluster Miras will obvious improve our statistics, but we are particularly hopeful about finding some longer period stars in the metal-rich bulge clusters.

Once there are theoretical models which deal effectively with mass-loss, and allow us to predict accurately the AGB tip luminosity for different populations, we will be able to use them to calibrate extragalactic

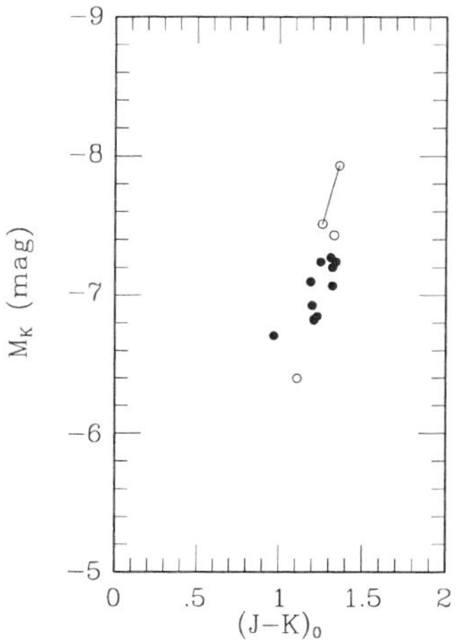

Figure 1 A colour-magnitude diagram for the Miras in globular clusters; open circles represent less certain luminosities. A luminosity of $K = -8.5$ is often assumed for the brightest stars in globular clusters, but $K = -7.5$ is actually a better estimate (see Feast et al. 2002).

systems. But, until that level of theoretical understanding is reached, those who wish to study extragalactic systems will make deductions based on a comparison with globular clusters or with the galactic bulge. The recent literature contains numerous studies of AGB populations in local group galaxies and beyond and it is interesting to see comparisons being made with galactic globular clusters and very different conclusions being drawn by different authors from essentially the same data. Figure 1 shows a colour magnitude diagram for cluster variables, of the kind typically used for comparison with extragalactic systems. The most luminous cluster star illustrated here is NGC 6553 v4, for which the reddening is uncertain; it is plotted in the figure as two connected open circles for two different reddenings. The lower mean luminosity seems more likely and this is one magnitude fainter than the $K = -8.5$ which some authors use.

It is also worth noting that comparing the luminosity of individual stars in extragalactic systems with those of AGB variables in the galactic bulge is even more fraught with uncertainty, because the shape of the bulge and the presence of significant numbers of foreground stars result in many stars having distances less than that of the centre and therefore luminosities that appear to be *much* brighter than they really are.

Variability is also a factor in comparing one system with another. The short period Miras, found in globular clusters, typically have peak-to-peak K amplitudes of around half a magnitude, so there is a high level of uncertainty associated with single measurements of the luminosity. Longer period stars have larger amplitudes, reaching over two magnitudes for the 1000 day variables discussed below.

2. Large Magellanic Cloud (LMC)

The PL relation for Mira variables was discovered for stars in the LMC, and Feast et al. (1989) refined earlier results to show that at K the O- and C-rich stars obeyed the same same PL relation. The bolometric luminosities seemed to show slightly different relations for the O- and C-rich stars, although there was always the suspicion that this was an artifact of the way that the bolometric magnitudes were calculated. Feast et al. also noted that O-rich stars with $P > 420$ days were significantly more luminous than the PL relation would predict.

Working independently, but at roughly the same time, Hughes & Wood (1990) came to similar conclusions; although they described the PL relation as having two linear parts, with a steeper slope over the long period ($P > 400$ days) range. Their derivation of a PL relation had considerable scatter, because their bolometric magnitudes were calculated from single observations rather than from the mean values.

More recently we have been working on much longer period stars in the LMC, which were discovered via their IRAS emission (e.g. Wood et al. 1992; Zijlstra et al. 1996). Most of these are obscured stars with high mass-loss rates. In globular cluster Miras the energy distribution peaks at a wavelength between 1 and 2 μm, while for these IRAS sources the energy peaks at longer wavelengths, $\lambda > 4\mu$m. There has, as yet, been little opportunity for systematic monitoring around the pulsation cycle at long wavelength, although a few repeated ISO observations allow us to estimate the bolometric amplitudes at around one magnitude. There are small systematic differences in the bolometric magnitudes obtained using ISO and IRAS. Furthermore, the way the colour corrections are treated can give rise to systematic differences between O- and C-rich stars. Thus we cannot yet claim to have accurate mean luminosities. Nevertheless, the overall impression given by all the available results is that most of the dusty Miras, with long periods, $420 < P < 1300$ days, fall on an extrapolation of the PL relation derived for O-rich stars with $P < 420$ days.

Looking in detail at the stars with bolometric magnitudes brighter than the PL extrapolation, we find that all those which have been studied show evidence for hot bottom burning (HBB), in particular they have very strong lithium lines (but see also Trams et al. 1999). Towards the end of the AGB evolution, in stars with initial masses in the range 4 to 6 $M_{\odot}$, the base of the H-rich convective-envelope can dip into the H-burning shell; the introduction of fresh H-rich material into the nuclear-burning shell allows the luminosity to go above the core-mass luminosity predictions (Blöcker & Schönberner 1991). Carbon is burned to nitrogen, and lithium can reach the surface via the beryllium transport mechanism (Sackmann & Boothroyd 1992). Smith et al. (1995) surveyed luminous AGB stars in the LMC and SMC for lithium - the clearest indication that HBB is taking place. Almost all stars with high lithium abundance lie above the PL relation, and we can now understand the change in slope of the PL, at around 400 to 420 days, as the effect of HBB in LPVs *without* particularly thick dust shells. Those stars with thick shells, like the ones for which we have ISO observations, are lower mass objects, lie near the PL, and probably never experienced HBB.

An important contribution is provided by the work of Nishida et al. (2000) who monitored 3 C-rich Miras, with thick dust shells, in SMC and LMC clusters, and were thus able to estimate pulsation periods and bolometric magnitudes for stars with known initial mass and metallicity. The periods are all around 500 days and the luminosities are very close to an extrapolation of the Feast et al. (1989) PL for short period O-rich stars. The turnoff masses for these clusters are around $1.5M_{\odot}$, and therefore much too low for their AGB stars to have undergone HBB.

3. Other Local Group Galaxies

Leaving the Magellanic Clouds, but staying in the local group we look at C-rich Miras in the Sagittarius dwarf spheroidal. This is the galaxy, discovered only in 1994, that is merging with the Milky Way on the far side of the bulge. It contains Miras with periods in the range 230 to 360 days, distinctly shorter than those in Magellanic Cloud clusters, as we might expect from this somewhat older population (Whitelock et al. 1999). The distance modulus derived from the Mira PL, $(m - M)_0 = 17.36 \pm 0.2$ mag, is in good agreement with that from RR Lyrae variables, $(m - M)_0 = 17.18 \pm 0.2$ mag.

There is a Mira in IC 1613 with a period of 641 days, the luminosity of which is considerably brighter than the prediction of the PL relation (Kurtev et al. 2001). With a spectral type of M3e it is clearly O-rich, and from what has been said above we must predict that it is undergoing HBB burning. It would be interesting to look for lithium in its spectrum.

The Leo I dwarf spheroidal contains several large amplitude variables (see Menzies, these proceedings). There is also evidence of very red C-stars from 2MASS observations of Fornax; these will almost certainly turn out to be C-Miras. We can look forward to similar discoveries in other local group galaxies and beyond, during the next few years.

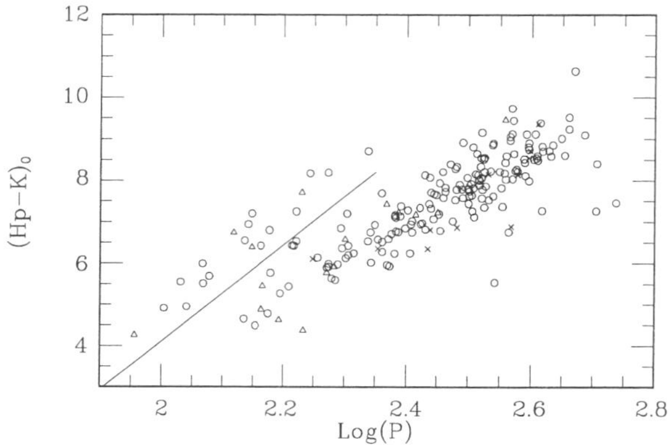

Figure 2 A combined Hipparcos, *Hp*, near-infrared, *K*, period-colour relation; circles represents Miras and triangles SRs, while Miras with S-type spectra are shown as crosses; the line separates SP-red and SP-blue stars (see Whitelock et al. 2000).

4. Solar Neighbourhood

Returning closer to home I want to finish by looking at what we know about luminosities from Hipparcos parallaxes (Whitelock et al. 2000; Whitelock & Feast 2000; Feast & Whitelock 2000a). It is worth noting that the stars selected for the Hipparcos input-catalogue had to be visually bright throughout their pulsation cycle. Therefore, this selection of stars have, of necessity, low mass-loss rates, $< 10^{-7} M_{\odot} yr^{-1}$; they are different from the LMC sample discussed above. The selection comprised 213 O-rich Mira-like variables with K magnitudes and pulsation periods.

Before discussing the parallax analysis I outline some of the characteristics of these stars, and their dependence on period and colour. Figure 2 shows the stars in a period-colour plot, where Hp is the Hipparcos broad-band magnitude. At short periods the variables divide into two sequences, a blue one that contains most of the stars and a parallel red one which contains a significant fraction of the shortest period stars; notice that both sequences contain SRs and Miras. The straight line in Fig 2 divides the stars into two groups: those above the line are the short period red (or SP-red) group, while those below, which form part of the sequence seen at longer periods, are the short-period blue (or SP-blue) group (Whitelock et al. 2000).

The pulsation amplitudes of the two groups are very similar, but the colour differences extend to their near-infrared colours - the mean spectral type of the red group, M4.5, is later than that of the SP-blues, M3. Furthermore, the kinematics and scale heights of the two groups are different. The SP-blue stars have larger scale heights, a greater velocity dispersion and a larger asymmetric drift than the SP-red stars. The SP-blue stars have similar characteristics to the Miras found in globular clusters, while the SP-reds seem to be rather different and apparently somewhat younger (see also Feast these proceedings). The differences

Table 1. PL Zero-Point from the Hipparcos Parallax.

No. Stars	β (mag)	σ_β	Stars Included
180	0.84	0.14	not SP-red; $\Delta Hp > 1.5$ mag
37	0.93	0.46	SP-blue only
18	0.40	0.24	SP-red only
38	0.90	0.31	Carbon stars

between these two groups justifies treating them differently in the parallax analysis.

Given the high uncertainties on the parallax measurements it is not practical to fit a PL relation to the data. Rather we assume the slope of the PL relation ($M_K = -3.47 \log P + \beta$), derived from the LMC work, and deduce β, the zero point, from the parallaxes. The PL relation is solved in the form:

$$10^{0.2\beta} = 0.01\pi 10^{0.2(3.47 \log P + K_0)},$$

where π is the parallax in mas. This allows us to use all the parallax data, and thus minimize the bias which would be introduced by selection. The right-hand side of the equation is weighted appropriately as described by Whitelock & Feast (2000a).

The equations were solved for various different subsets of the data, some of which are listed in Table 1 (it may be necessary to add a bias correction, ~ 0.04 mag to these values). The best result for the Miras is given by the group of 180 stars which excludes the SP-red group and the small amplitude variables, that should probably never have been considered as Mira-like. The value of the zero-point for that group, $\beta = 0.84 \pm 0.14$, corresponds to a distance modulus for the LMC of 18.64, which is comparable to values obtained in other ways. I should also draw your attention to the estimates of the zero point for the SP-red and SP-blue stars listed in the table. Given the large uncertainties the difference between them is not significant, but this difference between the SP-reds and the main Mira group (top row) is significant, and together with other evidence of differences, suggests that the SP-red stars may be brighter than the Mira PL relation would predict.

It is worth making a comparison with the various sequences in the PL diagram for the LMC, as discussed by Wood (2000) on the basis of Macho data. The SP-blue stars are presumably identical to the normal Miras, i.e. sequence C in Wood's nomenclature. According to Table 1 the SP-reds are about 0.5 mag brighter than the SP-blues, whereas Wood's B sequence is about 1.3 mag brighter. Note again that there is no difference in the mean pulsation amplitude of the two groups. It therefore seems that stars like these SP-reds have not yet been identified in the LMC. The kinematic difference between the SP-reds and SP-blues shows that they cannot be similar stars pulsating in different modes. The possibility remains that the SP-reds have a relationship to longer period Miras ($P > 235$ days) which do have similar kinematics. If that is the case then the SP-reds must be in a slightly earlier evolutionary phase than Miras on the PL relation.

To conclude my comments on the local Miras, it would seem that, with the exception of a small number with short periods and red colours, the large amplitude variables fall on the same PL relation as do Miras in the LMC and in globular clusters.

Finally, space has not permitted a discussion of the numerous OH/IR variables near the galactic centre, which prevailing wisdom suggests are faint, but which are difficult to measure accurately. More work is also need on C-star luminosities; the evidence points to their also obeying the PL, but dust shells often make luminosity estimates difficult.

Acknowledgments

I wish to thank my various colleagues, particularly Michael Feast, Jacco van Loon, and Albert Zijlstra for their patience in the face of my slow progress towards publication. My thanks also to Michael Feast and John Menzies for advice and critical reading of this manuscript.

References

Blöcker, T. & Schönberner, D. 1991, A&A, 244, L43
Feast, M.W. & Whitelock, P.A. 2000a, MNRAS, 317, 460
Feast, M.W. & Whitelock, P.A. 2000b, in The Evolution of the Milky Way, eds. F. Matteucci, & F. Giovannelli (Dordrecht: Kluwer), 229
Feast, M.W. et al. 1989, MNRAS, 241, 375
Feast, M.W., Whitelock, P.A. & Menzies, J.W. 2002, MNRAS, 329, L7
Hughes, S.M.G. & Wood, P.R. 1990, AJ, 99, 784
Kurtev, R. et al. 2001, A&A, 378, 449
Nishida, S. et al. 2000, MNRAS, 313, 136
Sackmann, I.-J. & Boothroyd, A. 1992, ApJ, 392, L71
Smith, V.V., Plez, B., Lambert, D.L. & Lubowich, D.A. 1995, ApJ, 441, 735
Trams, N.R. et al. 1999, A&A, 344, L17
Whitelock, P.A. & Feast, M.W. 2000, MNRAS, 319, 759
Whitelock, P.A., Marang, F. & Feast, M.W. 2000, MNRAS, 319, 728
Whitelock, P.A. et al. 1999, in IAU Symp. 192, The Stellar Content of Local Group Galaxies, eds. P.A. Whitelock & R.Cannon (San Fransisco: ASP), 136
Wood, P.R. 2000, PASA, 17, 18
Wood, P.R., Whiteoak, J.B., Hughes, S.M.G., et al. 1992, ApJ, 397, 552
Zijlstra, A.A., Loup, C., Waters, L.B.F.M., et al. 1996, MNRAS, 279, 32

PHOTOMETRIC PROPERTIES OF LONG-PERIOD VARIABLES IN THE LARGE MAGELLANIC CLOUD

Sachiyo Noda
National Astronomical Observatory, Osawa 2-21-1, Mitaka, Tokyo 181-8588 Japan
sachi.t.noda@nao.ac.jp

Mine Takeuti
Astronomical Institute, Tohoku University, Aoba, Sendai 980-8578 Japan
takeuti@astr.tohoku.ac.jp

Abstract Approximately four thousand light curves of red variable stars in the Large Magellanic Cloud (LMC) were selected from the 2.3-year duration MOA database by a period analysis using the Phase Dispersion Minimization method. Their optical features (amplitudes, periodicities, position in CMD) were investigated. Stars with large amplitudes and high periodicities were distributed on the only one strip amongst multiple structure on the LMC period-luminosity relation. In the CMD, the five strips were located in the order of the period. The stars with characterized light curves were also discussed.

Keywords: Red variables, period-luminosity relation, massive photometry

1. Introduction

The multiple and complicated period-luminosity relation for red variables in the LMC had been discovered using the microlensing database (Wood, Alcock, Allsman et al. 1999, Wood 2000). Although the Mira sequence (Feast, Glass, Whitelock, and Catchpole 1989, Hughes & Wood 1990) have been remarked as a distance indicator, such multiplicity is fatal for use as a distance indicator because their characteristics of each strip have not been revealed actually. The MOA (Abe, Allen, Banks et al. 1997; Hearnshaw, Bond, Rattenbury et al. 2000) database of the LMC obtained by large-scale photometry is quite appropriate to study the above problem. Not only related with the multiplicity of the period-

Y. Nakada et al. (eds.), Mass-Losing Pulsating Stars and their Circumstellar Matter, 27–34.

Table 1. The data series of MOA database

	Series 1	Series 2	Series 3
Period	1996 Jan.-1996 Dec.	1997 Jan.-1998 Jul.	1998 Aug.- present
Optics	f/13.5	f/6.25	f/6.25
FOV	$30' \times 30'$	$1^\circ \times 1^\circ$	$0.9^\circ \times 1.38^\circ$
CCD	$1K \times 1K$	$1K \times 1K$	$2K \times 4K$

luminosity relation, the photometric properties must be studied carefully to reveal the nature of AGB variables. Some interesting results of the study of the MOA database is presented.

2. The MOA project

The MOA is the microlensing research project, and the collaboration of about 30 astronomers from Japan and New Zealand. The observational targets are mainly the LMC and Galactic bulge. We observe every photometric night at the Mt. John University Observatory in the center of the South Island of New Zealand using small telescope (61-cm diameter) and large CCD (three chips of $2K \times 2K$ pixels) chips. There are three data series since 1996 (Table 1). The Series 1 is a period of test-drive, while the Series 3 is the current system. In the current Series, the 16 fields around the LMC bar are observed every night and approximately 4.4 million sources are included in the Series 3 LMC catalogue. The curves of transmission of the two color filters and the quantum efficiency of the CCD (SITe, $2K \times 4K$) are shown in Noda, Takeuti, Abe et al. (2002).

3. Red variables

Selection criteria. The selection of regular variables from the MOA database was carried out as the following. In the first criterion (Level 1), 313,706 stars of the MOA database identified with only one DENIS source (Deep Near-Infrared Southern Sky Survey; Epchtein, Deul, Derriere, et al. 1999) were selected. In the Level 2, 'non-photometric' data points were eliminated, and light curves of too small light variation were also removed in the Level 3. In the Level 4 to 8, careful period analysis (PDMM) were performed with necessary eye-estimate to exclude inappropriate samples, and finally, 4,858 red variables were obtained. Amongst the selected stars at the Level 1, K_S magnitudes (the effective wavelength $\approx 2.15\ \mu$m) were obtained for 67,107 stars. The histogram of K_S magnitudes is indicated in the left panel of Fig. 1.

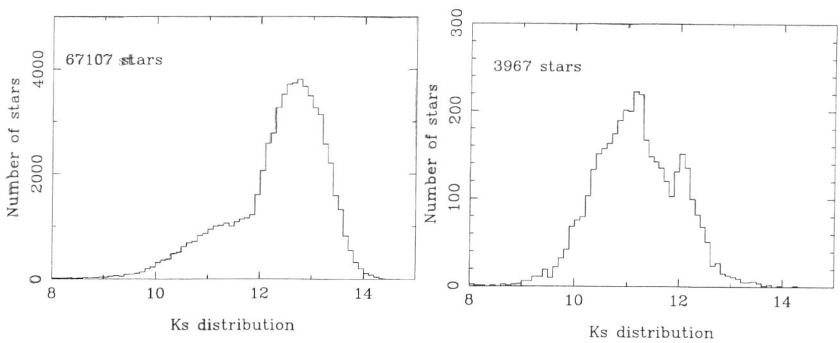

Figure 1. The distribution of the luminosity of non-variable (left panel) and variable (right panel) stars. Two clumps of variable stars are identified.

Two clumps of the red variables. The K_S magnitudes are tabulated for 3,967 stars among the finally selected variable stars. The histogram of K_S magnitudes of these stars is shown in the right panel of Fig. 1. The center of distribution of variable stars was $K_S \approx 11$ mag which corresponded to a bump at the brighter side of the peak in the left panel. The other small peak was found around $K_S \approx 12$ mag in the right panel. Because the large number of non-variables is found fainter than $K_S \approx 12$, it is no doubt about the existence of the fainter clump. The study of intrinsic properties of these two clumps will be important.

4. The period-luminosity diagram

Multiplicity. The K_S magnitudes as a function of $\log P$ for our 3,967 samples is presented in Fig. 2. The dashed line is the $\log P$-K relation for the oxygen-rich Mira sequence by the previous study (Hughes & Wood 1990), while the solid line is the same relation but shifted upwards by 0.29 mag. Some strips were found. In order to investigate the structure, the vertical separation of K_S magnitude from the PL relation by Hughes & Wood (1990) were examined, and six strips (groups $A \sim F$) were defined. The stars labeled with G are ignored because the periods are very short. Stars in the group F were likely long period Cepheids. 3,564 stars were included in the groups $A \sim E$. Note, the group C was the densest strip identified with the classical Mira sequence.

The distribution of amplitude. In Fig. 3, the amplitude histograms of the group $A \sim E$ are presented. The amplitude δR_m was defined as the difference of the magnitudes of the brightest and the faintest bins. In this figure, stars of large amplitude were mainly be-

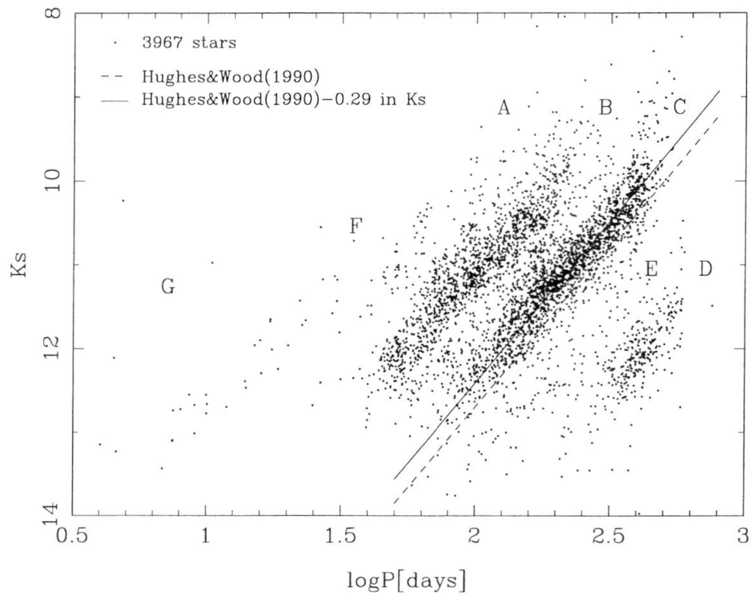

Figure 2. The period-luminosity diagram for 3.967 variables (see text).

longed the group C. For example, 95% of stars whose amplitude were larger than 0.9 were the member of group C.

Regularity of the period. The histograms for the periodicity, θ_r, histogram are presented in Fig. 4. θ_r $(0 < \theta_r < 1)$ is the relative parameter which is defined in the PDMM and indicates regularity of the light variation. Small θ_r indicates high regularity, while large θ_r indicates almost random variation. In this figure, the distribution of θ_r for group C showed it to be almost constant for $0.2 < \theta_r < 0.9$. It is obvious that stars with small θ_r, for example, 88% of stars with $\theta_r < 0.4$ were the member of group C.

The most remarkable sequence. Amongst stars which were satisfied the both condition of $\delta R_m > 1$ and $\theta_r < 0.4$, 96% were the group C component. This means the absolute magnitudes of long-period variables will be estimated properly when the sufficient number of light curves over several cycles are obtained. The luminosity of variables of large amplitude an d high regularity must be estimated by using the period-luminosity relation of group C.

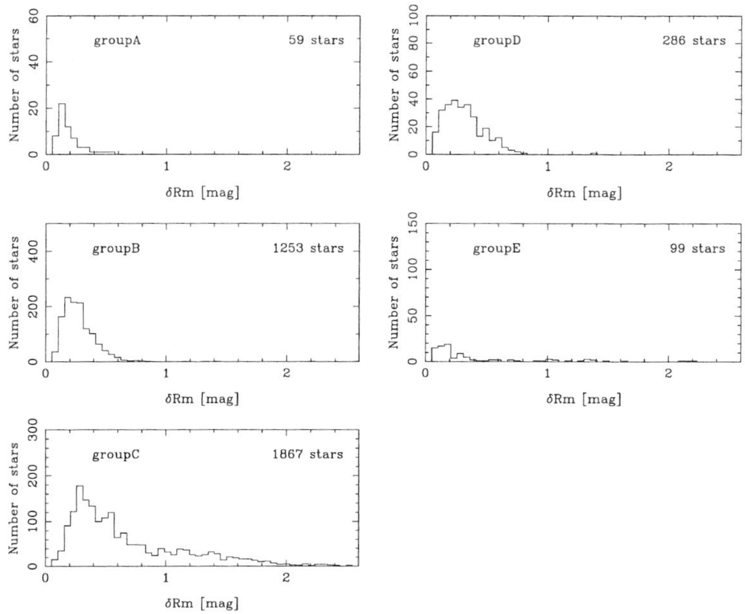

Figure 3. The histogram of amplitude, δR_m, for the five strips.

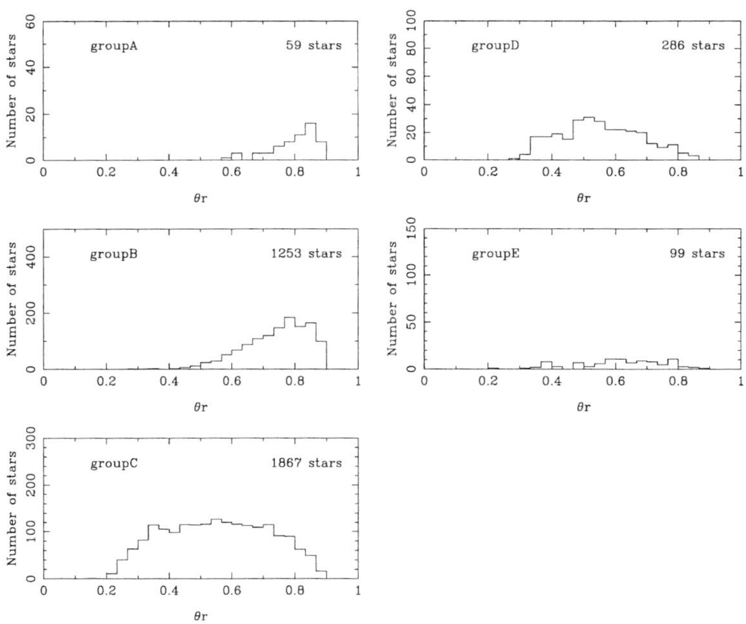

Figure 4. The histogram of periodicities, θ_r, for the five strips.

5. The $(\langle R_m \rangle - K_S, K_S)$ diagram

The color magnitude diagrams, $(\langle R_m \rangle - K_S, K_S)$, for each group are shown in Fig. 5. $\langle R_m \rangle$ indicates the mean magnitude in the MOA red light curve. In this figure, the group name and the number of sample are presented in the each panel. The bottom-right panel indicates the distribution of 67,107 red stars which were identified by DENIS and detected at least in K_S magnitude. Not only variables but also non-variables are plotted in this panel. As reported in the preliminary study by Noda et al. (2002), the variables were located in almost same domain on the CMD.

Because the present samples are very rich in the number, precise comparison of the properties of each group will be interesting. The mean K_S magnitudes of groups A - C correspond to the bright clump beside those of groups D and E are identical with the faint clump. The mean positions of each group on the CMD differ group to group too. Group A is brighter and bluer than group B, and the latter is also brighter and bluer than group C. The examined regression lines showed the slope of the lines for group A, B and C were arranged in the period order (The slope of group A is steeper than group B and so on). In the groups of the faint clump, group D is the fainter and redder one, and the slope of group E is slightly steeper than group D.

6. Other interesting results

RV Tauri-like feature. The light curves of RV Tauri stars are characterized by the alternative deep and shallow light minima. We have found that 13.5 % of the variable star samples show this typical RV Tauri type light curve. It is interesting that the majority of these stars (94 % !) belong to group C. Because the luminosity of these stars are typical to the Mira stars, it is clear that these stars are not the RV Tauri type. It will be mentioned that the RV Tauri-like light curve alone is not the characterized property for the RV Tauri stars.

Eclipsing variables. In the process of selection, we have found many stars showing the light curve typical to the eclipsing binary system. In 348 such stars, 159 stars are located on the same position as group D, and 103 stars are as group C, on the $(\log P,\ K_S)$ diagram. The existence of many eclipsing binaries at the same position as group D on the $(\log P,\ K_S)$ diagram is a new enigma about the AGB stars. It should be noted that group D of our paper is established after excluding the eclipsing binaries. The nature of group D stars will be studied without the connection of the eclipsing binaries.

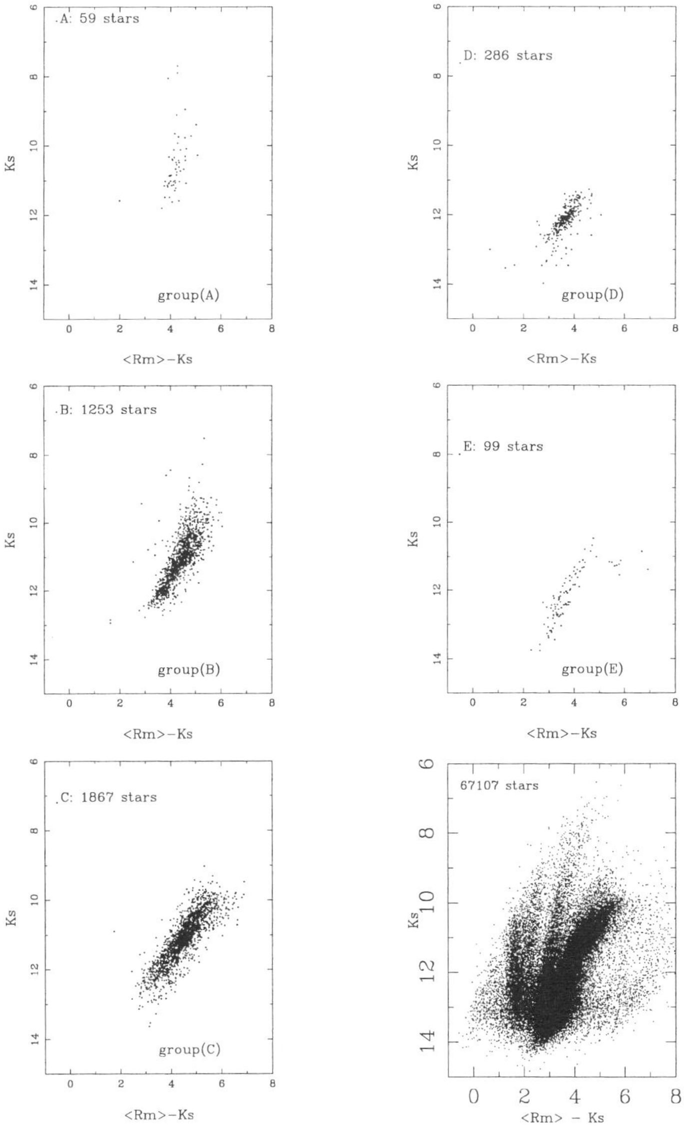

Figure 5. The ($\langle R_m \rangle$ - K_S, K_S) diagram. The bottom-right panel shows the distribution of 67,107 stars which were identified by DENIS and detected at least in K_S magnitude.

Period transition. Difference in the periods was found between the stars tabulated in Hughes & Wood (1990) and the present results. Such a difference was also found between the result of Noda et al. (2002) and the present one. It is found that the period transition of Galactic semi-regulars in the solar neighborhood was also common in the LMC red variables. Together with the RV Tauri-like light curves, these feature will be evidence for multi-mode behavior of the AGB variables.

7. Conclusion

Almost all of large amplitude or highly periodic stars in the LMC were the member of group C which is the most crowded and nearest strip to the classical Mira sequence. For example, the 94% stars of large amplitude ($\delta R_m > 1$) or 88% stars of inferior periodic ($\theta_r < 0.4$) were the group C component. If we require the both conditions, more than 90% were the member of group C strip. It is possible to utilize this type of sequence as a distance indicator, even if multiple relations are found in extra-galactic systems.

In the ($\langle R_m \rangle - K_S$, K_S) diagram, the stars of the five sequences show slightly different features in the order of period. Because all of the variables were found in a limited domain of the CMD, the intrinsic excitation mechanism will be the same, but the difference in the pulsation mode is suggested.

References

Abe, F., Allen, W., Banks, T., et al. 1997, in Variable Stars and the Astrophysical Returns of the Microlensing Surveys, ed. R. Ferlet, J.P. Maillard, & B. Raban, (Editions Frontières, Gif-sur-Yvette Cedex, France), 75

Cioni, M.-R., Marquette, J.-B., Loup, C., et al. 2001, A&A. 377, 945

Epchtein, N., Deul, E., Derriere, S., et al. 1999, A&A, 349, 236

Feast, M.W., Glass, I.S., Whitelock, P.A., & Catchpole, R.M. 1989, MNRAS, 241, 375

Hughes, S.M.G. & Wood, P.R., 1990, AJ, 99, 784

Hearnshaw, J.B., Bond, I.A., Rattenbury, N.J., et al. 2000, in IAU Colloquium No. 176, the Impact of Large-Scale Surveys on Pulsating Star Research, ed. L. Szabados & D.W. Kurtz (ASP Conf. Ser. 203), 31

Noda, S., Takeuti, M., Abe F., et al. 2002, MNRAS, 330, 137 (astro-ph/0111355)

Wood, P.R., Alcock, C., Allsman, R.A., et al. 1999, in IAU Symp. 191, Asymptotic Giant Branch Stars, ed. T. le Bertre, A. Lèbre, and C. Waelkens (San Francisco: ASP), 151

Wood, P.R. 2000, PASA, 17, 18

STELLAR SOURCES IN THE ISOGAL INTERMEDIATE BULGE FIELDS

D.K. Ojha[1], A. Omont[2], and the ISOGAL Team
[1] *T.I.F.R., Homi Bhabha Road, Colaba, Mumbai - 400 005, India*
[2] *IAP, CNRS, 98 Bis, Bd. Arago, F-75014, Paris, France*
ojha@tifr.res.in

Abstract We present a study of ISOGAL sources in the "intermediate" galactic bulge ($|l| < 2°$, $|b| \sim 1°$–$4°$), observed by ISOCAM at 7 and 15 μm (Ojha et al. 2002). In combination with near-infrared (IJK_s) data of DENIS survey, we discuss the nature of the ISOGAL sources, their luminosities, the interstellar extinction and the mass-loss rates. A large fraction of the detected sources at 15 μm are AGB stars above the RGB tip, a number of them show an excess in $([7]\text{-}[15])_0$ and $(K_s\text{-}[15])_0$ colors, characteristic of mass-loss. The latter, especially $(K_s\text{-}[15])_0$, allow estimation of the mass-loss rates ($\dot{M}$) and show their distribution in the range 10^{-8} to 10^{-5} $M_\odot$/yr.

Keywords: stars: AGB and post-AGB – stars: circumstellar matter - Galaxy: bulge

1. ISOGAL and DENIS Observations

ISOGAL observations with ISOCAM at $6''$ pixel field of view of nine intermediate bulge fields with good quality of 7 and 15 μm data are used in our study (see Ojha et al. 2002). For all the observed ISOGAL fields, we have used the data of the ISOGAL Point Source Catalogue (PSC, Version 1) described in Schuller et al. (2002). The near-infrared data used in this paper is acquired from the DENIS survey with special observations of the Galactic bulge (Simon et al., in preparation) in the three bands, I (0.80 μm), J (1.25 μm) and K_s (2.17 μm). Such DENIS observations cover the totality of the nine ISOGAL bulge fields.

2. Interstellar extinction

We have used the DENIS JK_s data to derive the JK interstellar reddening and hence the A_V for each star. The K_s/J-K_s magnitude-color diagrams of DENIS sources in the bulge fields show a well-defined red gi-

Y. Nakada et al. (eds.), Mass-Losing Pulsating Stars and their Circumstellar Matter, 35–38.

ant sequence shifted by fairly uniform extinction (see Ojha et al. 2002), with respect to the reference K_{s0} *vs* $(J-K_s)_0$ of Bertelli et al (1994) with Z=0.02 and a distance modulus of 14.5 (distance to the Galactic Center: 8 kpc). The near-infrared colors of this isochrone have been computed with an empirical T_{eff}–$(J-K)_0$ color relation built by making a fit through measurements for cool giants (see Schultheis et al. 1998). We have assumed that $A_J/A_V = 0.256$; $A_{Ks}/A_V = 0.089$ (Glass 1999). Fig. 1 shows the [7]-[15]/[15] color-magnitude diagram of ISOGAL sources.

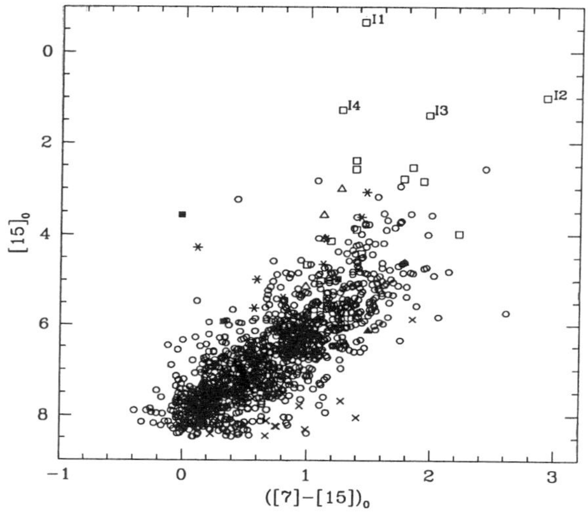

Figure 1. [15]/[7]-[15] magnitude-color diagram of ISOGAL sources in the ISOGAL fields. The open circles show the sources having 7/15 μm association quality flags of 3 & 4 (see Schuller et al. 2002). The crosses show the sources having association quality flags of 2. The asterisks display the associations with 2MASS (since the DENIS sources [$K_s \leq 6.5$] are saturated, they have been replaced by 2MASS). The remaining saturated DENIS sources (which have no 2MASS counterparts) are shown by filled squares. The bright ISOGAL sources associated with 2MASS sources (which have no DENIS counterparts) are shown by open triangles. Sources with rejected DENIS associations (quality flag 1 or $r>5''$) are denoted by filled circles. The filled triangles show the ISOGAL sources which have no available DENIS nor 2MASS counterparts, mostly because of saturation or non observation. The IRAS sources are shown by open squares and the brightest ones are denoted as "I" (see text).

3. Mass-loss rate ($\dot{M}$) in the bulge

In order to allow a rough calibration of the mass-loss rates, to discuss their relative distribution and to try to make a preliminary comparison

Table 1. Integrated mass-loss ($M_\odot$/yr/deg^2) in the "intermediate" bulge

	LBW98 ($\dot{M} > 10^{-8}$ $M_\odot$/yr)	WMF94 ($\dot{M} > 10^{-7}$ $M_\odot$/yr)
All fields with I1-4	4.0×10^{-4}	4.1×10^{-4}
All fields without I1-4	0.6×10^{-4}	2.4×10^{-4}
Fields with $\lvert b\rvert > 2$	0.6×10^{-5}	3.7×10^{-5}

with the solar neighbourhood, we have applied the results of previous work of Le Bertre & Winters 1998 (hereafter LBW98) and Whitelock et al. 1994 (hereafter WMF94), relating $\dot{M}$ and (K-[12])$_0$. We have used the relation between $\dot{M}$ and (K$_s$-[15])$_0$ with a slightly different calibration (see Ojha et al. 2002). Their results are significantly different because they rely on various assumptions about models for envelope structure and dust emissivity, dust-to-gas ratio, expansion velocity, luminosity, etc.

Fig. 2 shows the number distribution of mass-loss rates ($\dot{M}$) of ISOGAL sources in the bulge as a function of $\dot{M}$ for models of LBW98 and WMF94. It is seen in Fig. 2 that the contribution to the integrated mass-loss per bin of log $\dot{M}_{LBW}$ is important for all the considered bins, with large uncertainties (at least a factor of 2) because of incompleteness at low $\dot{M}$ and of the very small number of sources at high $\dot{M}$. However, especially with LBW98 model, there is a possibly dominant contribution from the four stars I1–4 with (K$_s$–[15])$_0 > 6$, $\dot{M} \geq 10^{-5}$ $M_\odot$/yr , whose contribution has been identified in Fig. 2b.

Numerical values for the integrated mass-loss are displayed in Table 1. Since it is difficult to decide whether the stars I1–4 belong to the central bulge/disk or not, we give also the same quantities for the three fields with $|b| > 2°$ where such stars with very large values of $\dot{M}$ ((K$_s$–[15])$_0 > 5$) are absent. The corresponding distribution of mass-loss rates are displayed in Figs. 2c and 2d. In the absence of stars with K$_s$–[15])$_0 > 5$, it is seen that all the considered magnitude bins provide an important contribution to the integrated mass-loss. However, since the total number of relevant stars with $|b| > 2°$ is about four times smaller than for the totality of the fields, where there are such four stars, one can not consider that their absence in fields with $|b| > 2°$ is statistically significant.

Such results show that the contribution of stars with relatively low mass-loss rates ($\dot{M} < 10^{-6}$ $M_\odot$/yr) is appreciable in the mass return to the interstellar medium for the old bulge population, and that even the stars with $\dot{M} < 10^{-7}$ $M_\odot$/yr have a significant contribution. However, it is not clear whether the stars with very large mass-loss rates,

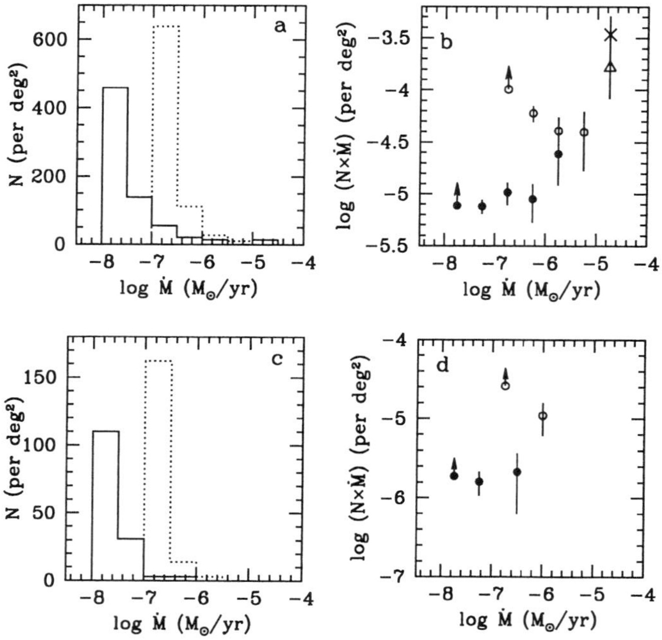

Figure 2. (a). Distribution of $\dot{M}$ of ISOGAL sources inferred from $(K_s\text{-}[15])_0$ in the bulge fields for models of LBW98 (full line) and WMF94 (dotted line). (b). The total mass-loss rate per deg^2 as a function of $\dot{M}$ for models of LBW98 (solid points) and WMF94 (open circles). The contributions of stars I1–4 are shown by a cross (LBW98) and an open triangle (WMF94). The error bars (with log $\dot{M}$ > -7.5 for LBW98 & log $\dot{M}$ > 7 for WMF94) show the Poisson noise and the arrows (log $\dot{M}$ < -7.5 for LBW98, log $\dot{M}$ < -6.5 for WMB94) show the incompleteness on the data. (c) & (d). Same as a & b, but limited to the fields at $|b| > 2$.

$\dot{M} \geq 10^{-5}$ $M_\odot$/yr, which are probably younger with a relatively large initial mass, may have a dominant contribution, as it seems the case for the solar neighbourhood (see e.g. Le Bertre et al. 2001).

References

Bertelli, G., Bressan, A., Chiosi, C., et al. 1994, A&AS, 106, 275

Glass, I.S. 1999, Handbook of Infrared Astronomy, (Cambridge: Cambridge Univ. Press)

Le Bertre, T., & Winters, J.M. 1998, A&A, 334, 173 (LBW98)

Le Bertre, T., Matsuura, M., Winters, J.M., et al. 2001, A&A, 376, 997

Ojha, D.K., Omont, A., Schuller, F., et al. 2002, A&A (submitted)

Schuller, F., Ganesh, S., Messineo, M., et al. 2002, A&A (submitted)

Schultheis, M., Ng., Y.K., Hron, J., & Kerschbaum, F. 1998, A&A, 338, 581

Whitelock, P.A., Menzies, J., Feast, M.W., et al. 1994, MNRAS, 267, 711 (WMF94)

GAIA: AGB STARS AS TRACERS OF STAR FORMATION HISTORIES IN THE GALAXY AND BEYOND

Arūnas Kučinskas
Lund Observatory, Box 43, SE-221 00, Lund, Sweden
Institute of Theoretical Physics and Astronomy, Goštauto 12, Vilnius 2600, Lithuania

Lennart Lindegren
Lund Observatory, Box 43, SE-221 00, Lund, Sweden

Toshihiko Tanabé
Institute of Astronomy, The University of Tokyo, Tokyo, 181-0015, Japan

Vladas Vansevičius
Institute of Physics, Goštauto 12, Vilnius 2600, Lithuania

Abstract

We discuss the tracing of star formation histories with ESA's space astrometry mission GAIA, emphasizing the advantages of AGB stars for the purpose. GAIA's microarcsecond-level astrometry, multi-band photometry and spectroscopy will provide individual distances, motions, $T_{\rm eff}$, $\log g$ and [M/H] for vast numbers of AGB stars in the Galaxy and beyond. Reliable ages of AGB stars can be determined to distances of $\sim$200 kpc in a wide range of ages and metallicities, allowing star formation histories to be studied in a diversity of astrophysical environments.

1. The GAIA mission, an astrometric and spectrophotometric survey of the Galaxy

The European Space Agency's GAIA mission, approved for launch in 2010–12, aims at surveying the Galaxy to 20th visual magnitude, using a combination of astrometric measurements (for trigonometric parallaxes and proper motions), multiband photometry (for basic stellar parameters like temperature and metallicity), and radial-velocity measurements. Targeted accuracies versus magnitude allow direct distances and motions to be obtained for large samples of intrinsically bright stars across the Galaxy and in some nearby Local Group galaxies. Expected typical accuracies are shown in Table 1. In total more than 1 billion stars

Y. Nakada et al. (eds.), Mass-Losing Pulsating Stars and their Circumstellar Matter, 39–42.

Table 1. Predicted accuracies, versus V magnitude, of individual AGB stars observed by GAIA. Standard errors estimated by Lindegren (unpublished, 2002) for parallax (π) and proper motion (μ), by Katz & Munari (2002) for radial velocity (v_r), and by Vansevičius et al. (2002) for photometrically derived $T_{\rm eff}$, $\log g$ and [M/H]. The last column is the maximum distance of an AGB star ($M_V \simeq -2$) at the given accuracies.

V mag	$\sigma(\pi)$ μas	$\sigma(\mu)$ μas yr^{-1}	$\sigma(v_r)$ km s^{-1}	$\sigma(\log T_{\rm eff})$	$\sigma(\log g)$	σ([M/H])	$d_{\rm max}$ kpc
15	13	8	1.1	0.007	0.20	0.24	25
17	32	18	6.3	0.01	0.27	0.32	60
19	90	50	–	0.04	0.60	0.63	150
20	160	90	–	0.13	1.1	1.3	250

will be observed, of which 50–100 million will obtain individual parallax distances to better than 5 per cent. A primary science goal is to study formation, evolution and structure of the Galaxy, for which large-scale mappings of star formation histories are essential. For a full description of the very broad range of science goals see Perryman et al. (2001).

In its present design GAIA comprises two astrometric instruments, with 1.4×0.5 m^2 apertures and a combined 0.5 deg^2 field of view, and a separate photometric/spectroscopic instrument with a 0.5×0.5 m^2 aperture. The latter performs photometry in $\sim$11 bands for astrophysical classification, and $R \sim 10^4$ spectroscopy in the 849–874 nm wavelength range, mainly for radial velocities. During its lifetime of at least 5 years, the satellite will scan the entire sky repeatedly, so that each object is observed at multiple epochs. The above numbers and accuracy predictions refer to the recently (May 2002) completed revised design, aiming at a substantially reduced mission cost compared with the previous baseline (Perryman et al. 2001), while preserving all science goals intact.

2. Tracing stellar populations using AGB stars

The availability of precise photometry is essential for age derivations using isochrone fitting to the main sequence turn-off (MSTO) point. Simulations of GAIA photometry demonstrate that this method may be successfully exploited with GAIA even in such distant stellar systems as the Magellanic Clouds (Kučinskas et al. 2002), but only for populations younger than $\sim$ 1 Gyr. In this paper we argue that GAIA observations of AGB stars can be used to determine star formation histories to even greater distances and for much older populations.

GAIA will provide a wealth of astrometric and spectrophotometric data on galactic and extragalactic AGB stars. Their uses are at least twofold: (a) as kinematic tracers, using distances and space motions obtained from the astrometric and radial-velocity data; (b) for age determinations, using basic stellar-atmosphere parameters ($T_{\rm eff}$, $\log g$ and [M/H]) derived from the spectrophotometric data, combined with distances and theoretical isochrones.

From the astrometric and radial-velocity accuracies in Table 1 it is obvious that GAIA will yield accurate distances ($< 10\%$) and full space velocities (< 1 km s^{-1}) for individual AGB stars up to distances of 10–15 kpc, if no interstellar extinction is present.

Extensive simulations by the Vilnius GAIA group (Vansevičius et al. 2002; Kučinskas et al. 2002) show that GAIA will also provide precise metallicities ($\sigma([\rm M/H]) \leq 0.3$) and gravities ($\sigma(\log g) \leq 0.3$) for AGB stars brighter than $V \sim 17$ (Table 1). Precise effective temperatures ($\sigma(\log T_{\rm eff}) \leq 0.04$) are derived down to $V \sim 19$. This holds within a broad range of metallicities ($[\rm M/H] > -2$) and ages (0.05–15 Gyr).

Metallicity estimates of intermediate age and old stellar populations can also be obtained from the slope of the red giant branch (e.g. Ferraro et al. 2000). Our simulations show that the method could provide an independent estimate of [M/H] with GAIA, effective up to distances of ~ 200 kpc, if no interstellar extinction is present (Kučinskas et al. 2002).

We have recently shown (Kučinskas et al. 2000) that reliable ages can be derived using isochrone fits to the AGB sequences on the observed HR diagram. It is essential for this procedure to have precise effective temperatures of the AGB stars, which can be derived by fitting synthetic spectral energy distributions to observed photometric fluxes (e.g., *BVRIJHK*). The method was successfully tested and compared with the MSTO method on a sample of populous star clusters in the Magellanic Clouds spanning a wide range of ages (Table 2 and Fig. 1). For galactic AGB stars, it is clear that the distance information needed to construct the observational HR diagrams will be available through GAIA. It thus appears that precise age estimates ($\sigma(\log t) < 0.3$) can be obtained for a wide range of ages (0.05–10 Gyr) and metallicities ($[\rm M/H] > -2$).

3. Conclusions

GAIA will provide unique astrometric and photometric data for studying individual and collective properties of stars in the Galaxy and its surroundings. AGB stars, being intrinsically bright, will provide precise individual distances, kinematics, $T_{\rm eff}$, $\log g$ and [M/H] up to distances of 10–15 kpc. Using isochrone fitting to the AGB stars will give reliable

Table 2. MSTO and AGB ages for a sample of LMC and SMC clusters. AGB ages (Kučinskas et al., in preparation) were derived using the same cluster metallicities (Col. 2) as for the MSTO estimates.

Cluster	[Fe/H]	MSTO	AGB
LMC:			
NGC 1783	−0.4	0.9 ± 0.4	0.8 ± 0.2
NGC 1846	−0.7	–	1.3 ± 0.3
NGC 1978	−0.7	2.0 ± 0.2	1.5 ± 0.5
NGC 1987	−0.4	–	1.0 ± 0.2
NGC 2121	−0.7	3.2 ± 0.5	3.5 ± 0.5
SMC:			
Kron 3	−1.3	8.0 ± 0.3	9 ± 3
NGC 152	−1.0	0.8	1.2
NGC 419	−0.7	1.2 ± 0.5	1.4 ± 0.2

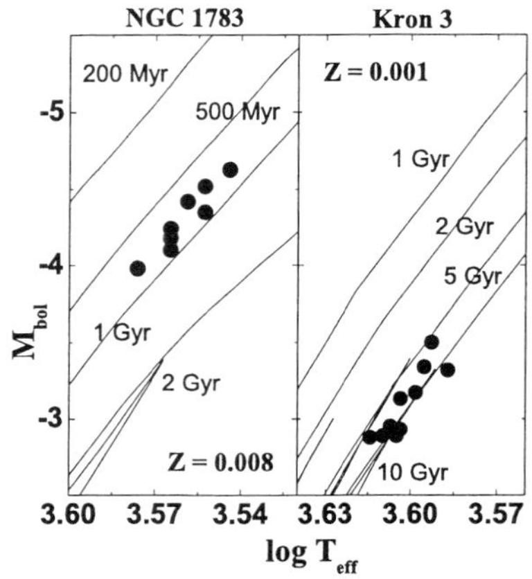

Figure 1. AGB sequences in NGC 1783 and Kron 3. Isochrones are from Bertelli et al. (1994).

ages ($\sigma(\log t) < 0.3$) for a wide range of ages and metallicities. If distances are known by other means (e.g. in distant clusters), the method can be used up to $\sim$ 200 kpc. Thus, AGB stars will allow the formation histories and kinematics of stellar populations to be probed in a diversity of astrophysical environments both in the Milky Way and in neighbouring galaxies.

Acknowledgments

The work was supported by a grant (NB00-NO30) of the Nordic Council of Ministers and by a grant of the Wenner-Gren Foundations. AK thanks the Workshop organisers for financial support to attend the event.

References

Bertelli, G., Bressan, A., Chiosi, C., Fagotto, F., & Nasi, E. 1994, A&AS, 106, 275

Ferraro, F.R., Montegriffo, P., Origlia, L., & Fusi Pecci, F. 2000, AJ, 119, 1282

Katz, D., & Munari, U. 2002, GAIA RVS Status Report, RVS-CoCo-004 (`http://wwwhip.obspm.fr/gaia/rvs/bibliography/RVS-CoCo-004.txt`).

Kučinskas, A., Vansevičius, V., Sauvage, M., & Tanabé, T. 2000, in ISAS Report SP 14, Mid- and Far-Infrared Astronomy and Future Space Missions, eds. T. Matsumoto & H. Shibai, 51

Kučinskas, A., Bridžius, A., & Vansevičius, V. 2002, Ap. Space Sci., 280, 159

Perryman, M. A. C., de Boer, K. S., Gilmore, G., Hog, E., Lattanzi, M. G., Lindegren, L., Luri, X., Mignard, F., Pace, O., & de Zeeuw, P. T. 2001, A&A, 369, 339

Vansevičius, V., Bridžius, A., & Drazdys, R. 2002, Ap. Space Sci., 280, 31

CHARACTERISTICS OF AGB STARS IN THE ISOGAL, MACHO AND OTHER DATABASES

I.S. Glass
South African Astronomical Observatory, PO Box 9, Observatory 7935, South Africa
isg@saao.ac.za

Abstract This contribution summarizes briefly some previous findings based on ISOGAL, MACHO and DENIS observations and goes on to include outlines of two projects now in progress:

(i) A complete sample of M stars in the Baade's Window NGC6522 field is being investigated using the MACHO, ISOGAL, DENIS and 2MASS databases to determine how variability and mass-loss are related to spectral sub-type.

(ii) The (MACHO) light curves of the defining stars of the infrared LMC mira period-luminosity relations are being analysed to check on their periods and to determine whether new information can reduce the scatter.

Keywords: Miras, semi-regular variables, ISOGAL, DENIS, MACHO, 2MASS

Introduction

The ISOGAL experiment surveyed many regions of the Galactic Plane and Inner Bulge at 7 and 15μm. MACHO has produced precise b- and r-band light-curve information for of order 10^7 stars in each of the Large Magellanic Cloud and the Galactic Bulge. Near-IR surveys have been made of these regions by 2MASS and DENIS. In addition, Hipparcos has produced light curves and distances for stars in the solar neighbourhood. The new data bases that have resulted offer new opportunities for studying the statistical properties of many types of stars at known distances.

In the region of the Baade's Windows of low absorption towards the Bulge, ISOGAL results have already demonstrated that many AGB stars show mass-loss rates comparable to those of short-period Miras (see *inter*

Y. Nakada et al. (eds.), Mass-Losing Pulsating Stars and their Circumstellar Matter, 43–46.

alia Glass et al, 1999)). Almost all the objects detected by ISOGAL appear as late M-type semi-regular variables (SRVs) in the MACHO database (Alard et al 2001, Glass and Alves, 2000). The variables are well-distributed in period from 10 to over 200 days and have amplitudes from 0.1 to 0.4 mag, too small to have been detected using the older photographic techniques. They are about twenty times as numerous as the Miras in the same field. Significant mass loss in the range 10^{-8} to 5 $\times$ 10^{-7} $M_\odot$ yr^{-1} may occur in variables with periods of 70 d or longer. Together with the Miras, they SRVs occupy the upper part of the K, $J-K$ colour-magnitude diagram (Schultheis & Glass, 2001). In the $J-H, H-K$ diagram, they form an extension to the sequence starting at earlier spectral types, but the miras stand out with significantly lower $J-H$, caused by water-vapour absorption.

1. Complete survey of M-giant stars

A complete sample consisting of all M-giants of subtypes from M1 to M8 within an annular region near the cluster NGC 6522 was defined by Blanco (1986). It contains 174 stars, positions for which were found from the original finding chart and the Digitized Sky Survey. All but two were identified in the DENIS Catalogue. Forty-seven are seen to be variables with amplitudes above $\sim$ 0.03 mag in the MACHO database.

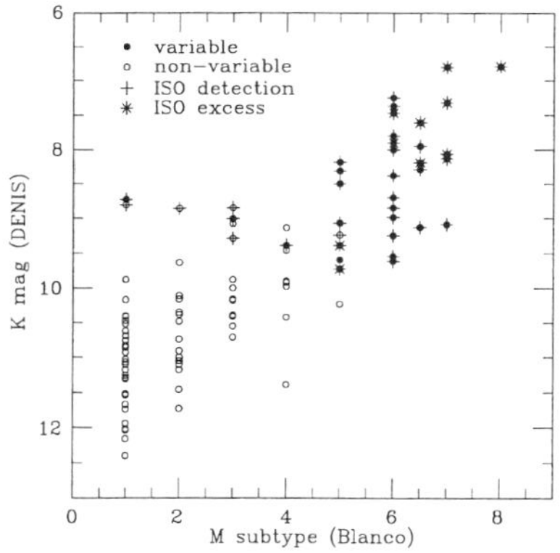

Figure 1 DENIS K mags of Blanco M-type giants from the region where the ISOGAL field overlaps the Blanco complete sample. Variables are distinguished from non-variables. ISOGAL detections are indicated as are stars with infrared excesses at 15μm.

The Blanco survey annulus is partly overlapped by the ISOGAL NGC 6522 field. In Fig 1 the DENIS K mags and Blanco spectral types for the overlap region are shown. This contains 101 of the 174 Blanco stars. The stars detected by ISOGAL, numbering 37, are indicated by superimposed crosses (+). These are predominantly of late spectral type, all those of M6 or later being detected. This is partly the effect of the ISOGAL faint end cut-off, but may also be affected by the increased K - [7] colour of later subtypes.

The ISOGAL detections showing [7] - [15] excesses, i.e., values > 0.5, have, in addition, superimposed × signs. Most of them are of late subtype and are very luminous.

The SRVs with single and double periods have been plotted on a K, log P diagram (Fig 2) for comparison with Wood's (2000) LMC SRV sequences (A, B & D). While they cover the same region of the diagram, they do not fall into Wood's neat sequences. The finite depth of the Bulge is one reason why scatter is expected (see also discussion below).

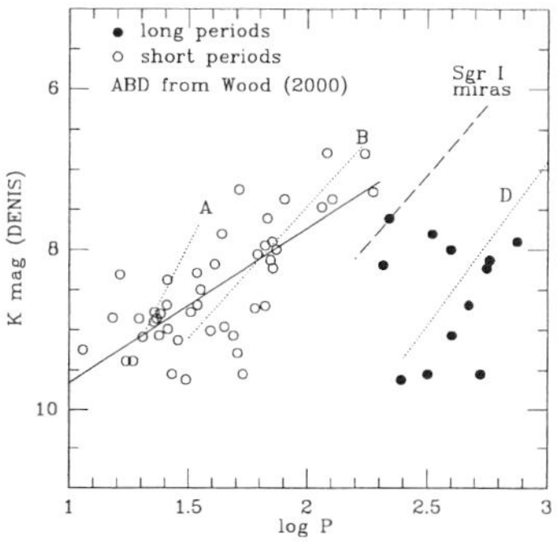

Figure 2 K, log$P$ diagram for SRVs in the NGC 6522 field. The solid line is a fit ($K = (-1.93 \log P + 11.6)$ to the data. The other lines are transformed SRV loci from Wood (2000) and miras from Glass et al (1985).

2. LMC mira P-L relation

The mira log P-L and log P-K relations (Glass & Lloyd Evans, 1981) are still best defined by the set of measurements reported by Feast et al (1989). The least dispersion of ±0.13 mag is observed for oxygen-rich miras in the K band. Somewhat larger values are found when carbon miras are included or the bolometric luminosities are considered. About 75% of the stars used in formulating these relations have b and r photometry in the MACHO database. Detailed light curves show that three of them are, in fact, SRVs, having amplitudes too low to be miras. However, the previous periods were found to be, in general, accurate enough that refining them (and elimating the SRVs) does not significantly alter the period-luminosity and period-K relations.

Most of the scatter in the relations arises from fluctuations in the average magnitude of a mira from cycle to cycle which remain within bounds estimated to be about ±10% from the long-term average at K. It could be the case that long-term monitoring will lead to an even tighter relation than exists at present.

The r amplitudes of the LMC carbon miras are less than those of the O-rich stars with comparable periods. The amplitudes of O-rich stars match those of similar period in the Sgr I Bulge field, but their period distribution is more extended towards the shorter end, where the amplitudes are relatively low, tending towards values associated with galactic SRVs.

Acknowledgments

The author thanks Dr M. Schultheis for his help and acknowledges the contributions of colleagues in the ISOGAL, MACHO and DENIS projects. He has made use of data from 2MASS and the Digitized Sky Survey.

References

Alard, C. et al, 2001, ApJ, 552, 289
Blanco, V.M. 1986, AJ, 91, 290
Feast, M.W., Glass, I.S., Whitelock, P.A., & Catchpole R.M. 1989, MNRAS, 241, 375
Glass, I.S., & Alves, D.R. 2000, in Lecture Notes in Physics 548, ISO Surveys of a Dusty Universe, eds. Lemke, D., Stickel, M. & Wilke, K. (Berlin: Springer), 363
Glass, I.S. et al., 1999, MNRAS, 308, 127
Glass, I.S., Whitelock, P.A., Catcpole, R.M., & Feast, M.W. 1985, MNRAS, 273, 383
Glass, I.S., & Lloyd Evans, T. 1981, Nature, 291, 303
Schultheis, M., & Glass, I.S. 2001, MNRAS, 327, 1193
Wood, P.R. 2000, Proc Astr Soc Australia, 17, 18

Discussion

Cioni: The red objects with long periods seem to lie on sequence D instead of on a vertical sequence in the PL plot as shown by Wood for the MSX sources. Do you interpret these sources on sequence D as Wood [does] or do you have a different explanation?
Reply: The MSX stars referred to have well-developed shells. The SRVs that I discuss have thin, undeveloped, shells; i.e., they are very different. Those near the D sequence are doubly-periodic objects as described by Wood at Montpellier for the LMC.
Chiang: How did you pick up the HBB stars: just because they are above the PL line?
Reply: Basically yes. However, some of them have spectroscopy and we see lithium enrichment and other evidence of HBB.
Feast: If part of the scatter in the mira P-L is due to to 'jitter' in the light curve then the 'real' PL must be very narrow.
Wood: Comment on above: The (K, log$P$) relations for sequences A and B, which are very small amplitude pulsators, show a large scatter. This seems at variance with a very small scatter in the mira (K, log$P$) relation. Indeed, we might expect a scatter due to variation in total stellar mass (and metal abundance).

II

PULSATION AND PROPERTIES OF MASS-LOSING STARS

STRUCTURE AND EVOLUTION OF AGB STARS: CURRENT UNDERSTANDINGS AND BEYOND

Masayuki Y. Fujimoto, Takuma Suda
Department of Physics, Hokkaido University
Sapporo, 060-0810 JAPAN
fujimoto@astro1.sci.hokudai.ac.jp: suda@astro1.sci.hokudai.ac.jp

Abstract

The present knowledge on the AGB stars is reviewed concerning the theory of internal structure, evolution and chemistry. The realization of AGB structure is elucidated from the viewpoint of gravothermal properties of gas spheres, including those with an ONeMg core. Recent developments in respect to the formation of carbon stars and the s-process nucleosynthesis are presented with focus on the extremely metal-poor stars, which may give insights into the formation and early evolution of our galaxy by these studies. Finally, we discuss the importance of non-standard effects in understanding the actual evolution of AGB stars and introduce some recent attempts to reveal their roles in the internal mixing and nucleosynthesis.

Keywords: Stellar evolution — asymptotic giant branch — carbon stars — s-process nucleosynthesis – elemental mixing

1. Introduction

The asymptotic giant branch (AGB) stars are composed of an electron-degenerate core and an extended envelope with deep surface convection, separated by the hydrogen and/or helium shell burnings. These shell burnings become active alternatively to supply a large luminosity near the Eddington limit of core, and in particular, the helium shell burning makes thermal runaways when it becomes active. Iben (1975) shows that after the helium shell flashes, the helium zone expands and the surface convection deepens to dredge up the nuclear products of flashes to the surface. This proves the formation mechanism of carbon stars and reveals the site of s-process elements for the first time. Since then, the

Y. Nakada et al. (eds.), Mass-Losing Pulsating Stars and their Circumstellar Matter, 51–58.

mixing and nucleosynthesis in the AGB stars has been attracting wide interest (e.g., Iben & Renzini 1983, Lattanzio & Boothroyd 1997).

The AGB stars are thought to end as white dwarfs, but not as supernovae, as the envelope is stripped off by the mass loss wind. This is inferred from the problem of the overproduction of iron, otherwise expected, although a reliable theory of mass loss is yet to be established. The mass losing AGB stars eject the nuclear products of helium shell flashes, dredged up to the envelope, and play important roles in the chemical evolution of CNO and light elements and the s-process elements. With low surface temperatures, they may be an substantial source of interstellar molecules and dusts. On the way to white dwarfs, an attractive spectacle of planetary nebula is provided as the final luminous stage.

The evolution of AGB stars has been one of the central subjects, and many reviews have been published on their characteristics and their roles in the evolution of galaxies (e.g., Wallerstein & Knapp 1998, Busso, Gallino & Wasserburg 1999). In this review, we restrict our discussion to the topics that have been seldom considered and are newly revealed. In particular, we present recent developments made in relation to the extremely metal-poor stars in the galactic halo, which are expected to give clues to understand the formation and evolution of our galaxy. On the other hand, there is growing evidence pointing to the importance of non-standard effects, presumably due to the stellar rotation, and we present an attempt to reveal their roles.

2. Characteristics of AGB Structure

The most striking feature of AGB stars is the inflation of envelope. Figure 1 compares the radius, r, of shell as a function of the mass, M_r, interior to it for a model star of mass $7M_\odot$ in various stages. As the star evolves to contract the core, the envelope expands greatly to be of radius much larger than that of main sequence star. This radial expansion has been a long standing issue in the theory of stellar evolution. A large radius itself can occur for stars with low central temperatures such as proto stars. Accordingly, the problem may be properly rephrased as why stars can expand despite the increase in the central value of P_c/ρ_c.

This aspect of stellar structure may be articulated by using a characteristic plane with homology variants U and V, defined as

$$U \equiv 4\pi r^3 \rho / M_r = d \ln M_r / d \ln r; \; V \equiv G M_r \rho / r P = -d \ln P / d \ln r \quad (1)$$

(Hayashi et al. 1962, Sugimoto & Nomoto 1980). The structures on the U-V plane are shown in Figure 2 for the same models as in Fig. 1, which are clearly distinguished from their morphology; the main sequence star, as well as the polytropes, display a monotonic variation from the center ($U = 3$ and $V = 0$) to the surface ($U = 0$ and $V = \infty$), while the stars of advanced stages have a loop on the way (Sugimoto & Fujimoto 2000). The existence of such solutions with different topologies is characteristic of the self-gravitating systems, and their structures may be classified by the number of loops along the U-V curve.

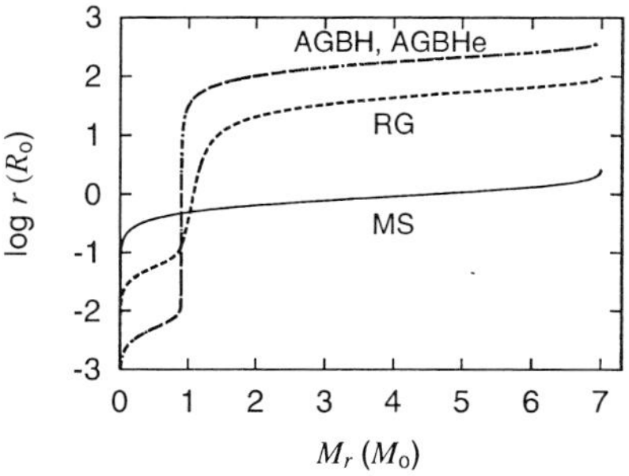

Figure 1. Radius, r, of shell against the mass, M_r, interior to it for the main sequence (MS), the red giant (RG), the AGB with hydrogen shell burning (AGBH) and the AGB with helium shell burning (AGBHe).

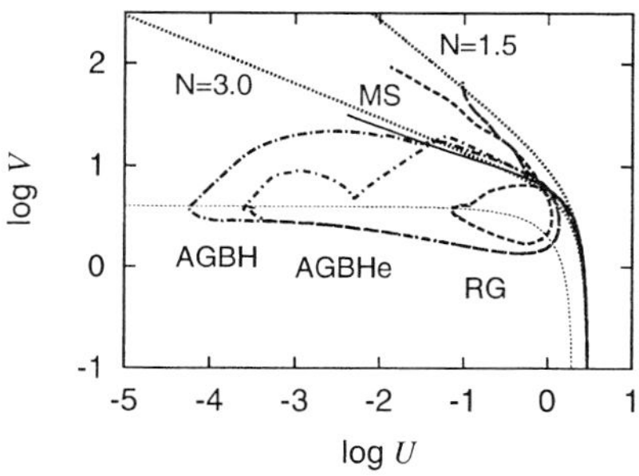

Figure 2. U-V curves for the same models as in Fig. 1. Thin and thick dotted lines denote the critical line $2U+V-4 = 0$ and two polytropes with the different indices N, respectively.

The transition from the dwarf-like (without a loop) to the giant-like (with a loop) structures is consequent upon the gravothermal contraction and the shell burning. The former augments P_c/ρ_c while the latter restrains changes in P_1/ρ_1 at the burning shell because of strong temperature dependence of nuclear burning rates. Furthermore, the shell burning demands the negative heat capacity in the shell for the stability, and hence, $V_1 \simeq N+1 \simeq 4$. Along with the gravitational contraction, therefore, the distance in P/ρ between the center and the burning shell is enlarged, which entails smaller U_1 and makes the loop bigger, as seen from the comparison between the red giant and AGB stars.

The change in the radial distance is described on the U-V plane as

$$d \ln r = d \ln(V/U)/(2U + V - 4). \tag{2}$$

Consequently, the excursion of U-V curves along the loop yields the radial inflation. The physical interpretation is as follows: The core contraction causes the decrease in the density and pressure at the burning shell, and hence, in order to support the weight of envelope, the shell burning is enhanced to expel the bulk of envelope into the region at larger distance from the center and at weaker gravity. Outside the loop, the U-V curves tend to trace those of dwarfs, which implies that the bulk of envelope in the red giant and AGB stars has configurations homologous to the dwarfs. In other words, the difference in the envelope characteristics between the AGB stars and the proto stars of same radius may stem from the inner boundary conditions.

3. Additional Evolutionary Path to AGBs

Most of the studies on the AGB star to date have been concerned with those with a CO core, evolved from the stars of low and intermediate masses ($M < 8M_\odot$). It has been shown, however, that the AGB configuration occurs with a ONeMg core and the carbon shell burning for stars of masses $9 \sim 11M_\odot$ (Garcia-Berro & Iben 1994, Ritossa, Garcia-Berro & Iben 1999 and references therein).

For these stars, carbon burns in non-explosive fashion and the ONeMg core contracts to get electron degenerate during the carbon shell burning. At the same time, the envelope exterior to the carbon burning shell expands and most of the helium zone is dredged up by the surface convection. When the carbon shell burning is extinguished, the core mass interior to the hydrogen burning shell reduces to $1.12 \sim 1.37M_\odot$, smaller than the Chandrasekhar mass limit, and the stars enter into the thermal pulsating AGB (TPAGB) phase with the hydrogen and helium double shell burnings.

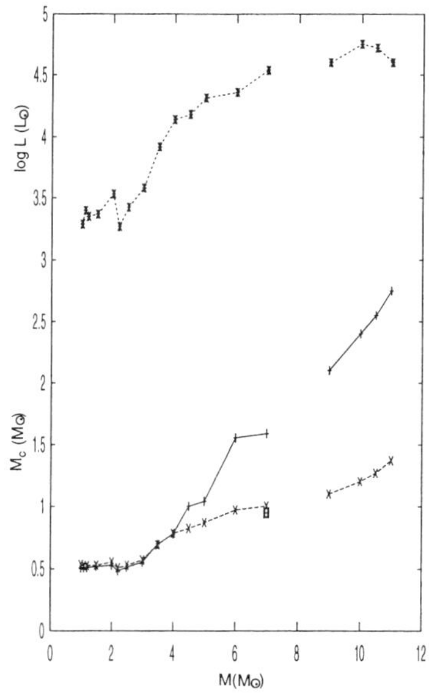

Figure 3 The luminosity (top) and the core mass (bottom) at the onset of TPAGB phase as a function of the initial mass stars of the solar metallicity (Iben 1975, Castellani et al. 1990, Straniero et al. 1997, Ritossa et al. 1999). Also plotted is the mass of helium core before the second dredge-up (middle).

The properties at the onset of TPAGB phase are summarized in Figure 3 as a function of initial masses, although they may be subject to uncertainties in the treatment of overshooting beyond the edge of convective core during the main sequence phase. The AGB stars with an ONeMg core have more massive core and larger luminosity than those with a CO core, which can be an alternative to the hot bottom burning in interpreting the luminous O-rich stars. If the envelope is ejected by the super wind, the former evolve to the planetary nebulae more massive and luminous with larger amount of mass and helium ejected.

4. Carbon Stars and s-Process Nucleosynthesis

It is well known that the carbon stars are formed by the third dredge-up during the TPAGB phase (Iben 1975); this works for stars of $M \geq 1.5 M_{\odot}$, which brings about large luminosities and large ratios of $^{12}C/^{13}C$ and is attendant with the s-process elements. The mass transfer from the companions in the close binary systems that have experienced the third dredge-up is also the accepted mechanism for the formation of Ba II stars and CH stars for the populations I and II.

Recently, a large number of carbon stars are discovered among the extremely metal-poor stars (EMPS) with the metallicity below [Fe/H] $\simeq -2.5$ in the HK survey (Beers et al. 1992). They exhibit extremely large carbon enrichments up to [C/Fe] ~ 2 and great enhancement in nitrogen also, while s-process elements are enriched in some, but not in all (see Aoki et al. 2002 and the references therein). These extremely metal-deficient carbon stars account for 25% of HK samples for [Fe/H] ≤ -3 (Rossi et al. 1999), while the CH stars do only for a few percents.

The EMPSs have been attracting wide interest as probes into the chemical and dynamical evolution of early Galaxy, for which critical is the understanding of their nature. These facts indicate that the predominant mechanism for carbon stars may be different from that for stars of younger generations. For EMPSs, another mechanism is known, triggered by the extension of convection during the helium core and/or shell flash through the overlying hydrogen-containing layers. This is made possible by the smaller entropy in the hydrogen burning shell for the smaller metallicity (Fujimoto et al. 1995). The formation mechanisms of carbon stars are classified according to the initial mass and the metallicity, as shown in Figure 4; the EMPSs develop carbon and nitrogen rich surface abundance during the helium core flash (case I) and at the very beginning of TPAGB phase (Case II and II′), earlier than the metal-richer stars (Fujimoto et al. 2000).

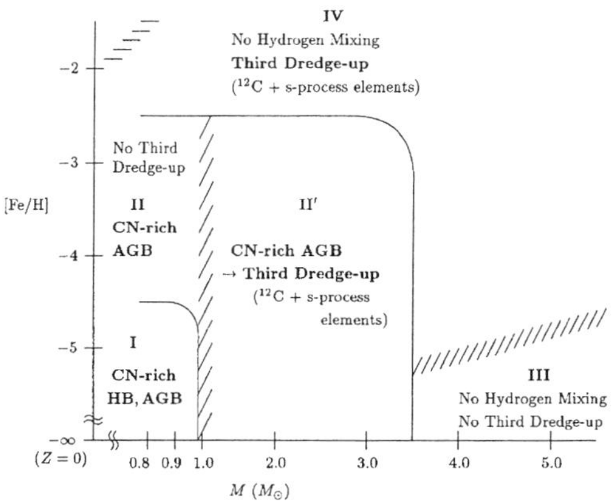

Figure 4 Evolution to carbon stars on the initial mass and metallicity plane (Fujimoto et al. 2000).

As for the s-process nucleosynthesis, $^{13}C(\alpha, n)^{16}O$ is regarded as the promising neutron source. But this source demands some kind of extra mixing operating at the base of convective envelope during the dredge-up phase to carry down hydrogen into the C-rich layer and form a ^{13}C pocket, as first suggested by Iben & Renzini (1982, 1983). Straniero et al. (1997) show that ^{13}C burns to ^{16}O under the radiative condition during the interpulse phases. Gallino et al. (1998) argue by parameterizing the size of the ^{13}C pocket that the radiative ^{13}C burning can reasonably reproduce the solar distribution of s-process element.

An important prediction from this radiative ^{13}C burning model is the metallicity dependence of s-process; the ratio of Pb/Ba has to increase for lower metallicity because of the decrease of seed nuclei (Gallino et al. 1998). Indeed Pb/Ba $\simeq 100$ is observed for three stars ([Fe/H] $= -1.67$, -1.7 and -2.45, Van Eck et al. 2000). But Aoki et al. (2000, 2001) find Pb/Ba $\simeq 1$ from two stars of [Fe/H] $= -2.71$, and -2.74. This indicates a different mechanism for the EPMSs. Indeed, Iwamoto et al. (2002 preprint) propose the convective ^{13}C burning, extending the above results by Fujimoto et al. (2000), in which engulfed hydrogen is diluted in the helium flash convection and provides right orders of neutron exposure to reproduce the observed patterns of s-process elements.

These evolutionary results, assisted with the binary mass transfer, give clues to understand the EMPSs, which exhibit large variations in the enhancement both for carbon and the neutron-capture elements (e.g., see Aoki et al. 2002).

5. Prevalence of Non-Standard Mixing

One of the central issues in the stellar evolution is the existence of non-standard mixing other than the thermal convection. There is the growing evidence pointing to the inward mixing of hydrogen in the radiative zone, such as the short recurrent periods of X-ray bursts, the enhancement of CNO and He in nova ejecta and the abundance anomalies of Mg and Al in globular cluster giants. The neutron source for radiative ^{13}C burning is also the case.

Among these are R-type carbon stars, which are low luminosities ($L \sim 100\ L_\odot$, Scalo 1976) but show no evidence for variations in the radial velocity, different from Ba II stars (McClure 1997). They have rather low carbon isotopic ratio ($^{12}C/^{13}C \leq 10$) and lack the enhancement of s-process elements, typical of the Ba II and CH stars (Dominy 1984). These properties cannot be accommodated to the mechanisms, discussed above, within the standard framework of stellar evolution, suggesting that the non-standard mixing may play some part in their formation.

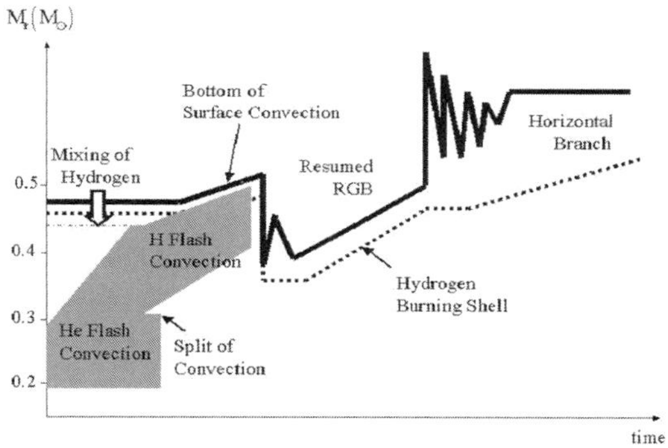

Figure 5. A scenario for the R-type carbon stars, postulating the engulfment of hydrogen by the helium flash convection.

We propose a new scenario for the origin of R-type carbon stars, postulating the inward mixing of hydrogen by some extra mechanism into the upper helium core (Suda & Fujimoto 2002, preprint). The progress of evolution is illustrated in Figure 5. At the tip of red giant branch, the helium core flash is ignites to drive a convection outward to erode hydrogen mixed down over a few pressure scaleheights. Hydrogen engulfed is carried down to the middle of helium convection to burn mainly via $^{18}O(p,\alpha)^{15}N(p,\alpha)^{12}C$. This powers the hydrogen shell flash and splits the convective zone into two. The upper convection, driven by the hydrogen shell flash, elevates the entropy much higher, which makes it possible for the surface convection to penetrate into their site and to dredge up carbon and nuclear products to the surface. Solid and dotted lines in the figure denote the loci of the bottom of surface convection and of the hydrogen burning shell, respectively. Because of an appreciable decrease of core mass, the star climbs the red giant branch again as a carbon star. This scenario can well reproduce the observed properties of R-type carbon stars. These results indicate the importance of the non-standard mixing in the evolution of AGB stars, and further investigations are necessary with their effects included for the proper understandings of AGB stars.

6. Summary

The structures of AGB stars can be understand in terms of gravothermal nature of gas sphere and shell burning. Stars of mass $9 \sim 11 M_{\odot}$ evolve to the AGB and undergo the TPAGB phase after the central shell burning. Because of larger core masses, the AGB stars with ONeMg core can be alternative interpretation to the hot bottom burning. The standard picture is now available for the mixing and nucleosynthesis in AGBs, and yet, for the proper understandings, it is necessary to investigate the non-standard effects, e.g., by the stellar rotation. The peculiarities of AGBs in their chemistry and dynamics may provide a probe to explore the mass loss , the structure of circumstellar shells, and the parts in the evolution of galaxies.

References

Aoki, W., Norris, J.E., Ryan, S., Beers, T.C., & Ando, H. 2000, ApJ, 536, L97

Aoki, W., Ryan, S., Norris, J.E., Beers, T.C., et al. 2001, ApJ, 561, 346

Beers, T.C., Preston, G.W., & Shectman, A. 1992, AJ, 103, 1987

Busso, M., Gallino, R., & Wasserburg 1999, Ann. Rev. Astron. Astrophys, 37, 239

Carcía-Berro, E., & Iben, I. Jr. 1994, ApJ, 434, 306

Dominy, J. F. 1984, ApJS, 55, 27

Fujimoto, M.Y., Sugiyama, K., Iben, I. Jr., & Hollowell, D. 1995, ApJ, 444, 175

Fujimoto, M.Y., Ikeda, Y, & Iben, I. Jr. 2000, ApJ, 529, L25

Gallino, R., Arlandini, C., Busso, M., et al. 1998, ApJ, 497, 388

Hayashi, C., Hoshi, R. & Sugimoto, D. 1962, Prog. Theor. Phys. Suppl, 22, 1

Iben, I. Jr. 1975, ApJ, 196, 525

Iben, I. Jr., & Renzini, A. 1982a, ApJ, 259, L79

Iben, I. Jr., & Renzini, A. 1982b, ApJ, 263, L23

Iben, I. Jr., & Renzini, A. 1983, Ann, Rev. Astron. Astrophys, 21, 271

Lattanzio, J.C., & Boothroyd, A.I. 1997, in Astrophysical Implications of the Laboratory Study of Presolar Materials. Eds. Bernatowicz & Zinner (AIP), p.85

McClure, R. D. 1997, PASP, 109, 256

Ritossa, C., García-Berro, E., & Iben, I. Jr. 1999, ApJ, 515, 381

Rossi, S., Beers, T.C., & Sneden, C. 1999, in The Third Stromlo Symposium: The Galactic Halo, eds. B.K. Gibson etal. (ASP Conf. Ser. 165), 264

Scalo, J. M. 1976, ApJ, 474, 489

Straniero, O., Gallino, R., Busso, M., Chieffi, A., et al. 1995, ApJ, 229, 624

Sugimoto, D., & Nomoto, K. 1980, Space. Sci. Rev., 25, 155

Sugimoto, D., & Fujimoto, M. 2000, ApJ, 538, 837

Van Eck, S., Goriely, S., Jorissen, A., & Plez, B. 2001, Nature, 412, 793

Wallerstein, G., & Knapp, G.R. 1998, Ann. Rev. Astron. Astrophys, 36, 369

NONLINEAR PROPERTIES OF THE SEMI-REGULAR VARIABLE STARS

J. Robert Buchler
University of Florida, USA
buchler@phys.ufl.edu

Z. Kolláth
Konkoly Observatory, Budapest, HUNGARY
kollath@konkoly.hu

Abstract We demonstrate how, with a purely empirical analysis of the irregular lightcurve data, one can extract a great deal of information about the stellar pulsation mechanism. An application to R Sct thus shows that the irregular lightcurve is the result of the nonlinear interaction of two highly nonadiabatic pulsation modes, namely a linearly unstable, low frequency mode, and the second mode that, although linearly stable, gets entrained through a 2:1 resonance. In the parlance of nonlinear dynamics the pulsation is the result of a 4 dimensional chaotic dynamics.

Keywords: Stellar pulsations, Variable stars, Chaos

In the following we lump together under the label 'semiregular' *largo sensu*, the stars of RV Tau type, the Semi-Regular stars and some of the Mira variables. All of these stars have lightcurves of varying degrees of irregularity. Most of our information about them comes from amateur astronomer data bases (AAVSO, AFOEV, BAAVSS and VSOLJ) which contain data on a large number of bright stars spanning almost a century. Many have at least several decades of good temporal sampling. Unfortunately, the data are very noisy, especially around the lightcurve minima, and for our analyses we therefore have to bin, smooth and filter the data to form a time-series with equal time-intervals. Figures 1 and 2, on top, show sections of lightcurves for 4 selected stars, *viz.* R Sct (RV Tau type), R UMi (SR type), R Cyg and X Aur (both Mira type).

Y. Nakada et al. (eds.), Mass-Losing Pulsating Stars and their Circumstellar Matter, 59–66.

The questions that we are asking here are: What are the nature and cause of the irregularities? Is there one or several underlying mechanisms? Do the the lightcurve data contain any quantitative information that can be extracted and exploited? If so, how does it correlate with luminosity, and can it be used for distance measurements? Most past work (*cf.* these conference proceedings) has concentrated on extracting period-luminosity relations which sheds light on the evolutionary and pulsational status and on the stellar structure of these stars. But is there additional information in the data?

Figures 1 and 2, on the right side, display the Fourier spectra of the 4 sample stars. One might be tempted to classify R UMi and X Aur as monoperiodic and R Sct and R Cyg as biperiodic as in Kiss *et al.* (2002). However, these stars are *not multiperiodic*. Neither the amplitudes, nor the phases, nor even the frequencies are constant in time. This can best be seen in the corresponding time-frequency plots. Instead of the more common wavelet or Gabor transform we have made our time-frequency plots with a Choi-Williams kernel (Cohen 1994) which has the property of sharpening features (*cf.* also Kolláth & Buchler 1997). It is not astonishing that the *instantaneous amplitude* in the dominant peak (f_0) is seen to vary (dark corresponding to higher values on the adopted greyscale), but very interestingly the *instantaneous frequency* varies as well! In order to make the structure of the 'harmonic' region ($\approx 2f_0$) visible on the same greyscale we have scaled up the amplitudes in that region. Remarkably, the harmonic frequency does not move synchronously with the dominant frequency. Furthermore, for R Sct, the $2f_0$ power seems to switch back and forth between $2f_0$ and $2f_0+\Delta f$. A similar behavior occurs in X Aur.

None of these features of the time-frequency plots, nor the irregularity of the lightcurves for that matter, can be explained by evolution, by dust or spots, by binarity or by stochasticity, even though all of these effects can be present and influence the lightcurve. Instead, the time-frequency plots suggest a low dimensional underlying chaotic dynamics consisting of the nonlinear interaction of a few modes. We suggest that multimode might be a better label for these stars than multiperiodic.

In those cases where the amateur data and the Cadmus (private communication, and Buchler *et al.* 2002) data overlap, the time-frequency analysis gives essentially the same results - *e.g.* these fingerprints of nonlinear mode interactions are insensitive to observational noise.

We wish to stress that our nonlinear approach to the study of the pulsations of the semiregular stars, namely the *global flow reconstruction* is fully *empirical*, *i.e.* devoid of theoretical modelling. Only two working assumptions are made:

1. *the lightcurve is produced by the (deterministic) nonlinear interaction of a small number of pulsation modes*,

2. *the system is autonomous*, *i.e.* we ignore time dependence such as evolution over the span of the data.

For additional details we refer to Buchler & Kolláth (2000) and Buchler *et al.* (1996). Our assumptions imply that the star's behavior is describable by a differential system in a **physical phase space** of *a priori* unknown dimension d.

$$\frac{d\mathbf{x}}{dt} = \mathbf{F}(\mathbf{x}) \tag{1}$$

where $\mathbf{x}$ is a d-dimensional vector whose components are the phasespace variables (which could be modal amplitudes and phases, for example). For a single oscillatory mode, *e.g.* d would be equal to 2, for 2 coupled oscillatory modes $d = 4$. The involvement of a secular mode would add 1 to the dimension.

In parallel we now introduce a **reconstruction space** of dimension d_e, in which we construct successive position vectors

$$\mathbf{X}^{\mathbf{n}} = \left(s_n, s_{n-\tau}, s_{n-2\tau}, \ldots, s_{n-(d_e-1)\tau}\right),$$

using the observational data $s_n = s(t_n)$, the magnitude in our case. The quantity τ is called the delay parameter. If our assumptions are satisfied, then the temporal behavior should be captured by an evolution equation

$$\mathbf{X}^{n+1} = \mathcal{M}(\mathbf{X}^n), \tag{2}$$

provided d_e is large enough. We could also have introduced a differential system akin to 1 in this space – our $\mathbf{X}^n$ is merely a stroboscopic description of the dynamics. The map $\mathcal{M}$ is assumed to be a sum of all the multivariate monomials up to some order (usually 4) and the unknown coefficients are determined by a least squares fit from the data. (We minimize $||\mathbf{X}^{n+1} - \mathcal{M}(\mathbf{X}^n)||$ over the data set). A powerful embedding theorem assures us that the dynamics in the physical phasespace and in the reconstruction space are the same provided that d_e is large enough. Consequently, *from the study the behavior of the reconstructed system we can infer otherwise unknown properties of the physical phase space.*

Once we have constructed a map from the data set we can iterate that map and generate '*synthetic signals*' which are much longer than the observational data set. From the latter we can then compute Lyapunov exponents and the fractal dimension d_L. We consider a reconstruction successful when the synthetic signals are robust with respect to a range of smoothing parameters and a range of delay parameters τ, are stable, and when the results are independent of the embedding dimension d_e, as long as the latter is large enough. The lowest value for which we obtain robust results will be called d_e^{min}.

In Buchler *et al.* (1996) we analyzed the AAVSO data of R Sct. Here we have repeated our analysis with a richer data base obtained by combining the AAVSO, AFOEV, BAAVSS and VSOLJ data and extending the basis to date. In Fig. 4 we display the R Sct lightcurve together with a typical synthetic signal in 4D. The synthetic signal is clearly seen to capture the nature of the lightcurve. This becomes even more evident when one looks at the lightcurve data over the last 150 years (Kolláth 1990). With the extended data basis our results do not change and remain very interesting.

1. In 3D no robust reconstructions are possible. The results are not changed by going to $d_e = 5$ and 6. They are also independent of the delay parameter τ within a broad range. We conclude that $d_e^{min} = 4$.

2. One of the Lyapunov exponents is always positive, implying that the pulsation is chaotic. The fractal dimension of the attractor is $d_L \sim 3.2$. The values of the exponents and of d_L is largely independent of d_e.

3. Clearly the (Euclidean) dimension of the physical phasespace is sandwiched between $d_L = 3.2$ and $d_e^{min} = 4$. *We therefore can infer that* $d = 4$. This suggests that the lightcurve is generated by the nonlinear interaction of two vibrational modes, consistently with the time-frequency analysis.

4. When the map is linearized around its fixed point one obtains 2 spiral stability roots
$\lambda_1 = \pm i\, 0.0068 \times 2\pi + 0.0044 d^{-1}$, and $\lambda_2 = \pm i\, 0.0145 \times 2\pi - 0.0062 d^{-1}$. Because the fixed point of the map corresponds to the equilibrium state of the star, the two spiral roots corroborate the presence and excitation of two vibrational modes. Furthermore it tells us that there is a first, *linearly unstable mode* of frequency $f_0 = 0.0068$ and a second mode, *linearly stable* one, with frequency slightly greater than $2f_0$. This is in agreement with the time-frequency analysis. Note that these modal properties come from a map which was obtained through a fit to the data. None of these properties were imposed.

We conclude that there is no need for a *deus ex machina*, such as irregular convective overshoots, to explain the nature of the irregular pulsations. From our empirical data analysis we have arrived at a useful physical picture: *The irregular pulsation is the result of the nonlinear interaction of two strongly nonadiabatic pulsation modes.* A lower frequency, linearly unstable mode entrains a stable, higher frequency one through a 2:1 resonance. The unstable mode wants to grow, but shares kinetic energy with the stable one which then dissipates it, and the cycle repeats. Because of the strongly nonadiabatic nature of these modes the motion is irregular (chaotic).

The reader may wonder why the same resonant scenario gives rise to a synchronized, periodic pulsation in the classical bump Cepheids (Buchler 1993). The physical reason is that the latter are only weakly nonadiabatic, the ratio η of growth rate to frequency is of the order of 0.01. In the semiregulars the ratio of luminosity to mass is more than ten times larger, and η is of order unity. The fact that the amplitude can vary on the timescale of the period is of course a necessary (but not sufficient) condition for chaotic behavior.

The amateur astronomer data bases contain a number of semiregulars with sufficient coverage to allow the same approach. We have embarked on the analysis of these stars to see if this mechanism of a resonant interaction is shared by other (most?) semiregular stars. Preliminary results for stars such as R UMi are very encouraging and again indicate a low dimensional chaotic nature of the pulsations. We also note that similar results have been obtained with high quality data (Buchler, Kolláth and Cadmus 2002). However, because our goal is to extract quantitative information, such as dimensions, Lyapunov exponents, etc., at this stage we feel that more work is necessary to establish the robustness of the results that we have obtained.

We wish to thank the organizers for their generous support which made our participation possible. Our thanks also go to the AAVSO, AFOEV, BAAVSS and VSOLJ for allowing us to use their data. This work has been supported by NSF (AST9819608) and OTKA (T038440).

References

Buchler, J. R. 1993, in *Nonlinear Phenomena in Stellar Variability*,Eds. M. Takeuti & J.R. Buchler (Kluwer: Dordrecht), repr. from ApSS 210, 1

Buchler, J. R., Kolláth, Z., Serre, T. & Mattei, J. 1996, ApJ, 462, 489

Buchler, J.R. & Kolláth, Z., 2001, "Nonlinear Analysis of Irregular Variables", in *Nonlinear Studies of Stellar Pulsation*, Eds. M. Takeuti & D.D. Sasselov, ASS Libr. Ser., 257, 185 [http://xxx.lanl.gov/abs/astro-ph/0003341].

Buchler, J.R., Kolláth, Z. & Cadmus, R. 2002, *Chaos in the Music of the Spheres*, Proceedings of CHAOS 2001, Potsdam, Germany, (in press); [http://xxx.lanl.gov/abs/astro-ph/0106329]

Cohen, L. 1994, Time-Frequency Analysis. Prentice-Hall PTR. Englewood Cliffs, NJ

Kolláth Z., 1990, MNRAS 247, 377

Kolláth, Z. & Buchler, J.R. (1997), *Time-Frequency Analysis of Variable Star Light Curves* – in *Nonlinear Signal and Image Analysis*, *Ann. NY Acad. Sci.* 808, 116.

Kiss, L.L., Szatmary, K., Cadmus, R. R. & Mattei, J.A., 1999, A&A 346, 542

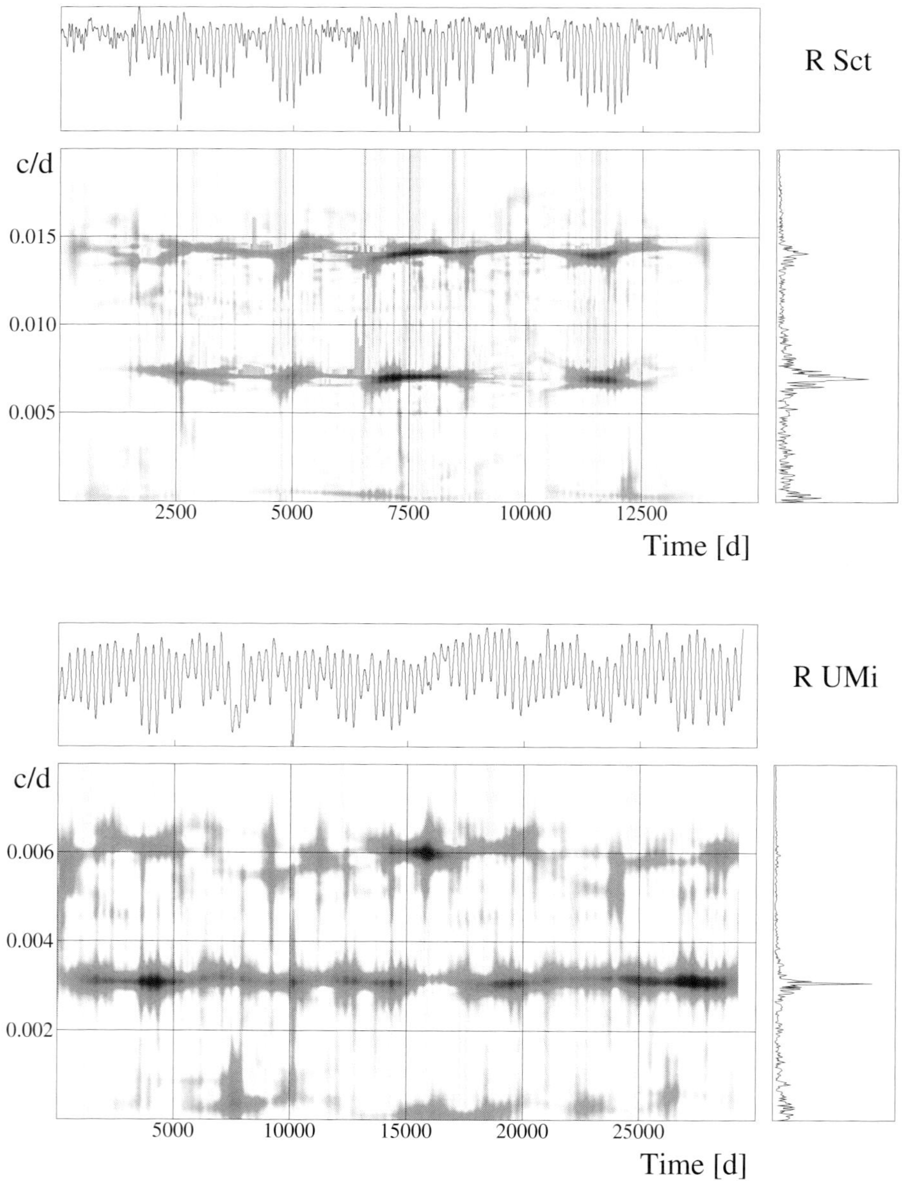

Figure 1. Time-frequency plots for R Sct and R UMi; see text for details.

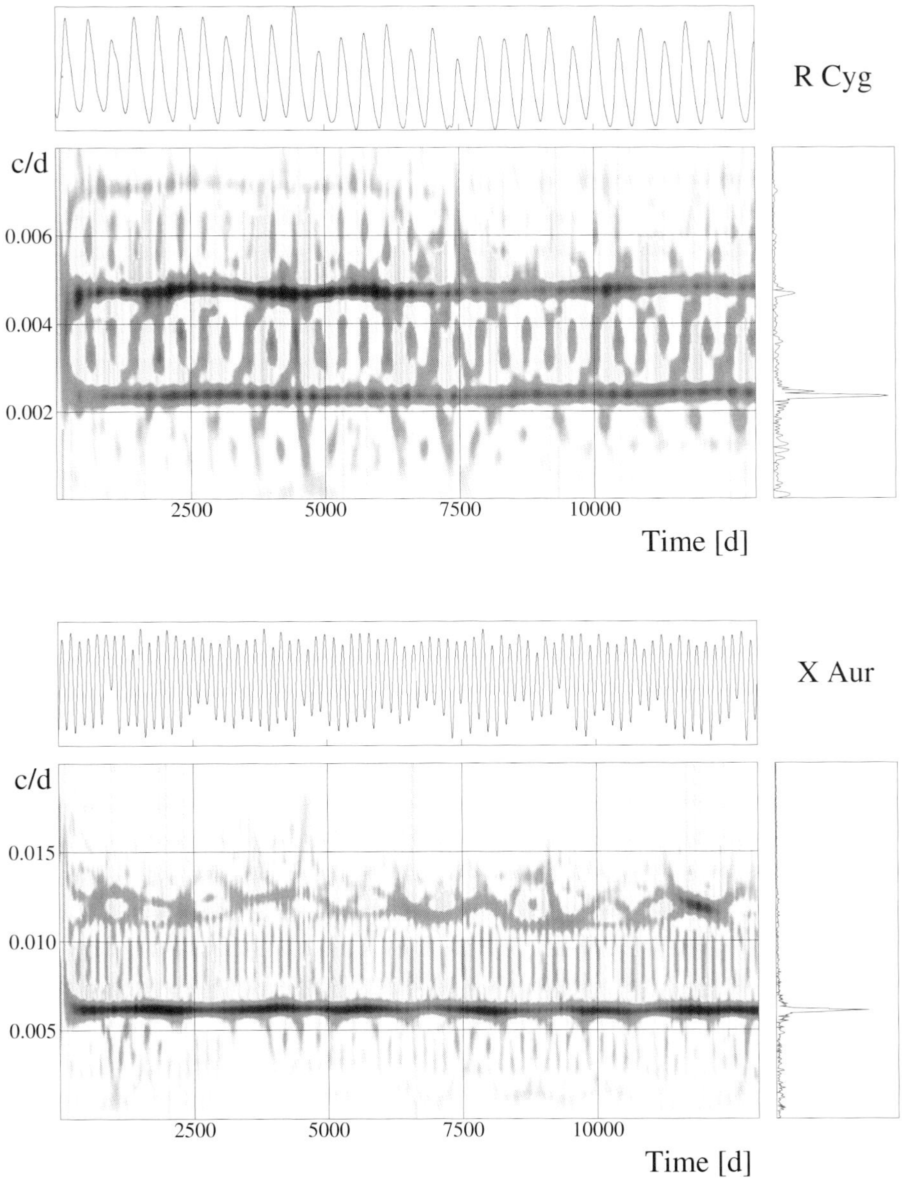

Figure 2. Time-frequency plots for R Cyg and X Aur.

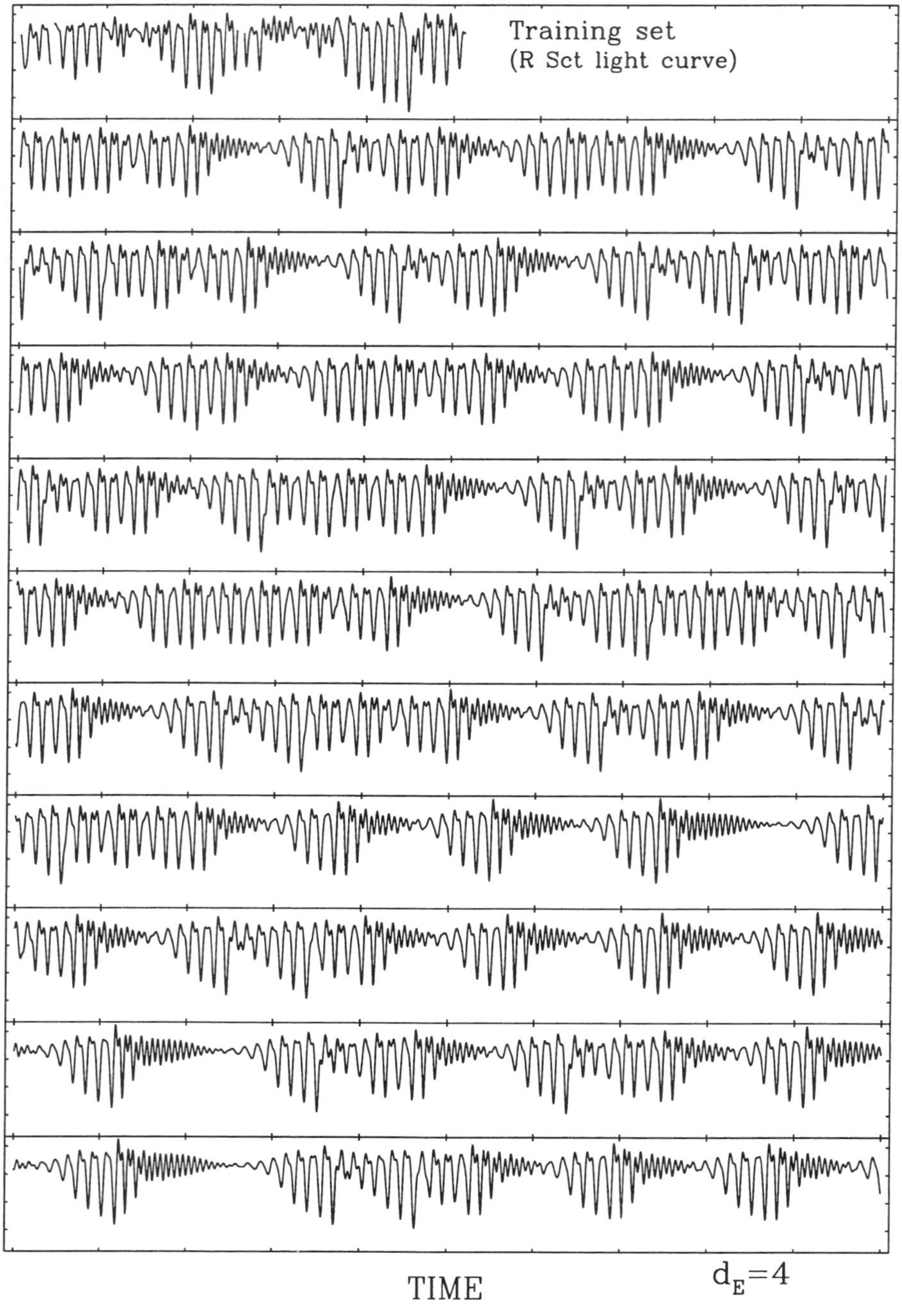

Figure 3. R Sct: *top:* section of R Sct light curve; *below:* long synthetic signal generated with a 4D map.

REGULAR AND IRREGULAR PULSATIONS OF LUMINOUS BLUE VARIABLES

Ernst A. Dorfi
Institut für Astronomie, Universität Wien,
Türkenschanzstr. 17, A-1180 Wien, Austria
dorfi@astro.univie.ac.at

Alfred Gautschy
ETH-Bibliothek, Rämistr. 101,
CH-8092 Zürich, Switzerland
gautschy@library.ethz.ch

Abstract The evolution of massive stars in the range from $20\,M_\odot$ to about $60\,M_\odot$ leads through oscillatory unstable phases during their core hydrogen as well as during the early core helium burning stages. Large-scale, non-linear pulsations with periods around one day can develop and lead to shock waves travelling through the stellar atmosphere. Ballistic motions of the outermost layers on the dynamic time scale of the atmosphere are interrupted by strong compressional shock waves. The deposition of energy and momentum leads to atmospheric stratifications that differ significantly from what prevails if hydrostatic stratification is assumed.

Keywords: Stellar pulsations, radiation hydrodynamics, variable stars

Introduction

Luminous blue variables (LBVs) are stars of great interest because they undergo the most dramatic changes in spectral type, brightness, and color of any class of single non-explosive stars (see e.g. the conference proceedings edited by A. Nota and H.J.G.L.M. Lamers (1997) on various reviews of LBVs properties). Numerous non-variable very high luminosity objects ($L \simeq 10^6\,L_\odot$) are surrounded by nebulosities hinting at phases with considerable mass loss prior to the LBV stage. The LBV class of stars includes the S Dor, P Cgyni and Hubble-Sandage variables. The non-variable as well as variable objects can be found at the same

Y. Nakada et al. (eds.), Mass-Losing Pulsating Stars and their Circumstellar Matter, 67–74.

location on the HR Diagram. Due to the short LBV phase of a few 10^4 years (based on statistical arguments of LBVs in LMC and SMC, e.g. Bohannan 1997) it is still unclear how the stars appear during their quiet phases.

The nature of LBV variability is not yet sufficiently well documented observationally. On time scales of years and months, eruption-like outbursts have been observed. However, on the time scale of a few days, also small-amplitude, quasi-regular photometric variations have been detected (e.g. Nota & Lamers (1997) for a summary on LBV properties). The small number of known LBVs – about 30 objects – together with an insufficient observational coverage makes it difficult to deduce the physical nature of either the micro-variability or the outbursts.

Also the theoretical picture of the LBV-phase is neither complete nor understood. The suggested models range from strange-mode instabilities (Kiriakidis et al. 1993) to oscillatory dynamically unstable pulsations of stars near to the Eddington limit (Stothers & Chin 1993). However, none of these models is able to explain the micro-variability, the eruptive outbursts or the stellar winds as observed during the LBV phase.

The observationally deduced mass-loss rates of LBVs not being in a major outburst range from $10^{-4}\mathrm{M}_\odot/\mathrm{yr}$ for η Car to $6 \cdot 10^{-7}\mathrm{M}_\odot/\mathrm{yr}$ for R71 in the LMC (e.g. Leitherer 1997). The terminal outflow velocities are typically around 200 km/s.

1. Pulsational computations

Starting with hydrostatic initial models calculated with a standard stellar evolution code, we have performed radial non-linear pulsation computations. For several stages of either core hydrogen burning or early core helium burning the full set of the radiation hydrodynamics equations (RHD, e.g. Mihalas & Mihalas 1984) was solved to model the dynamical evolution of the stellar envelope. The numerical method used for these non-linear pulsation simulations was described in detail by Dorfi (1998). In the dynamical computations the radiative transfer was solved with the Rosseland-mean OPAL92-opacity data for the radiative flux and the Planck-mean opacity for the corresponding radiation energy equation. The calculated structural profiles (density, temperature, etc.) served as the input for a frequency-dependent radiative transfer. After folding the emergent spectra with the filter passbands we obtained the UBVRI-light curves which will, however, not be further discussed in the current context (see Dorfi & Gautschy 2000 for more details).

Since stellar-evolution scenarios predict larger stellar masses to evolve through a LBV phase we concentrate here on two typical pulsationally

Table 1. Table of LBV model parameters

Name	M [$M_\odot$]	L [$10^3 L_\odot$]	$T_{\rm eff}$ [K]	P [days]	$\Delta R/R$	Δu [km/s]	$u_{\rm s}$ [km/s]	$\Delta m_{\rm V}$ [mag]
M45T1	45	718	17 800	-	1.543	601	213	0.29
M60H	60	933	34 680	1.611	0.368	598	211	0.27

unstable models with $45\,M_\odot$ and $60\,M_\odot$, respectively. The properties of two LBV models are summarized in Table 1. One model M60H exhibits rather regular pulsations (cf. Fig. 1) with a period P given in column 5. In contrast, the second model M45T1 develops a rather complicated pulsational pattern (cf. Fig. 2) with no clear period. Both examples are discussed in more detail in the following sections. The relative changes of the photospheric radius $\Delta R/R$ are given in column 6. Only masses higher $30\,M_\odot$ show radial oscillations with large velocity variations Δu (column 7). These motions go along with strong shock waves running through the atmosphere; the shock speed $u_{\rm s}$ is given in column 8. Finally, the last column gives the changes in the visual magnitude $\Delta m_{\rm V}$ which allows a direct comparison with observed LBVs. According to the results we emphasize the non-adiabatic nature of the pulsations where large velocity amplitudes are accompanied by small luminosity variations of the order of about 0.1 mag. For light curves in various passbands we refer to Dorfi & Gautschy (2000). Due to the dominance of the radiation pressure and the very high luminosity-to-mass ratios these peculiar LBV oscillations are essentially confined to the outer layers. From a comparison with linear stability analysis they seem to be rather regular radial modes (for a review e.g. Gautschy & Saio 1995)

2. Regular pulsations

The temporal evolution of the pulsating star M60H with initial stellar parameters of $M = 60\,M_\odot$, $L = 933\,000\,L_\odot$ and $T_{\rm eff} = 34\,680$ K is plotted in Fig. 1. The motion of different mass shells over two pulsational cycles shows that the period of the outer (observable) shell is $P = 1.611$ days, the period of variability changes as one looks deeper into the stellar interior. In contrast to this outer periodicity the deep-lying shells oscillate with half of that period, namely $P = 0.805$ days which corresponds exactly to the pulsation period obtained from a linear stability analysis. The non-linear reactions of the outermost layers induce the longer periodicity that is relevant for an observer because the motions of the upper atmosphere are dominated by two strong shock waves seen at phases 0.35 and 0.7. The outward running waves travel at speeds of $u_s = 136\,\mathrm{km/s}$ and $u_s = 211\,\mathrm{km/s}$. The corresponding Mach numbers vary between 5 and 8. After an acceleration event the mass shells move basically along ballistic trajectories that are determined by the dynamical time scale of the stellar envelope. The sound travel time through the initial hydrostatic model is 1.502 days, which is shorter than the observed period because the pulsating star has a larger mean radius.

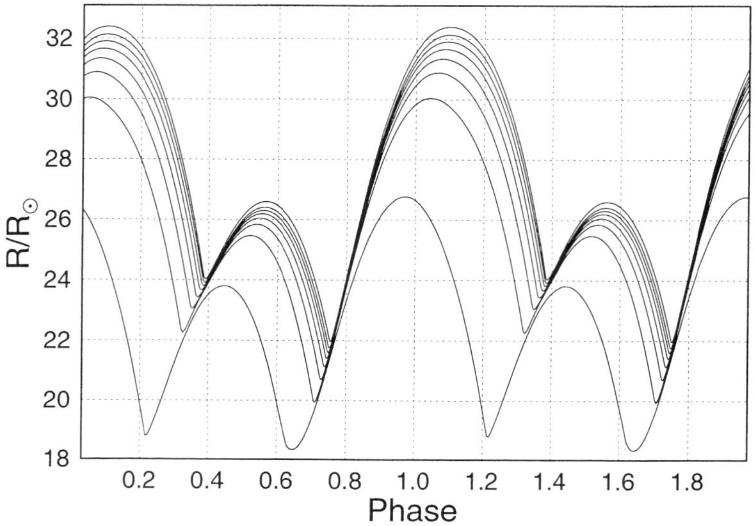

Figure 1. The non-linear oscillation of M60H with $60\,M_\odot$ is dominated by strong shock waves running through the stellar atmosphere shown by the paths of different mass shells as a function of the pulsation period of $P = 1.611$ days (see text for more details).

The strong shock compressions heat the thermal gas and lead to a subsequent expansion of the atmosphere. The photospheric radius averaged over a pulsational cycle increased from its hydrostatic value of $R_{\rm ph} = 21.3\,\mathrm{R}_\odot$ to $\bar{R}_{\rm dyn} = 25.8\,\mathrm{R}_\odot$. The relative size of the atmosphere of the initial model is $\Delta R/R_{\rm ph} = 7.8 \cdot 10^{-3}$. Due to using a Lagrangian outer boundary condition, we could easily compute the average size of the *pulsating* atmosphere to be $\Delta R/\bar{R}_{\rm dyn} = 8 \cdot 10^{-2}$. Hence, the typical scale length of the atmosphere increased by more than a factor of 10. The radiative features associated with the overall expansion are imprinted in the light curves and in the spectral line formation. In addition to this average effect, the shock waves cause additional light variations at phases 0.38 and 0.76. However, due to the low optical depth of the outermost layers, the travelling shock features do not inflict phase shifts on the light curves as observed in different passbands.

3. Irregular pulsations

As seen in the previous section, the non-linear pulsations of massive stars can been characterized by stellar atmospheres that are largely decoupled from the oscillating stellar body. The following example of a $45\,M_\odot$ star with an effective temperature of $T_{\rm eff} = 17\,800\,$K and a luminosity of $L = 718\,300\,L_\odot$ illustrates the such a behavior in an irregular pulsation.

Figure 2 plots the temporal variation of the stellar parameters over almost one year. The full lines show the physical variables at the photosphere whereas the dashed lines correspond to the outermost layers of our computations. We used an Lagrangian boundary condition to detect any possibly occurring mass loss. The stellar oscillation follows a rather complicated pattern due to the non-linear interaction of three unstable modes found by a linear stability analysis. As inferred from the figure, a large-scale motion of the outermost layers is triggered by several shock waves depositing their energy. These large-scale flows expand the atmosphere up to a maximum value of $533\,R_\odot$ starting from an initial hydrostatic stellar radius of $R_{\rm ph} = 91.5\,R_\odot$ with a a relative size of the atmosphere of $\Delta R/R_{\rm ph} = 1.8 \cdot 10^{-2}$. During the depicted time period spanning almost one year, the mean photospheric dynamical radius $\bar{R}_{\rm dyn}$ increases to $213\,R_\odot$ whereas the outermost atmospheric layer lies in the mean at $315\,R_\odot$. This corresponds a mean relative atmospheric size that balloons to $\Delta R/\bar{R}_{\rm dyn} = 0.48$, with a maximum value that tops at 1.51. The average atmospherical layers have been stretched by a factor of 27.3 or by 86 when the maximum extension is measured. The mean photospheric temperature decreases to $12\,800\,$K which is about 1300 K higher than the value given by a simple atmospheric equilibrium scaling according to $T_{\rm dyn} = T_{\rm eff}\sqrt{R_{\rm ph}/\bar{R}_{\rm dyn}}$ indicating that the atmospheric stratification is not determined by such simple equilibrium conditions. We emphasize that although the underlaying pulsation pattern is dominated by more irregular motions, the outermost stellar layers show a more regular large-scale flow given by the dynamical time scale of the atmosphere. Not every shock wave emerging from the stellar surface leads directly to a radial extension of the whole stellar atmosphere. This property is most clearly visualized by comparing the photospheric velocity (full line in Fig. 2b) with the velocity at the outer computational boundary (dashed line in Fig. 2b). The low optical depth of the atmosphere manifests itself in the comparison of the photospheric (full line) and the external temperature evolution (dashed line) shown in Fig. 2c. Both lines deviate from each other when the atmosphere allows the shock waves to pass through its larger extension. This overall fea-

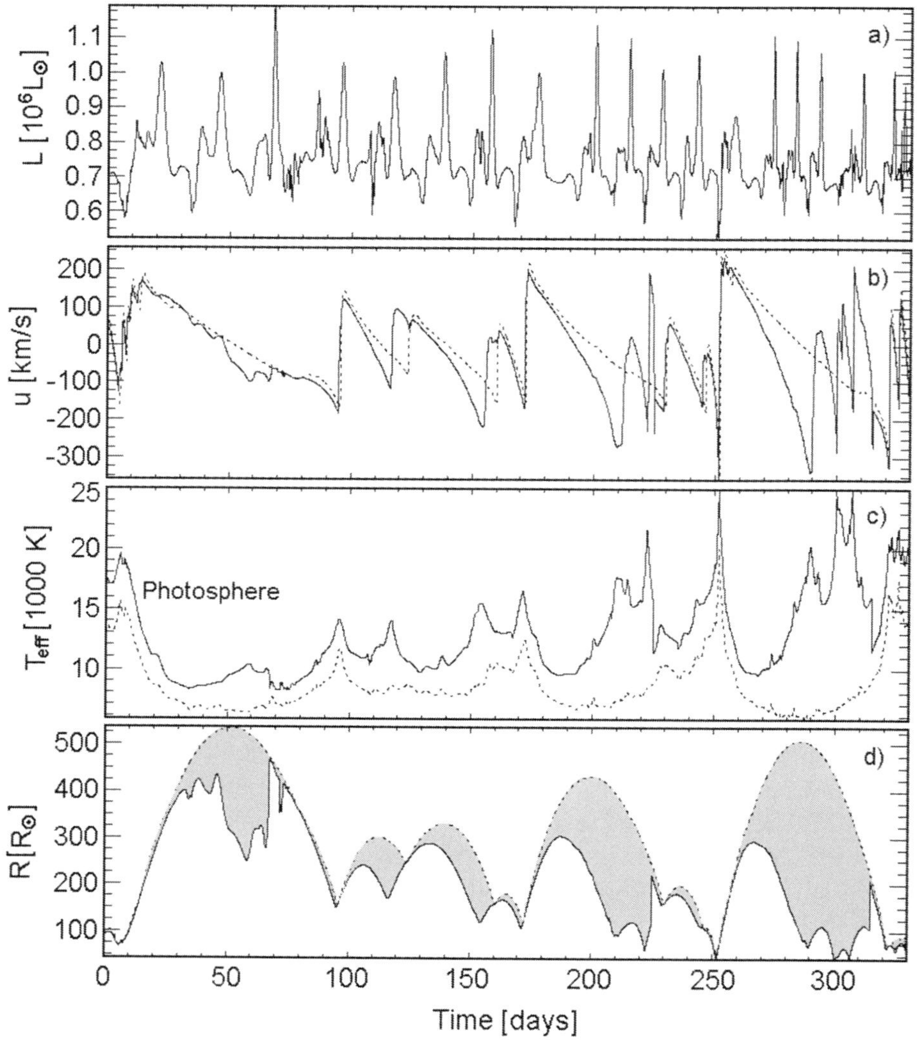

Figure 2. The non-linear oscillation of a 45 $M_\odot$ star shows the interaction of several modes and the corresponding reaction of the outer stellar layers (dashed lines). The short-term changes in the luminosity (a), in units of 10^6 $L_\odot$, are caused by the radiative shock waves. The velocity (b) varies between an outflow of 200 km/s and an infall velocity of −300 km/s. The temperatures (c) at the photosphere and the outermost layer points at the decoupling of the stellar atmosphere. The grey shaded areas (d) in the radius (in units of $R_\odot$) illustrate the difference between the location of the photosphere and the almost ballistic behavior of the extended atmosphere.

ture of almost independent atmospherical dynamics is also seen in the more regular pulsation of the $60\,M_\odot$ model (cf. Fig. 1) where only every second wave results in a noticeable levitation of the atmospheric layers.

4. Concluding remarks

As seen from the presented selection of non-linear computations, observational properties of LBVs might only loosely couple to the global stellar parameters since the atmosphere (determining the observables) reacts on the underlying stellar pulsations on its own intrinsic time scale. Several shock waves shape the atmosphere and lead to changes of the luminosity. However, the dominance of the radiation pressure acts on the non-adiabaticity of the pulsations as well as on the developing atmospheric stratification. In general, we expect a rather complicated behavior during these pulsationally unstable LBV phases since both effects are responsible for the emerging light curves. However, the large transparency of the atmosphere allows the radiation to be less influenced by the large scale gas motions and – as seen from Table 1 – the observable V-band luminosity changes remain below 0.3 mag. Due to the action of the waves, the temperature and density stratification of dynamical atmospheres differ strongly from hydrostatic counterparts. Both effects of temporal variations as well as of radial dependence may have a profound influence on the observationally determined stellar parameters. In particular, the effective temperatures can be off by more than 1000 K in both directions, depending on the details of the atmospheric changes.

Our attempts to generate a pulsationally-driven mass loss of LBVs failed up to now. The overall structure of the resulting stellar atmospheres is much closer to a dynamical unstable situation than to a hydrostatic atmosphere. For stellar masses below $30\,M_\odot$, however, the dynamical atmospheres remain close to hydrostatic ones. For higher masses, the atmospheres of pulsating LBVs gradually decouple from the deeper interior even for regular pulsations. For both, regular and irregular oscillations, the gas motions evolve into radiating shock waves which travel with typical speeds of around 200 km/s. They dissipate enough energy to strongly levitate and accelerate the stellar atmosphere up to 30% of the corresponding escape speed. Since the velocity amplitude in the atmospheres for stars above $45\,M_\odot$ are in the range of 600 km/s, Doppler-shifts in a large number of expected spectral lines provide an additional source of radiation pressure which was not yet taken into account. Further on, the frequently observed bipolar structure of LBV nebulosities hints at significant contribution from rotation which opens another avenue to enhance the stellar mass loss.

Acknowledgments

This work has been supported by the Österreichischer Fonds zur Förderung der wissenschaftlichen Forschung (FWF) under project number P15506.

References

Bohannan, B., 1997, in ASP Conf. Ser. 120, Luminous Blue Variables: Massive Stars in Transition, eds. A. Nota & H.J.G.L.M. Lamers (San Francisco: ASP), 3

Dorfi, E.A., 1998, in 27th Saas Fee Course (Berlin: Springer), 263

Dorfi, E.A., & Gautschy, A. 2000, ApJ, 545, 982

Gautschy, A., & Saio, H. 1995, ARA&A, 33, 75

Kiriakidis, M., Fricke, K.J., & Glatzel, W. 1993, MNRAS, 264, 50

Leitherer, C. 1997, in ASP Conf. Ser. 120, Luminous Blue Variables: Massive Stars in Transition, eds. A. Nota & H.J.G.L.M. Lamers (San Francisco: ASP), 58

Mihalas, D., & Mihalas, B.W. 1984, Foundations of Radiation Hydrodynamics (New York: Oxford Univ. Press)

Stothers, R.B., & Chin, C.-W. 1993, ApJ, 408, L85

Discussion

Scholz: Why does the pulsation in the $60\,M_\odot$-plot look so *regular* although the atmosphere is noticeably decoupled from the interior and the two time scales (travelling time of the shock front through the atmosphere and the pulsation time) are different?

Dorfi: For one the models presented the pulsation time scale from a linear stability analysis is 0.8 days and the sound crossing time of the initial stellar model is 1.5 days, which is almost twice the fundamental period. Due to this coincidence the resulting pulsation develops a very regular pattern but as seen in the computation of a $45\,M_\odot$ star such regular pulsations may not be the general case.

Elizur: Since the instability is driven by radiation pressure - isn't it important to do the gas opacity more accurately?

Dorfi: Concerning the opacity we have two effects. First, the pulsational instability is driven deep in the stellar envelope where the optical depths are so high that the Rosseland-mean provides the appropriate description of the opacity. Secondly, the outer most layers of the atmosphere are strongly influenced by the opacity and there a frequency-dependent *and* velocity-dependent treatment of the gas opacity could change the dynamics of the stellar atmosphere leading also to a significant mass loss. These effects are currently investigated.

MASS-LOSING AND PULSATING AGB STARS IN THE MAGELLANIC CLOUDS

Martin Groenewegen
Instituut voor Sterrenkunde, Celestijnenlaan 200B, B-3001 Leuven, Belgium
groen@ster.kuleuven.ac.be

Abstract

I present some new results on mass-losing and pulsating stars in the LMC and SMC, based on an analysis of OGLE/MACHO light curves and DENIS/2MASS/ISO infrared data, for samples of AGB stars detected by IRAS, and by the recent ISOCAM "minisurvey" of the SMC.

The conclusions are that (1) the scatter in the PL-relation at long periods is (much) larger than that for $P < 400$ days, (2) the SEDs of some carbon stars are not well fitted in the blue, indicating possibly some missing opacity problem in the model atmospheres, or that the effective temperatures are below 2600 K, the lowest temperature in the available model grid, and (3) the period-mass loss rate relation for C-rich SMC stars seems to follow that derived for Galactic carbon-rich Miras, suggesting no explicit dependence of mass loss rate on metallicity.

Keywords: Stars: AGB and post-AGB, Stars: mass-loss, Magellanic Clouds

Introduction

The mass-loss phenomenon in cool stars manifests itself most clearly through infrared emission. Although ground-based near-infrared (NIR) observations of Asymptotic Giant Branch (AGB) in the Magellanic Clouds (MCs) exist in the literature of the 1980s, they are mostly of individual objects (since observations at that time were obtained with single-channel photometers) and are not very deep. The realisation that "obscured" stars, that is stars with a very high mass loss rates that are faint or invisible in the optical, were in fact detected in the IRAS survey (Whitelock et al. 1989, Reid et al. 1990, Loup et al. 1997), has sparked a lot of follow-up observations on these IRAS detected AGB stars, ranging from single-epoch NIR observations (Reid 1991, Zijlstra et al. 1996, van Loon et al. 1997, 1998, Groenewegen & Blommaert 1998), NIR

Y. Nakada et al. (eds.), Mass-Losing Pulsating Stars and their Circumstellar Matter, 75–82.

monitoring (Wood et al. 1992; Wood 1998; Whitelock, this proc.), and later on, follow-up ISO observations (Trams et al. 1999; Blommaert et al. 2002, in preparation) in the mid-IR.

However, the sample of IRAS detected AGB stars in the MCs is in many ways biased, as it represents a sample of AGB stars with the highest mass loss rates and/or objects that were close to maximum light in their pulsation cycle, as the observed IRAS fluxes were very close to the IRAS detection limit.

In recent years, the DENIS (Cioni et al. 2000), and 2MASS surveys have made it possibly to study red AGB stars in the MCs in a way that is less biased. The same will be possible soon using the data obtained with a small unbiased survey of the MCs with the ISOCAM instrument (Loup et al. 2002, in preparation).

Naturally, long-period variables are long known to exist in the Magellanic Clouds (MCs) (e.g. Payne-Gaposhkin 1971). Large numbers of Semi-Regular and Mira variables have been identified by comparing Schmidt plates over a several epochs (e.g. Hughes 1989, Reid et al. 1995, and references therein).

In recent years the surveys searching for micro-lensing events (MACHO, EROS, OGLE, MOA) have resulted in the detection of innumerable variable objects, many of which are long-period variables. Wood et al. (1999) and Wood (2000), using MACHO (see also Wood in these proceedings), and confirmed by Cioni et al. (2001), using EROS data, have shown that in a magnitude - period diagram (at least) 5 sequences can be identified, four of which seem to be indisputably related to the Mira variables, pulsating in the fundamental mode, and semi-regular variables pulsating in the fundamental, first, second and third overtone mode.

In the present paper, I take data from the recently available mirco-lensing and IR surveys to discuss some aspects of variable stars in the MCs.

1. The samples and results

The following samples of stars have been considered. Firstly, all (as far as I am aware off) 19 AGB stars detected by IRAS in the LMC, with known pulsation periods (from Wood et al. 1992, Wood 1998, and some preliminary results dated 1998 from the now finalised IR monitoring observations of LMC AGB stars discussed by Whitelock during this proceedings), as I previously discussed in Groenewegen et al. (2000). For these stars ISO data was gathered from Trams et al. (1999). As

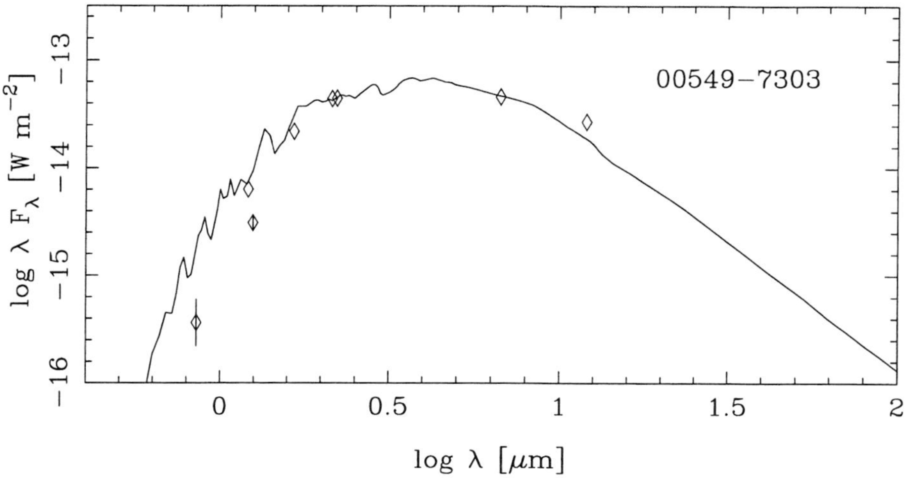

Figure 1. Fit to the SED of a star using amorphous carbon dust and a carbon star model atmosphere. Parameters are L= 13 700 $L_\odot$, $\dot{M}$= 2.8 $\cdot 10^{-6}$ $M_\odot yr^{-1}$. Note the lack of agreement in the optical.

pulsation periods are already available, no search in the OGLE or MACHO database was made for this sample.

The second sample considers the AGB stars detected by IRAS in the SMC as discussed by Groenewegen & Blommaert (1998). ISO observations (mostly 12 and 25 μm data, and some CAM-CVF spectra) have been obtained for a large subset of these stars (Blommaert et al. 2002). The ground-based IR positions of the stars have been correlated with the DENIS (Cioni et al. 2000) and 2MASS database, in order to obtain improved positions and IJK and/or JHK photometry. The improved position or the ground-based IR position was then correlated with the OGLE and MACHO databases. In many cases there was no match as the stars are too red to have been detected in the optical. Sometimes there was more than one match within a few arcsec, and then it was checked which source showed long period, large amplitude behaviour. Seven objects of this sample of 28 have an counterpart in OGLE and/or MACHO.

The third sample considered contains a selected sub-sample of 15 objects of the over 1000 objects detected in the ISOCAM "minisurvey" of the SMC (Loup et al. 2002), with well determined ISOCAM fluxes and with DENIS and 2MASS counterparts. Ten of these have an counterpart in OGLE and/or MACHO.

The OGLE and MACHO light curves were downloaded, and analysed with the interactive program Period98 (Sperl 1996). Basically, a Fourier

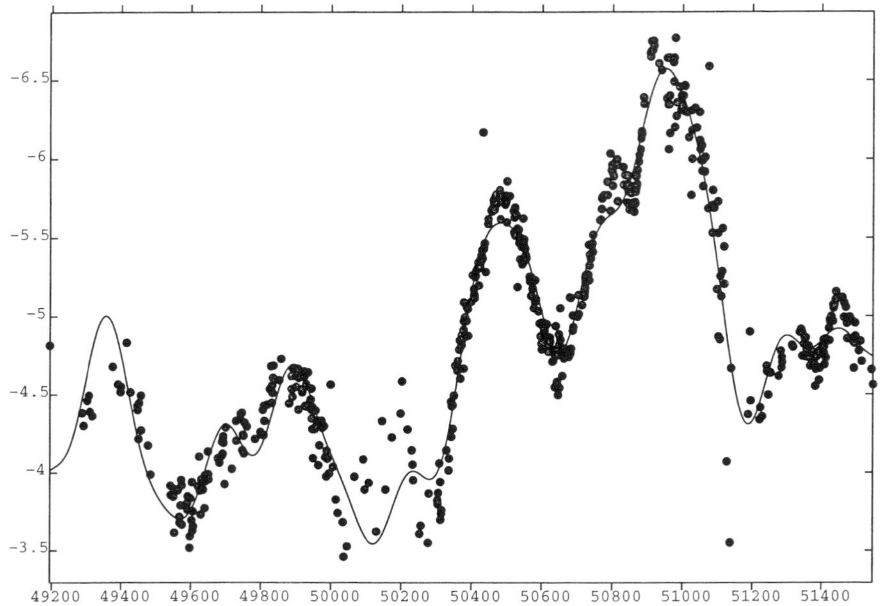

Figure 2. MACHO Red-channel light curve of the object in Fig. 1. MACHO Red-channel magnitude is plotted versus Julian date.

analysis is performed on the data and estimates of the frequency with the most power, and the amplitude are returned. Then, a sinusoidal fit is made to the data. A Fourier analysis can then be performed on the residuals, and so on, until no significant decrease in the rms is obtained.

The IR (2MASS, DENIS and previous single-epoch or monitoring) data, IRAS and ISO results are gathered to construct the Spectral Energy Distributions (SEDs). The SEDs are fitted with a dust radiative transfer code (Groenewegen 1993), basically to derive luminosity and total mass loss rate. In all calculations a gas-to-dust ratio of 200, an expansion velocity 10 km/s, and distances to the LMC and SMC of respectively 50 and 63 kpc are assumed. The central stars are represented by model atmospheres of M-stars (Fluks et al. 1994), or carbon stars (Loidl et al. 2001). The type of dust is either silicate or amorphous carbon dust, respectively. In most cases the chemical type of the object is not known. However, the thermal emission of carbon and silicate dust are so different, that, in many cases, a good fit to the SED is only obtained for one type of dust.

Figures 1-3 give examples of a fit to the SED, and the MACHO and OGLE light curves for the same object. The fit is good at longer wave-

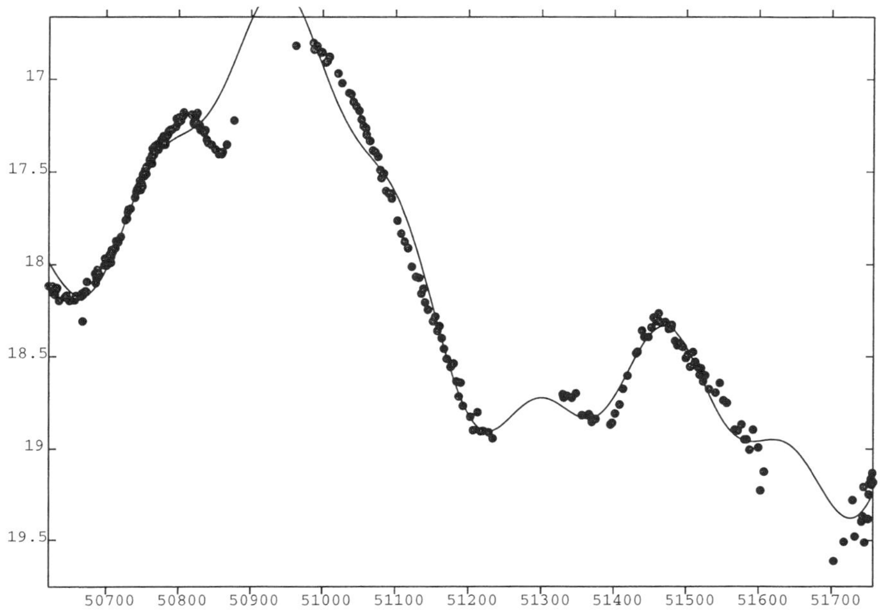

Figure 3. OGLE I-band light curve of the object in Fig. 1. The I-magnitude is plotted versus Julian date. Note the good correspondence in the shape of the light curve compared to Fig. 2.

lengths but poor short ward of 1 μm. Although the example presented is extreme, I often found a disagreement in the optical between data and model, possibly indicating a missing opacity problem in the carbon star model atmospheres, or, simply, that the effective temperatures of the carbon stars in the sample are generally below 2600 K, the lowest available temperature in the model grid.

Figures 4-5 combine the results of analysing the light curves and fitting the SEDs for all 36 stars considered here. For illustrative purposes, the observations are compared to the predictions of a synthetic AGB evolution code. The code was written by Dr. J. Wagenhuber in 1998 just before he left astronomy, and basically implements the recipes described in Wagenhuber & Groenewegen (1998). The output of the code is the time evolution of various stellar quantities. The luminosity and effective temperature are first combined to calculate the stellar radius, and then with the current stellar mass to estimate the fundamental pulsation period, taking the pulsation constant from Wood (1990). The model shown in Fig. 4 is for a 4 $M_\odot$ star, and the output of the model is sampled every 10^3 years. It is assumed that pulsation (in the fundamental mode) takes place the entire AGB phase.

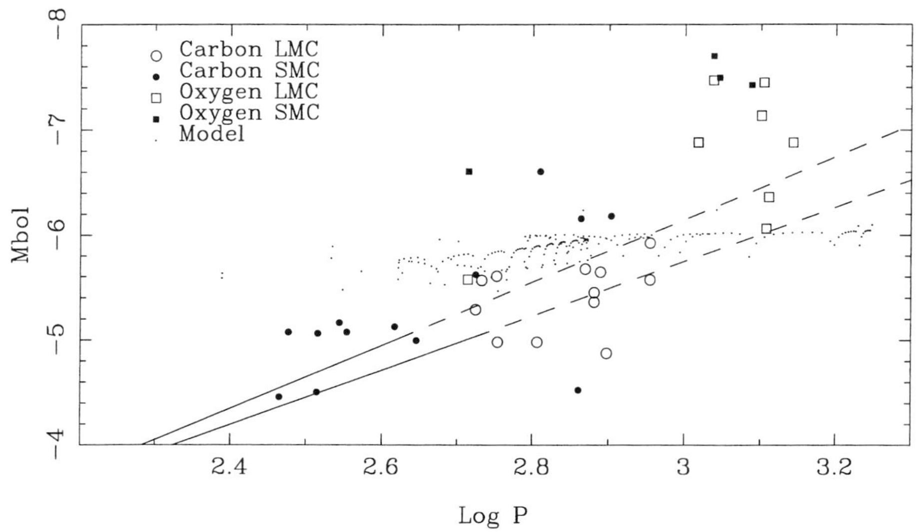

Figure 4. The *PL*-diagram from the stars studied here. Galaxy and chemical type are indicated by different symbols. The two straight lines represent the *PL*-relation for oxygen-rich and carbon-rich Miras in the LMC (respectively the upper line from Feast et al. 1989, and the lower line from Groenewegen & Whitelock 1996), the solid part indicating the period range over which these relations were derived, and the dashed part indicating the extension. The dots represent the evolution of a 4 $M_\odot$ star from its first thermal pulse to the Post-AGB phase, pulsating in the fundamental mode. Each dot corresponds to 10^3 years. Details of the evolution will depend on the assumed mass loss rate law.

Figure 4 shows the period-luminosity diagram, including previously derived relations (see caption). The scatter is considerable. The determination of the luminosities are not corrected for the phase of the light curve – and doing so would lead to less scatter – but it seems very unlikely that the true dispersion is as small as the about 0.15 mag that is typically found in *PL*-relations derived for $P \lesssim 400$ days (Feast et al. 1989, Groenewegen & Whitelock 1996).

Figure 5 shows the relation between mass loss rate and pulsation period. The carbon stars seem to follow quite closely the relation derived in Groenewegen et al. (1998) for Galactic Carbon Miras. In order words: no evidence is seen for a clear explicit dependence of mass loss rate on metallicity. However, a detailed comparison is not straightforward as it involves poorly known quantities as the dust-to-gas ratio and dust opacities in the two galaxies.

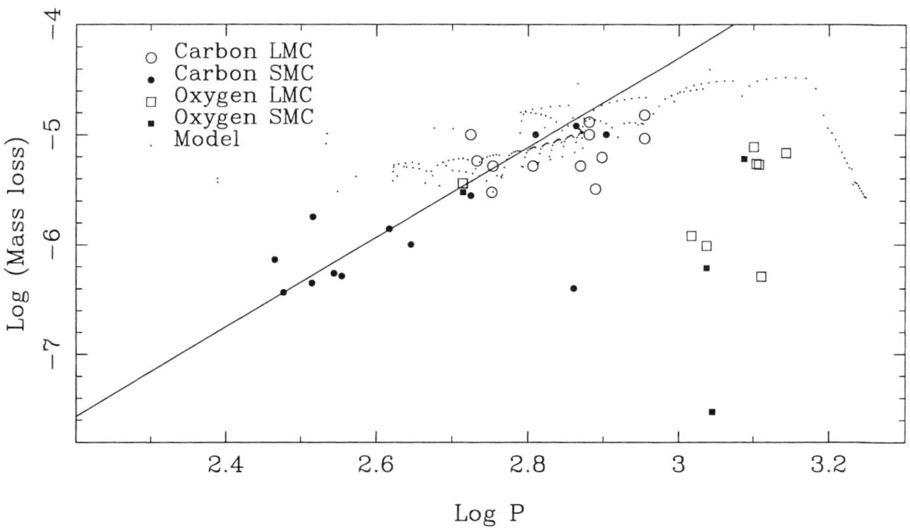

Figure 5. The relation between period and mass loss rate. Symbols as in the previous figure. The line represents the relation derived in Groenewegen et al. (1998) for Galactic carbon-rich Miras.

Acknowledgments

I would like to thank Adam Frankowski (Torun, Poland) for "reviving" the numerical code from Josef Wagenhuber's scientific legacy, during a 3-week visit to the astronomical institute in Leuven in April 2002, and Cecile Loup (IAP, Paris) for providing the list of objects in the SMC "minisurvey".

This publication makes use of data products from the Two Micron All Sky Survey, which is a joint project of the University of Massachusetts and the Infrared Processing and Analysis Center/California Institute of Technology, funded by NASA and the NSF.

This paper utilizes public domain data obtained by the MACHO Project, jointly funded by the US Department of Energy through the University of California, Lawrence Livermore National Laboratory under contract No. W-7405-Eng-48, by the National Science Foundation through the Center for Particle Astrophysics of the University of California under cooperative agreement AST-8809616, and by the Mount Stromlo and Siding Spring Observatory, part of the Australian National University.

References

Cioni, M.-R., Loup, C., Habing, H.J., et al. 2000, A&AS, 144, 235
Cioni, M.-R., Marquette, J.-B., Loup, C., et al. 2001, A&A, 377, 495
Feast, M.W., Glass, I.S., Whitelock, P.A., & Catchpole, R.M. 1989, MNRAS, 241, 375
Fluks, M.A., Plez, B., Thé, P.S., et al. 1994, A&AS, 105, 311
Groenewegen, M.A.T. 1993, Ph.D. Thesis, Chapter 5, University of Amsterdam
Groenewegen, M.A.T., & Whitelock P.A. 1996, MNRAS, 281, 1347
Groenewegen, M.A.T., & Blommaert, J.A.D.L. 1998, A&A, 332, 25
Groenewegen, M.A.T., & Whitelock, P.A., Smith, C.H., & Kerschbaum, F. 1998, MNRAS, 293, 18
Groenewegen, M.A.T., van Loon, J.Th., Whitelock, P.A., Wood, P.R., & Zijlstra, A.A. 2000, in IAU Symp. 177, The Carbon Star Phenomenon, ed. R.F. Wing, (Dordrecht: Kluwer), 385
Hughes, S.M.G., 1989, AJ, 97, 1634
Loidl, R., Lançon, A., & Jorgensen, U.G. 2001, A&A, 371, 1065
Loup, C., Zijlstra, A.A., Waters, L.B.F.M., & Groenewegen, M.A.T. 1997, A&AS, 125, 419
Payne-Gaposchkin, C.H. 1971, Smithsonian Contrib. Astrophys, 13, 1
Reid, N., Tinney, C., & Mould, J. 1990, ApJ, 348, 98
Reid, N. 1991, ApJ, 382, 143
Reid, I.N., Hughes, S.M.G., & Glass, I.S. 1995, MNRAS, 275, 331
Sperl, M. 1996, DSSN, 10, 5
Trams, N.R., van Loon, J.Th., Waters, L.B.F.M., et al. 1999, A&A, 346, 843
van Loon, J.Th., Zijlstra, A.A., Whitelock, P.A., et al. 1997, A&A, 325, 600
van Loon, J.Th., Zijlstra, A.A., Whitelock, P.A., et al. 1998, A&A, 329, 169
van Loon, J.Th., Groenewegen, M.A.T., de Koter, A., et al. 2000, A&A, 351, 559
Wagenhuber, J., & Groenewegen, M.A.T. 1998, A&A, 340, 183
Whitelock, P.A., Feast, M.W., Menzies, J.W., & Catchpole, R.M. 1989, MNRAS, 238, 769
Wood, P.R. 1990, in From Miras to Planetary Nebulae, 67
Wood, P.R., Whiteoak, J.B., Hughes, S.M.G., et al. 1992, ApJ, 397, 552
Wood, P.R. 1998, A&A, 338, 592
Wood, P.R., Alcock, C., Allsman, R.A., et al. 1999, in IAU Symp. 191, Asymptotic Giant Branch Stars, eds. T. Le Bertre, A. Lebre, & C. Waelkens, (San Francisco: ASP), 151
Wood, P.R. 2000, PASA, 17, 18
Zijlstra, A.A., Loup, C., Waters, L.B.F.M., et al. 1996, MNRAS, 279, 32

THE GALACTIC KINEMATICS OF MIRA VARIABLES

Michael Feast
Astronomy Department, University of Cape Town, Rondebosch,7701, South Africa
mwf@artemisia.ast.uct.ac.za

Abstract The galactic kinematics of Mira variables derived from radial velocities, Hipparcos proper motions and an infrared period-luminosity relation are reviewed. Local Miras in the 145-200day period range show a large asymmetric drift and a high net outward motion in the Galaxy. Interpretations of this phenomenon are considered and (following Feast and Whitelock 2000) it is suggested that they are outlying members of the bulge-bar population and indicate that this bar extends beyond the solar circle.

Keywords: stars: AGB: - Galaxy: kinematics and dynamics - Galaxy: structure

1. Introduction

The galactic kinematics of Mira variables have for a long while been of importance in helping us understand both the nature and evolution of Miras as well as the structure of our own Galaxy. This paper is primarily concerned with the second point - what do we learn from Miras about galactic structure? The paper also concentrates on local kinematics, leaving out a detailed discussion of the kinematics of the galactic bulge, much work on which has of course been done here in Japan. It has been possible to take a fresh look at the local kinematics of Miras using Hipparcos astrometry, extensive new infrared photometry and published radial velocities. This was done in a series of three papers (Whitelock, Marang and Feast (2000), paper I: Whitelock and Feast (2000), paper II: Feast and Whitelock (2000a), paper III). The present paper summarizes some of the relevant results from these papers and extends the discussion of the kinematics.

Y. Nakada et al. (eds.), Mass-Losing Pulsating Stars and their Circumstellar Matter, 83–89.

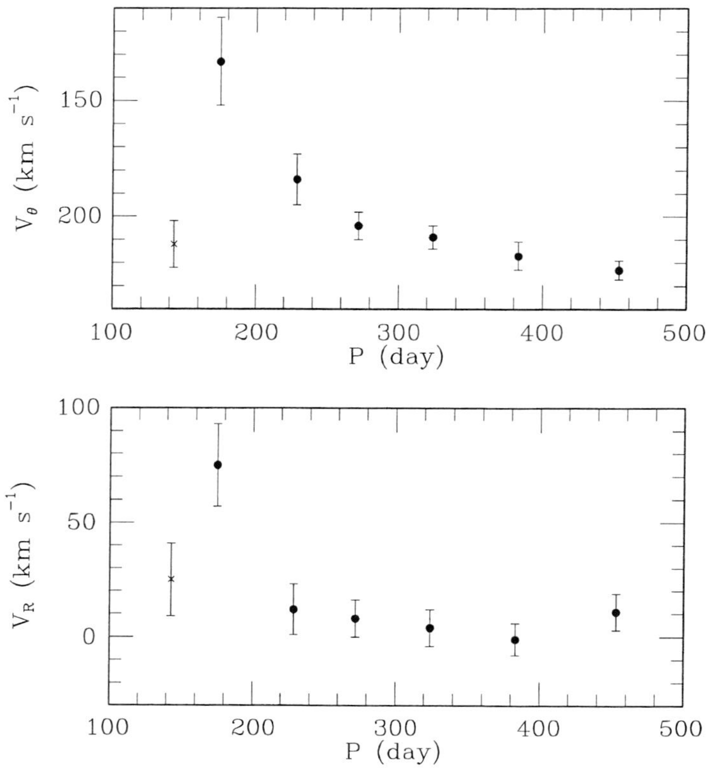

Figure 1. (a) The mean V_θ in various period groups plotted against the mean period of the group. (b) The mean V_R in various period groups plotted against the mean period of the group. The cross represents the SP-red variables in both figures.

2. The Period-Kinematic Sequence

It has been known for many years that there is a relation between the kinematics and the periods of Mira variables in the general solar neighbourhood. In particular the asymmetric drift and the velocity dispersion increase as one goes from longer to shorter periods. There was however an anomaly. The shortest period group (periods less than $\sim$ 150 days) differed strongly from the general trend (Feast 1963). Following an initial study by Hron (1991), it was shown in paper I (see also Whitelock 2002) that combining infrared and Hipparcos magnitudes one sees two sequences of Miras at shorter periods in the period - colour plane, the SP-red and SP-blue stars, and this must be taken into account when discussing the kinematics.

Figure 1a shows a plot of mean velocity of Miras in the direction of galactic rotation, V_θ, in period groups. This plot (data in paper III) is a distinct improvement on earlier work. It depends on space motions rather than just radial velocities and the SP-red stars have been grouped separately and are denoted by a cross. The figure shows that at the longest periods the stars of the main Mira sequence are moving at close to the circular velocity of galactic rotation, which is taken as 231 $km.s^{-1}$ (Feast and Whitelock 1997). There is a clear dependence of mean rotational velocity on period for this main Mira sequence with the short period (145-200 day) Miras having a large asymmetric drift ($-97 \pm 20 km.s^{-1}$).

The main Mira sequence now seems reasonably well understood. The infrared colour-period relation for this sequence is the same as that for Miras in globular clusters (paper I and Whitelock 2002), strongly suggesting a similarity between the field and cluster Miras over the period range in common. Not only do the clusters show that their Miras lie at the tip of the AGB, they also show that there is a period-metallicity relation (e.g. Feast and Whitelock 2000b). This allows us to study the galactic kinematics of old populations as a function of metallicity (and possibly also age) over a range of ages and metallicities where there are few if any useful, precise, tracers. For instance globular clusters containing Miras are often classed as "disc" clusters and treated together in kinematic discussions. However the field Miras show that there is a considerable variation of kinematic properties in the period/metallicity range of relevance to these clusters.

Where do the SP-red stars fit in? A full answer to this question may come when more members of the class are identified and when good individual parallaxes are available for a significant number. The best guess at present (see e.g. Whitelock 2002) is that, unlike the stars of the main Mira sequence, they are not at the end of their AGB lives but that they will evolve into longer period Miras of the main Mira sequence.

3. Local Miras and a Galactic Bar

A particularly interesting result in paper III is the evidence of a net radial, outward motion (V_R) for Miras in the solar neighbourhood. This is shown as a function of period in figure 1b There is evidence of a small outward motion over most of the period range of the Miras. However it is very marked in the group with periods in the range 145 to 200 days where one finds, $V_R = +75 \pm 18 km.s^{-1}$ for a mean period of 173 days. Note that in view of the discussion in section 2, the SP-red Miras are omitted in the discussion of the present section.

There has been evidence in the literature for many years that some groups of old stars in the solar neighbourhood show a predominant, but modest, outward motion in the Galaxy, though this has not always been recognized. The Hipparcos results, giving parallaxes and proper motions for many thousands of common stars have enabled this to be studied in greater detail. It is now clear that there is a group of old stars with a small asymmetric drift and a modest outward velocity ($V_\theta =\sim 190\text{km.s}^{-1}$, $V_R =\sim +30\text{km.s}^{-1}$ (Dehnen 1998, 1999, 2000, Fux 2001). Fux has referred to these stars as forming the "Hercules Stream". Dehnen, Fux and also Quillen (2002) each suggest that the stars of the Hercules Stream were originally in circular orbits which have been perturbed into their present state by the influence of a (pre-existing) galactic bar. However, the detailed dynamic processes envisaged differ from author to author.

Fux has suggested that the short period Mira group just discussed belongs to the Hercules Stream. However whilst the Hercules Stream consists of old stars with a wide range of metallicities (Raboud et al. 1998) the large local outward motion in the case of Miras is confined to a restricted range in period, and therefore also in metallicity. The effect is also more extreme than in the Hercules group. This is shown in Figure 2 which plots V_θ against V_R for individual Miras in the 145 to 200 day group.

The dotted oval shows the extent of the Hercules Stream as delimited by Dehnen (2000, fig 9). Evidently these Miras do not concentrate in the region of the Hercules Stream.

It is clear that some of the stars in figure 2 are on highly eccentric orbits. About half the sample have perigalactic distances less than about 2kpc and some at least will pass through the galactic bulge. Despite the small number of stars in this diagram, there is also evidence of a relation between V_θ and V_R. In paper III we derived the curve shown in the fig 3 which makes the first order assumption that the stars are moving in simple elliptical orbits of different eccentricities but with their major axes aligned. There is then only one free parameter; the position of this major axis with respect to the sun-centre line. The curve is drawn for the best fitting value of this position angle (17^{+11}_{-4} degrees). It provides a rather good fit to the data and since this angle is, within the uncertainties, similar to that proposed for the bar in the galactic bulge, we suggested that these 145-200 day Miras in the general solar neighbourhood were outlying members of the bar population itself. If this is the case the bar population extends out to beyond the solar circle.

This discussion depends on the Miras in the general solar neighbourhood and is thus limited in number. In a recent paper Kharchenko et al. (2002) have suggested that over a larger volume of space, the mean V_R

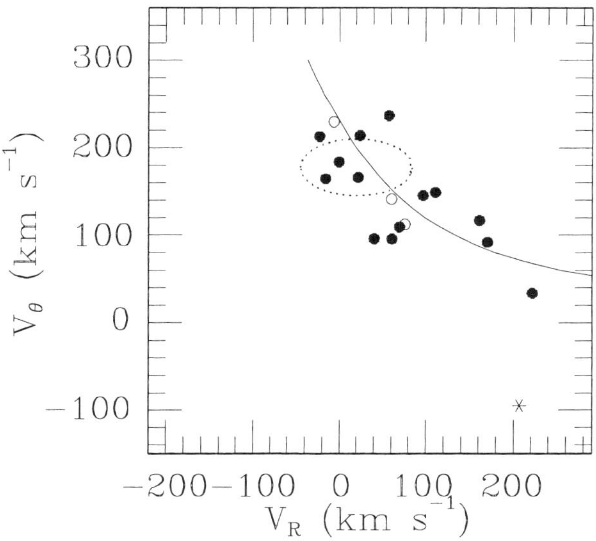

Figure 2. A plot of V_θ against V_R for local Miras with periods in the range 145 to 200 days. The open circles denote stars for which the standard error of a velocity component is greater than 20km.s^{-1}. The asterisk show the position of S Car which is on a highly eccentric retrograde orbit. The curve and dotted oval are discussed in the text.

for 145-200 day Miras is near zero. More details of their work are need for a proper discussion. However the following points are worth noting. (1) the distances adopted by Kharchenko et al. are obtained from visual magnitudes and the reddenings from a model. They must thus be rather uncertain. (2) A certain amount of trimming is carried out (velocities further from means than 3σ are rejected). (3) Complications in the analysis will arise when one goes to a large volume if the Galaxy is not axi-symmetrical. (4) No distinction is made in Kharchenko et al. between the SP-red and SP-blue stars discussed above and this is particularly important in the period range under discussion.

Whilst therefore the large scale picture remains unclear, it seems rather remarkable that the nearby Miras in the period range 145-200 days with radial velocities, Hipparcos proper motions and infrared photometry show a marked asymmetry in V_R. As figure 2 shows, all the stars in this group with ($V_\theta \lesssim 160$km.s^{-1}) have positive values of V_R. For an axi-symmetrical galaxy there should be a symmetrical distribution of V_R about zero in this figure at any given V_θ. The deviation from such a distribution is sufficiently striking that it seems difficult to attribute it entirely to chance.

Acknowledgments

I would like to thank Professor Nakada (University of Tokyo) and the organizers of this meeting for making my attendance possible. This paper depends on work done in collaboration with Patricia Whitelock.

References

Dehnen, W. 1998, AJ, 115, 2384
Dehnen, W. 1999, ApJ, 524, L35
Dehnen, W. 2000, AJ, 119, 800
Feast, M.W. 1963, MNRAS, 125, 367
Feast, M.W. & Whitelock, P.A. 1997, MNRAS, 291, 683
Feast, M.W & Whitelock, P.A. 2000a, MNRAS, 317, 460 (paper III)
Feast, M.W. & Whitelock, P.A. 2000b, in The Evolution of the Milky Way, eds. Matteucci, F. & Giovannelli, F. (Dordrecht: Kluwer), 229
Fux, R. 2001, A&A, 373, 511
Hron, J. 1991, A&A, 252, 583
Kharchenko, N., Kilpio, E., Malkov, O. & Schilbach, E. 2002, A&A, 384, 925
Quillen, A.C. 2002, astro-ph, 0204040
Raboud, D., Grenon, M., Martinet, L., Fux, R. & Urdy, S. 1998, A&A, 335, L61
Whitelock, P.A. 2002, this volume
Whitelock, P.A., Marang, F. & Feast, M.W. 2000, MNRAS, 319, 728 (paper I)
Whitelock, P.A. & Feast, M.W. 2000, MNRAS, 319, 759 (paper II)

Discussion

Habing
In your calculations of the galactic orbit, did you assume the bar is stationary? It may rotate.
Feast
For a rotating bar one needs to assume the simple elliptical orbits precess. One is then concerned with the present orientations of the Mira orbits and the bar. So the conclusions are not affected
van Langevelde
1. You find a best fit of ϕ, the angle between the major axis of the Mira orbits and the bar, but there must be a whole range of eccentricities. What is the range of perigalactic distances?
2. What makes the 145-200 day group special? Would that indicate something about the population/age of the bar?
Feast
1. The distribution in figure 2 is essentially a distribution in eccentricity. About half the stars in that plot go within 2 kpc of the centre. The exact perigalactic distance depends of course on the mass model.
2. That is not entirely clear. There are Miras with a range of periods in

the bulge. At periods longer than about 200 days, the local population is dominated by variables on much more nearly circular orbits. Possibly this is because the galactic density gradient of Miras is a function of period.

Nakada

Do the bulge Miras dominate the short period (P 145-200 days) group in the solar neighbourhood?

Feast

About half the Miras in this group (SP-red stars omitted) have perigalactic distances sufficiently small that we can probably say they belong to a bulge population. However there is no evidence at present that the stars in this period group which are on more nearly circular orbits are different physically from those on highly eccentric orbits. In that sense they can perhaps all be regarded as a bulge type population.

PERIOD EVOLUTION IN MIRA VARIABLES

Albert A. Zijlstra, Mikako Matsuura
UMIST, department of Physics, Manchester M60 1QD, UK

Timothy R. Bedding
University of Sydney, School of Physics, Sydney 2006, Australia

Abstract We investigate a number of Mira variables which show evolution in their periods. Three different type of period changes are found: continuous evolution, sudden changes and meandering periods. Of the order of 1% of Miras show evidence for period changes, but unstable periods may be common among the longest period Miras. In the case of R Hya, the period change was accompanied by a dramatic change in mass-loss rate, consistent with the mass-loss formalism of Vassiliadis & Wood. Their formalism then predicts that the mass-loss rates of the longest period Miras ($P > 400$days) may fluctuate by a factor of 2–3 over time scales of a century.

1. Introduction

Long-term monitoring of Mira variables has revealed a few cases of period evolution, the best known being that of R Hya. The pulsation equation, $\log P = 1.5 \log R - 0.5 \log M + \log Q$, shows that period evolution can be driven by a change in radius of the star. The most popular model for radius changes, that of a thermal pulse, suggests that $\sim 1\%$ of Miras may show period evolution. In this paper we discuss period changes in Miras. Three different types are found, one of which appears to be found in a large fraction of long-period Miras.

2. A case study: R Hya

R Hya, the third-brightest Mira variable in the sky, has shown a steadily declining period, first discovered by Olbers (1841). Wood & Zarro (1981) attributed its changing period to a thermal pulse, which should have occurred around the time of the discovery of R Hya, in AD 1662.

Y. Nakada et al. (eds.), Mass-Losing Pulsating Stars and their Circumstellar Matter, 91–94.

The star has been extensively monitored by, e.g., VSOLJ, BAAVSS, and AAVSO observers since about 1900. Previous observations are available mostly as dates of maximum (and more rarely, minimum), and are described in Müller & Hartwig (1918) and Argelander (1869). We analyzed all data using the weighted wavelet Z-transform, which is well suited to search for time-dependent frequencies. Before AD 1850 the data is to patchy for such analysis, and we obtain periods from individual sets of observations.

The resulting period evolution is shown in Figure 1. The earliest time at which the period is well determined is AD 1705. Before this, there are two observations close to maximum, which are in phase with the same period. Figure 1 shows that the main decline occurred between c. 1770 and 1950. After 1950, the period remained stable at 385 days. Before 1770 the period was 495 days, with no evidence for rapid change. Wood & Zarro (1981) discuss a possible period increase during the oldest observations, but we find do not confirm this.

The evolution of the pulsation amplitude (only available from 1900 onwards) mirrors the period change, declining until 1950 and remaining constant since.

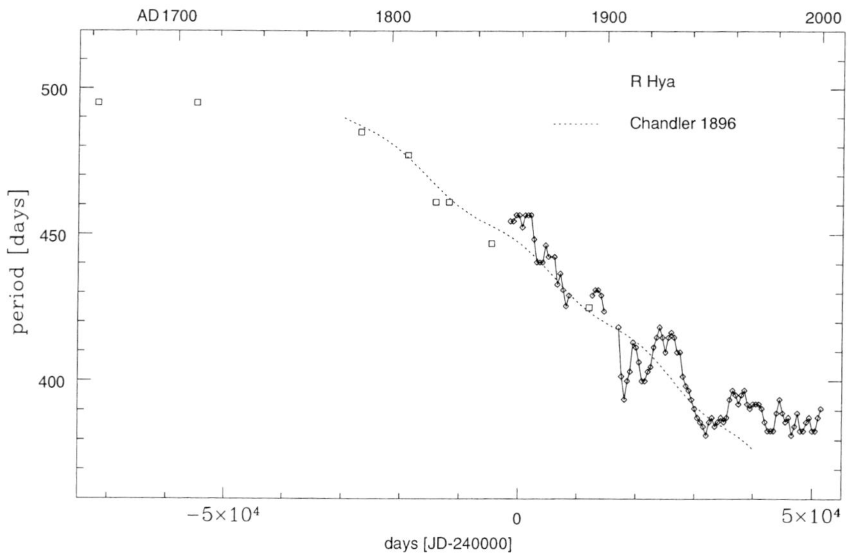

Figure 1. The period evolution of R Hya between 1662 and 2001. The first (dashed) point is uncertain; the period is well determined from 1704 onwards. Extrapolation of the linear decline suggests the decline began around 1770. The dashed line is the fit proposed by Chandler (1896).

3. Three types of period variations

C-type R Hya is a (rare) example of a continuously or monotonically changing period. In spite of many searches, very few similar stars have been found. We have analyzed all suspected cases using the same wavelet transform method, and confirm three cases. One of these, R Aql, was already well established, with a period declining from 365 days at its discovery (around 1850), to 275 days at present. The decline has been continuous, with no indications for epochs of stable periods. (But the observations cover a time span shorter than that of the R Hya decline phase, of 200 years). We also find a decreasing period in RU Vul, from 160 to 110 days over 50 years. This star is a small amplitude semi-regular, rather than a Mira. Finally, W Dra shows a period *increase* from 155 to 180 days over 90 years of observations.

S-type A few Miras show sudden and fast period changes after a long phase of stable periods. The best case is T UMi, which for 60 years behaved as a normal Mira with a period of 320 days. Since 1980 its period has declined rapidly to 240 days at present. Such rapid change is unprecedented among Miras. BH Cru may be a similar case. It was discovered as a variable in 1965, and its period has increased since 1975 from 420 to 530 days. It is not known whether its period was stable before, but we include it with T UMi because of its similar extreme rate of change. BH Cru has attracted attention because its chemical type changed from SC to CS during its evolution, which has been attributed to carbon dredge-up. We find that this change, due to the appearance of C_2 bands, can be explained by a lower stellar temperature and does not imply an increase in carbon abundance.

M-type A number of long-period Miras show evidence for meandering or fluctuating periods. Cases include W Hya, S Ori, T Cep and R Nor. Their periods change by about 10% over 50 years, followed by a return to the previous period. This may be due to a true instability in the extended stars. All these stars have periods of 400 days or longer.

All these Miras show total period changes of between 10% and 25%. The change per cycle is between 0.1% and 1%, being largest for the suddenly evolving stars. In almost all cases, the amplitude follows the period evolution closely with very similar fraction change. The time scale for the evolution $(P/\dot{P})$ is 100–1000 years.

An important question is what fraction of Miras show unstable periods. Not all known Miras have been observed sufficiently extensively that changing periods can be determined, and strictly speaking only a lower limit to the fraction can be determined. Most of the Miras listed above are bright (V< 8 at maximum) and we therefore determine the fraction relative to the bright and (mostly) well studied Miras.

We find that the fraction of Miras showing continuous changes is 1–2%. They are however more common, 10% or more, among the brightest (V< 6) Miras. Sudden changes are found in fewer than 1% of Miras. In contrast, meandering periods are found in $\sim$ 15% of Miras with periods longer than 400 days. The data suggests that most or even all Miras with periods longer than 500 days have unstable periods.

4. Mass loss variability

Two main mass-loss parametrizations have been proposed: Vassiliadis & Wood (193) suggest that $\log \dot{M} \propto P$, whilst the Blöcker (1995) law depends on L and R of the star. Blöcker's law is mostly insensitive to periods changes (a small effect enters because the period changes indicate a change in radius), but the VW formula suggests a large effect.

R Hya provides a test case. Its IRAS LRS spectrum (Hashimoto et al. 1998) shows the absence of a silicate feature, which can be modelled with a detached dust shell. The derived inner radius of the shell is consistent with an interruption to the mass loss starting 200 yr ago. The mass-loss rate declined by at least a factor of 10. This agrees with the VW formalism, but is much larger than predicted by Blöcker's law. R Hya supports a strong dependency of the mass-loss rate on period.

If this holds for other Miras, than all period changes discussed here will have given rise to large changes in the mass-loss rates. Even for the smaller changes in the meandering Miras, mass-loss rates would vary by a factor of a few. There is some evidence that mass-loss rates in Miras can vary on time scale of the order of 100 yr, e.g. as shown by the rings seen around post-AGB stars (Kwok et al. 2001). Fluctuating periods may be one factor contributing to the formation of such rings and other radial inhomogeneities.

References

Argelander, F.W.A. 1869, Astron. Beobachtungen zur Sternwarte Bonn, 7, 315

Blöcker, T. 1995, A&A, 297, 727

Chandler, S.C. 1896, AJ, 16, 145

Hashimoto, O., Izumiura, H., Kester, D.J.M., & Bontekoe, Tj.R. 1998, A&A, 329, 213

Kwok, S., Su, K.Y.L., & Stoesz, J.A. 2001, ApSS, 265, 115

Müller, G., & Hartwig, E. 1918, Geschichte und Literatur des Lichtwechsels (Poeschel, Leipzig)

Olbers, H. 1841, Schumachers Jahrbuch für 1841, 98

Vassiliadis, E., & Wood, P.R. 1993, ApJ, 413, 641

Wood, P.R., & Zarro, D.M. 1981, ApJ, 247, 247

CHAOTIC PULSATIONS IN IRREGULAR VARIABLES

G. Ambika
Department of Physics, Maharajas College, Cochin-682 011 India
nalanda4@satyam.net.in

M. Takeuti
Astronomical Institute, Tohoku University, Aoba, Sendai 980-8578 Japan
takeuti@astr.tohoku.ac.jp

A. K. Kembhavi
IUCAA, Ganeshkhind, Pune - 411 007 India
akk@iucaa.ernet.in

Abstract It is now well established that the irregularities in the light intensities of some variables are chaotic in origin due to the inherent nonlinearities in the underlying dynamics. By taking typical stars, we present an analysis of their light intensities using time series analysis emphasizing aspects of nonlinear dynamics. The fractal dimensions and other indices are calculated to establish the strength of chaos and the minimum modes involved in their pulsations. In CY Aqr we have found high dimensional chaos.

Keywords: Variable stars, stellar pulsations, nonlinear dynamics, CY Aqr

1. Introduction

Recently, the observational data of the light curves of several irregular variable stars has been subjected to systematic studies using the technique of nonlinear time series analysis (Buchler and Eichborn 1987; Takeuti and Buchler 1993; Buchler 1990; Takeuti and Sasselov 2001). The motivation behind such studies has been to distinguish cases where the irregularity arises from pure stochastic processes from those that are chaotic due to the underlying nonlinearity in the dynamics of the pulsations. There have also been cases where the irregularity is found

Y. Nakada et al. (eds.), Mass-Losing Pulsating Stars and their Circumstellar Matter, 95–98.

to be noise or stochasticity (either inherent or external) superposed on periodic pulsations. Thus the emphasis is to identify some pattern in the irregularity or complexity of the pulsations and then to quantify or index this pattern by computing fractal dimensions and Lyapunov exponents (Hilborn 1994).

The approach has its own limitations in that it is highly data dependent viz. the purity and stationarity of the available data, the choice of parameters in the analysis etc. Even though the indices characterizing chaos like dimensions and Lyapunov exponents are dynamical invariants, it is a computational hazard to try various combinations of the parameters like embedding dimensions delay time etc. until an optimum choice is arrived at. The most appropriate filter for the data is also at times crucial for the analysis.

2. Time series analysis and chaos

The available observational data consists of the magnitudes of the light intensities measured over a certain time span, which should typically at least 20 times the period of the star. To recapture the underlying dynamics, this scalar series $s(t)$ is split into a vector series $\mathbf{y}(t)$ in a reconstructed space of higher dimension m (called the embedding dimension space) formally equivalent to the actual space. These delay vectors are constructed as Abarbanel (1996).

$$\mathbf{y}(t) = \{s(t), s(t+d), s(t+2d), \ldots s(t+(m-1)d)\}. \quad (1)$$

Here d is the delay time and m is the embedding dimension. The first part of the analysis is to choose the optimum values for d and m for the particular data. d should be large enough so that temporal correlations are almost absent and m should be the minimum required for a proper unfolding of the geometry of the reconstructed attractor or trajectory. Then from Takens' theorem, the number of degrees of freedom or the phase space dimension q is such that $m \geq 2q+1$ (Takens 1980). For sufficiently small q values, we conclude the system has low dimensional chaos. In such cases, we proceed to calculate the fractal dimensions and Lyapunov exponents.

3. Chaos in CY Aqr

CY Aquarii (V = 10.42 - 11.16) is one of the most regularly observed SX Phoenix type stars. Coates, Fernley, Sekiguchi and et al. (1994) concluded that the star is singly periodic with the period of 0.061 d with clear changes of the times of maximum light. McNamara, Powell and Joner (1996) studied the effective temperature and the surface gravity with the $ubvy\beta$ and $V(RI)_C$ photometry and obtained the physical parameters, $M = 1.06M_\odot$, $R = 1.59R_\odot$, and $M_V = 2.47$. The star is metal-poor [Fe/H] $= -1.5$ evolved star but not LVPs or SRs. Because of the existence of high quality photometric observations, the results of nonlinear study is interesting to reveal the nature of the stellar pulsation.

The data used in the present analysis is from the IAU Archives of unpublished variable star photometry with 3503 entries covering the time between JD 2447392.7466 and 2447397.7638 containing about 85 periods of the star. The details of data acquisition was described in Coates et al. (1994). For the nonlinear time series analysis, we use the TISEAN package (Hegger, Kantz and Schreiber 1996). The data after a preliminary filtering is subjected to the stationarity test using a space time separation plot (stp). The saturation and the horizontal nature of the curves indicate stationarity of the data (Kantz and Schreiber 2000). The periodic nature is found but small deviation indicating nonlinearity is also found.

We therefore try to reconstruct the dynamics corresponding to our data. The first step is to isolate a good estimate for d and m. For $d = 11$, the orbits are almost fully resolved. Since further increase of d does not produce much change we fix d to be between 9 - 12. We calculate the correlation $C(d)$ (Abarbanel 1996) for different d values and find that its value crosses zero first at $d = 40$. So also the time delayed mutual information exhibits its first minimum at $d = 47$. Since these give varying results, to get the optimum choice of d and m, we vary both simultaneously and apply the method of false nearest neighbours (fnn) (Hegger et al. 1996). We get $d = 11$ and $m = 17$ as the minimum required values.

The trajectories in the reconstructed space has a complex distribution of points on the attractor. They are then said to process a fractal structure and to resolve the geometry, we compute the fractal dimensions. The most useful and easily obtainable from a time series data from the numerical point of view is the correlation dimension d_2. We obtain d_2 as 2.3827 ± 0.0531 from repeated calculations. The corresponding value of m lies in the range 15 - 17.

The dynamical complexity of the attractor is quantified through the Lyapunov exponents, that define the average rate of stretching of two nearby points during the dynamical evolution of the system. To establish chaos, it is enough to calculate the maximal Lyapunov exponent (MLE). For this, we use the Kantz algorithm (Kantz 1994), and obtain 0.02307 ± 0.00268 for the MLE with $d = 11$.

4. Concluding remarks

In the present work, we have analysed the light curves of CY Aqr and have found that the fractional value of d_2 and the positive nature of MLE. The irregularities in the light curves of this star are chaotic in nature. Ambika, Kembhavi and Mattei (n. d.) have found d_2 as 3.3 - 3.9 for $m = 6$ and $d = 7$ in an SRd variable SX Her. Buchler et al. (1996) has established chaos in R Sct and AC Her with $d_2 = 3.2$ and $m = 4$. Recently Buchler et al. (2001) have similar results for AGB variables like V CVn, R UMi, and RS Cyg. However, recent work on a symbiotic star Z And by Gallego et al. (1999) gave a higher value for $d_2 = 9.33 \pm 0.06$ and $m > 80$. Analysis of three LPVs, o Ceti, R Leo and V Boot (Kollath and Szeidl 1993; Cannizzo, Goodings and Mattei 1990), do not reveal strong evidence for chaos. More work on a large number of stars before any general pattern of behaviour can be conclusively arrived at.

References

Abarbanel, H.D.I. 1996, Analysis of Observed Chaotic Data (Springer Verlag NY)
Ambika, G., Kembhavi, A.K., & Mattei, J.A, presented at the workshop on Celestial Mechanics and Dynamical Systems, held at IUCAA, Pune, 2001.
Broomhead, D.S., & King, G.P. 1986, Physica, 20 D, 217
Buchler, J.R. 1990, The Numerical Modelling of Nonlinear Stellar Pulsations, NATO ASI Series C (Kluwer Academic Publishers), vol. 302
Buchler, J.R., & Eichhorn, H. 1987, Chaotic Phenomena in Astrophysics, Ann. New York Academy of Sciences, 497
Buchler, J.R., Kolláth, Z., & Cadmus, R 2001, astro-ph/0106329
Buchler, J.R., Kolláth, Z., Serre, T., & Mattei, J. 1996, ApJ, 462, 489
Cannizzo, J.K., Goodings, D.A., & Mattei, J.A. 1990, ApJ, 357, 235
Coates, D.W., Fernley, J.A., Sekiguchi, K., Barnes, T.G., & Frueh ,M.L. 1994, MNRAS, 266, 1
Gallego, M.C., Gareía, J.A., Vaquero, J.M., & Bajo, E.S. 1999, Acta Astron., 49, 171
Hegger, R., Kantz, H., & Schreiber, T. 1996, Chaos, 9, 143
Hilborn, R.C. 1994, Chaos and Nonlinear dynamics (Oxford University Press)
Knatz, H. 1994, Phys. Lett, A185, 77
Kantz, H., & Schreiber, T. 2000, Nonlinear Time Series Analysis (Cambridge University Press)
Kollath, Z., & Szeidl, B. 1993, A&A, 277, 62
McNamara, D.H., Powell, J.M., & Joner, M.D. 1996, PASP, 108, 1098
Takens, F. 1980, in Dynamical Systems and Turbulence, ed. Rand, D.A., & Young, L.S. (Springer Verlag, Berlin), 366
Takeuti, M., & Buchler, J.R. (eds) 1993, Nonlinear Phenomena in Stellar Variability, Ap&SS, 210
Takeuti, M., & Sasselov , D.D.(eds) 2001, Stellar Pulsation – Nonlinear Studies, ASSL Series (Kluwer Academic Publishers), Vol. 257

TIME DEPENDENT MODELLING OF MASS LOSS ON THE AGB

Yvonne Simis

Astrophysikalisches Institut Potsdam, An der Sternwarte 16, 14482 Potsdam, Germany

simis@aip.de

Abstract Previous numerical hydrodynamics calculations (Simis et al. 2001) indicated that the concentric shells around post-AGB objects and Planetary Nebulae may form due to an instability in the AGB wind. In this presentation I show the results of a series of calculations for various stellar parameters (T_*, L_*) that should enable to determine when this instability takes place. Also, I present first results of calculations that follow the outflow on a timescale longer than that of the mass loss modulation (which is typically of the order of a hundred years). This enables to study the effect of stellar evolution on the wind.

1. Mass loss on the AGB: when and how?

A time dependent, two-fluid (gas and dust) numerical hydrodynamics code for the calculation of dust forming AGB winds was developed. The chemical composition of the gas and the nucleation, growth and evaporations are taken into account consistently, but only pure carbon grains are considered. Stellar pulsation in the form of a piston at the inner boundary can be added. A detailed description of the code can be found in Simis et al. (2001,2002). Two new aspects of the code have been added recently. Firstly, a routine for optically thick radiative transfer was included. This routine uses the TRANSPHERE code for radiative transfer in spherical symmetry (C.P. Dullemond, priv. comm.). The second new feature is the possibility to modify the stellar luminosity and temperature, in order to follow stellar evolution.

The first result obtained with the hydrodynamics code was the reproduction of the concentric shells around IRC +10216 (Mauron & Huggins, 1999). To ensure that the modulation in the mass loss, observed as the shell structure, truly originates in the flow, this particular calculation was performed without stellar pulsation. We found that, when allowing grain drift, an instability in the momentum transfer from grains to gas

Y. Nakada et al. (eds.), Mass-Losing Pulsating Stars and their Circumstellar Matter, 99–102.

could lead to modulation in the mass loss rate. This instability is a result of the fact that two outflow modes are possible: a high mass loss outflow occurs when gas and grains are strongly coupled because the grains are large and hence drift slowly through the gas. The mass loss rate is lower if gas and grains decouple, i.e. when the grains are small and drift rapidly through the gas. Both outflow modes are unstable: small grains tend to remain small, because they drift rapidly to the gas and therefore do not have the opportunity to grow, conversely, large grains are likely to grow efficiently because their drift velocity is low. Due to the quadratic dependence of the drag force on the drift velocity, this force can increase quite sudden, causing a switch from the low mass loss mode to the high mass loss mode. The sudden onset of the drag force causes an outward moving shock and simultaneously, an inward moving rarefaction wave. This enforces a decrease of the density in the subsonic region, leading to both higher grain drift speeds and smaller grain sizes again. The time the rarefaction wave needs to cross the subsonic, dust forming region seems to set the time scale for the mass loss modulation. For more details see Simis et al. (2001).

In order to investigate when on the AGB (i.e. for which T_*, L_*, M_*) the instability can occur, a grid of 20 models was calculated. Each model can be characterised by the density scale height at the photosphere, $\mathcal{H}_0$, which is a function of T_*, L_*, M_*. It was found that large values of the scale height correspond to fairly stationary outflows, for the lowest scale heights the density decrease is too steep to produce a dust driven wind. For intermediate scale heights, mass loss modulation at a time scale of the order of a hundred years could take place. Figure 1 shows the evolution of $\mathcal{H}_0$ for the 5 $M_\odot$, solar metallicity, theoretical AGB track by Vassiliadis & Wood (1993, hereafter VW93). Overplotted, in grey-scales, are the "stability regimes" that follow from the grid of calculations. Instability of the outflow increases with darkness. Note that I do not want to claim that these calculations define exactly the "instability region" on the AGB. In order to do so, many more calculations covering a larger part parameter space, are required. Rather, Fig. 1 should be interpreted as an indication that such a region can exist. For an extended report on these results see Simis (2002).

According to the mechanism proposed here, the shells are formed while the star is still on the AGB. This, or any other mechanism that produces the shells while the star is on the AGB, can only explain the presence of the shells if these do not damp out on a time scale that is shorter than the time between formation and the end of the AGB. Figure 2 shows a calculation by Meijerink & Mellema (2002) of the evolution of the shells at large distance. They applied the density and velocity mod-

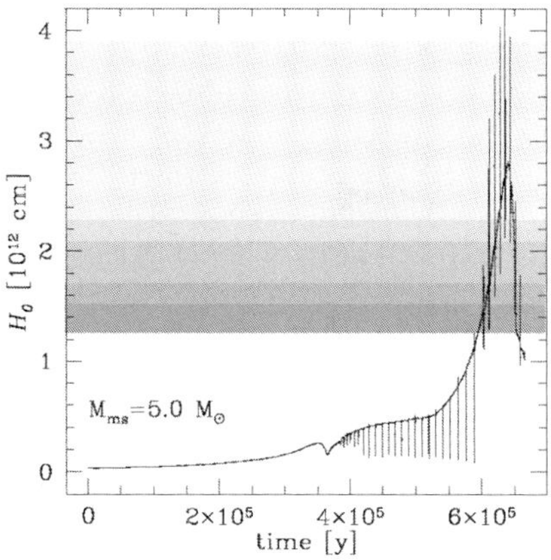

Figure 1. Density scale height near the photosphere for the 5 $M_\odot$ solar track by Vassaliadis & Wood (1993). Grey-scales indicate the "stability regions" retrieved from a grid of hydrodynamic models. Light color indicates fairly stationary outflow, darker regions indicate modulated outflow.

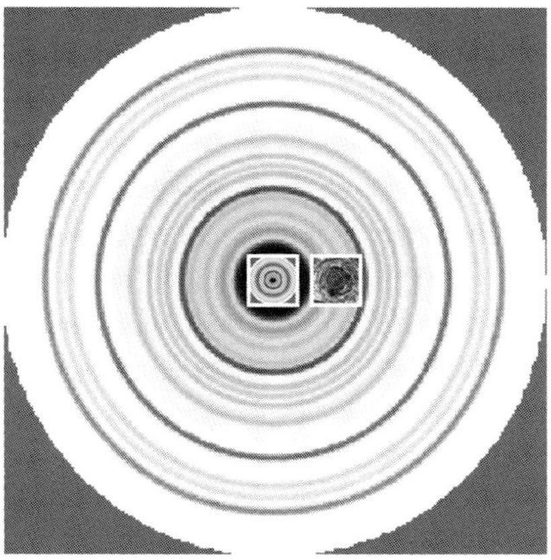

Figure 2. Gas column density of large scale calculation for IRC +10216 (Meijerink & Mellema 2002). Plotted on the same scale: small scale calculation (dust) (Simis et al. 2001) and observed shells (B+V band, scattered light) (Mauron & Huggins 1999).

ulations of the gas at the outer boundary of the IRC +10126 calculation (Simis et al. 2001) as a time dependent inner boundary condition and calculated the long term evolution of the shells. The figure shows the gas density with a average profile subtracted after roughly 30000 years. The outermost shells at that time are at a distance of 0.225 pc. This is comparable to the size of the outermost shells around the Cat's eye Nebula (NGC 6543), they have radii 0.17 pc (assuming a distance to NGC 6543 of 1.0 kpc) (Balick et al. 2001). This calculation shows that mass loss modulation on the AGB does not damp out on a relatively short time scale, but is likely to survive at least until the ionisation by the young WD.

2. Long term AGB calculations

As a first step towards combined stellar evolution and hydrodynamical wind calculations, I have incorporated the time dependent temperature and luminosity from the theoretical tracks by VW93 in the wind calculations. Figure 3 shows the evolution of the stellar temperature, luminosity, mass and the mass loss rate during the last few thermal pulse cycles of a star with initial mass of 3.5 $M_\odot$ and solar metallicity. The temperature evolution of the calculation is prescribed to follow exactly the theoretical track. The variations in the luminosity are due to motion the stellar radius as a result of the stellar pulsation, implemented using

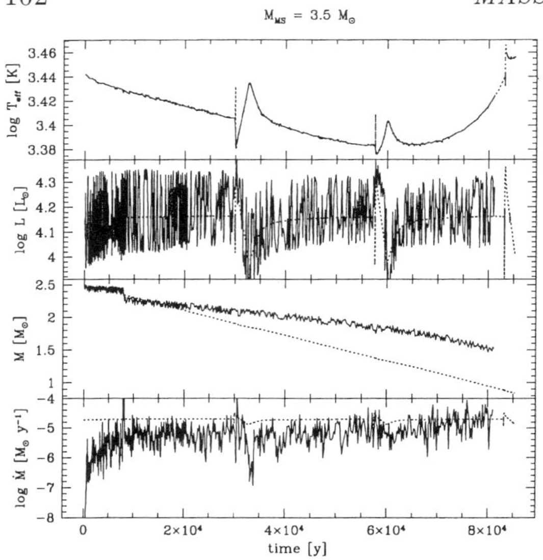

Figure 3 Effective temperature, luminosity, stellar mass and mass loss rate as a function of time during the final three thermal pulse cycles. The dotted lines are the tracks calculated by VW93 for a star with main sequence mass of 3.5 times solar. The full lines are the result from a the model calculation described here.

a piston inner boundary with period 650 days and amplitude 8 $km\ s^{-1}$. Note that the time between two subsequent output events produced by the calculations is much longer than the pulsation period, so that the variation of the luminosity in the plot does not represent the luminosity variation due to the pulsation in detail. The mass loss rate in the hydrodynamical wind calculation is initially smaller than the rate applied to calculated the mass loss in the track from VW93. This illustrates the need for consistently coupling stellar evolution and wind/shell calculations, which is a current topic of research. In such a coupled calculation the stellar evolution calculation provides the time-dependent temperature and luminosity for the wind calculation which then produces a mass loss rate that is needed as an outer boundary condition for the stellar evolution calculation.

Acknowledgments

I thank Kees Dullemond, Rowin Meijerink and Garrelt Mellema for input to this work and the LOC and DFG grant SCHO 394/23-1 for travel support.

References

Balick, B., Wilson, J., & Hajian, A. 2001, AJ, 121, 354
Mauron, N., & Huggins, P. J. 1999, A&A, 349, 203
Meijerink, R., & Mellema, G. 2002, in preparation
Simis, Y. J. W., Icke, V., & Dominik, C. 2001, A&A, 371, 205
Simis, Y. J. W., Dominik, C., & Icke, V. 2002, A&A, submitted
Simis, Y. J. W. 2002, A&A, in preparation
Vassiliadis, E., & Wood, P. R. 1993, ApJ, 413, 641

ASYMMETRIC MASS-LOSS ON THE AGB: EXAMPLES FROM IRAS DATA

R. Weinberger, B. Aryal
Institut für Astrophysik, Universität Innsbruck
Technikerstr. 25, A-6020 Innsbruck, Austria
ronald.weinberger@uibk.ac.at; binil.aryal@uibk.ac.at

Abstract A wide-angle search for extended structures (arcs, filaments, cavities etc.) around ca. 400 Galactic planetary nebulae on IRAS SkyView maps revealed several candidates. Some of these dust structures can hardly be explained by chance superpositions of otherwise unrelated objects. They show clear signs of bi- or quadru-polar shape and will, due to their huge dimensions, probably have started to form already in the AGB phase. If so, then the mass-loss in the AGB phase is not spherically symmetric in a number of cases.

Keywords: AGB stars, asymmetrical mass-loss, IRAS data

1. Introduction

The mass-loss on the AGB and in the subsequent phases are of considerable importance for the understanding of the late stages in the evolution of not too massive stars. Planetary nebulae (PNe) are particularly important examples for these phases: their morphologies and physics are clearly governed by the preceeding mass-loss processes.

It is usually assumed that the mass-loss on the AGB was spherically symmetric and not earlier than in the proto-planetary phase changed to a bipolar mass outflow (see e.g. Hrivnak et al. 2001). Details of this change are however widely unknown, but some kind of disk or ring collimating the outflow appears to exist, possibly created by a companion orbiting the former AGB star.

As to the search for dust in the vicinity of ABG stars and (young) PNe, several papers can be found which show that there are a number of structures like filled shells, detached shells, and arcs that can carry the imprints of the mass-loss history and the structure and density of

Y. Nakada et al. (eds.), Mass-Losing Pulsating Stars and their Circumstellar Matter, 103–106.

the interstellar matter (ISM). For example, Speck et al. (2000) reported the detection of parsec-sized dust shells around two very young PNe. Particularly revealing is an extensive work based on IRAS data by Young et al. (1993) around ~500 red giants, where only 4 shells were found that are very large with D >4 pc. That is, giant structures which would not only be of significance for the late stellar evolution but also for the dynamics of the ISM were hardly found up to now; the reason is that obviously no systematic search was conducted within many degrees around AGB stars and PNe. As to PNe we have carried out such a systematic search based on the IRAS NASA SkyView maps.

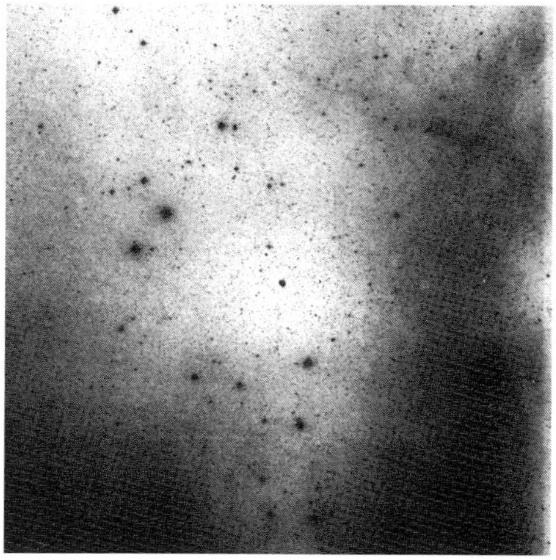

Figure 1 An ultrahigh-contrast copy of a part of a Palomar Observatory Sky Survey red print with a field size of $3^o \times 3^o$ around the high-latitude planetary nebula NGC 4361 (black centrally located spot).

2. Motivation and Search

The original motivation for our search for huge dust structures around PNe was our detection of a very large 'cavity' around the high-latitude PN NGC 4361 on prints of the Palomar Sky Survey; since the exceedingly faint wide-spread emission in this region is reflected light from thin ('cirrus') dust clouds, we considered this as a hint that the PN or its precursor might have (partly) blown off the ISM around this object. This cavity is also evident at 100 μm IRAS SkyView maps. However, a chance superposition of NGC 4361 and a hole in the ISM cannot be ruled out. Therefore a couple of years ago we have started a systematical search on IRAS SkyView maps for huge structures around PNe in order to find additional, more convincing, candidates.

Regions with sizes of 2°, 5°, 10°, and 15° around ~400 PNe were examined at 12, 25, 60, and 100 μm. The ~400 PNe represent ~1/4 of the registered Galactic PNe population and comprise all NGC, IC, and Abell objects as well as other extended PNe.

3. Results

Huge dust structures around one dozen PNe were found by us. A few of them look quite convincing and thus may serve as proofs that the phenomenon of these giant features, i.e. arcs and/or holes etc. indeed exists.

NGC 6826 and its neighborhood serves as the first example. At 60 μm, in a $3.5^o \times 3.5^o$ region, one can see (image scaling: histogram equalization) two large infrared emitting arcs spaced by about 30 pc (assuming a distance of d = 1.1 kpc for the PN). The western arc, displayed as a contour map, clearly shows that it is connected with the PN. The position angle connecting both arcs is practically identical with an axis connecting two deviations from circular symmetry in the halo of the PN - probably an effect of interaction with the matter of the arcs. The arc's mass is a few solar masses. We believe that these arcs are mainly swept-up ISM, and if the (constant) wind velocity was 15 km/sec, they were created about $1\,10^6$ years ago. That means, this bipolar mass-loss has commenced in the AGB phase.

Another example, this time for a huge cavity, is the vicinity of NGC 2899. This object probably stems from a star of several solar masses. Here, at 60 μm, we discovered a distinct cavity in infrared emission, 14 pc wide, which turned out to be actually quadru-polar at higher resolution in the infrared. The axes of this quadru-polar hole coincide with the axes of the strikingly butterfly-like PN. Other features - like a dust arc in the south-west - support our interpretation that the cavity is related to NGC 2899. Using the standard formulae for the evolution of bubbles in the ISM, and assuming a mass-loss rate of 10^{-7} solar masses per year for the precursor of the PN, an ambient gas density of 0.1 per cm^3, and wind velocites of 150 and 15 km/sec respectively, the formation of the hole must have started at least $5.5\,10^5$ years ago, i.e. in the AGB phase.

A third example - this time representing a shell and two bipolar dust blobs - is NGC 1514. The location of the midpoints of the lines joining these two pairs of blobs fits to the proper motion vector of the PN. Again, by assuming a constant wind velocity of 15 km/sec, the whole phenomenon has commenced in the AGB phase.

Further, the PN Lo 4 shows two pairs of blobs; the inner pair, at higher resolution (100 μm) looks as if a precessing jet is at work. Again, we believe that already in the AGB phase the whole phenomenon has begun to form.

Several other examples exist, but some of them will be not real, as we think, but represent chance projections.

4. Discussion

If - at least some - of the huge bi-polar dust structures and/or cavities are real, then

- spherically symmetric mass-loss on the AGB is not mandatory
- huge dust structures are much more frequent, since many are hidden by foreground/background dust, and others are partly destroyed by interaction with clouds in the ISM
- such ejected matter from stars of moderate/small mass has long-range effects, i.e. it may dynamically influence the ISM over many pc, representing a new process shaping the ISM
- the concept of companions to AGB stars gains further support

Our next goals will be to search for huge dust structures also around young white dwarfs, to attempt optical spectroscopy (perhaps we may find signs of mild shocks), and to try to make models of these phenomena.

Acknowledgments

The authors thank the Austrian Fonds zur Förderung der wissenschaftlichen Forschung (FWF) for the support of the project P15316.

References

Hrivnak, B.J., Kwok, S., & Su, K.Y.L. 2001, ApJ, 121, 2775
Speck, A.K., Meixner, M., & Knapp, G.R. 2000, ApJL, 545, L145
Young, K., Phillips, T.G., & Knapp, G.R. 1993, ApJ, 409, 725

THE EXTENDED ATMOSPHERE AND EVOLUTION OF THE RV TAURI STAR, R SCUTI

M. Matsuura, A.A. Zijlstra
Department of Physics, UMIST, P.O. Box 88, Manchester M60 1QD, UK
(M.M.) m.matsuura@umist.ac.uk

I. Yamamura
Institute of Space and Astronautical Science, Kanagawa, 229-8510, Japan

T.R. Bedding
School of Physics, University of Sydney 2006, Australia

Abstract We found molecular bands in absorption and in emission in the infrared spectra of the RV Tau stars, R Sct. This is a unique property among the RV Tau stars, but rather commonly seen in Mira variables. The period evolution of this star does not fit with an ordinary post-AGB scenario. Therefore, we argue an alternative choice of the evolutional stage; R Sct is not a post-AGB star but an AGB star in the helium-burning phase of the thermal-pulse cycle.

Keywords: stars: AGB – stars: atmospheres – stars: individual: R Sct

1. Introduction

RV Tau stars are pulsating variables characterized by alternating deep and shallow minima in their light curves. They are generally considered to be post-AGB stars with low initial masses (Jura 1986).

R Sct (HR 7066; HIP 92202) has a reported period of 147 days (Kholopov et al. 1988). The effective temperature varies from 4750 to 5250 K (Shenton et al. 1994); the spectral type may vary as late as M3 at minimum phase (Kholopov et al. 1988). This is the one of the latest spectral type among the RV Tau stars. From the analysis of sodium lines (Lèbre & Gillet 1991), pulsation activity is expected in this star,

Y. Nakada et al. (eds.), Mass-Losing Pulsating Stars and their Circumstellar Matter, 107–110.

which is not common in RV Tau stars. Therefore, it is interesting to see the nature of R Sct.

2. The infrared spectra

We obtained the SWS data for R Sct from the ISO Data Archive, and we analyse spectra from 2.4 to 17 μm. R Sct was observed on 10 March, 1996. The spectral resolution is $\lambda/\Delta\lambda$ =300–500, depending on the wavelength.

We calculate molecular spectra using a circular slab model (Matsuura et al. 2002 and Yamamura et al. 1999) and fit the observed spectra. The fitted results are shown in Fig. 1. The observed features are identified with H_2O, SiO, CO, and CO_2 bands either in absorption or in emission. H_2O and CO_2 are unstable in the photosphere of a star with an effective temperature above ~3500 K. The excitation temperatures of molecules are 4000 K for CO, 1200–2200 K for H_2O, and about 800 K for CO_2. The excitation temperatures are lower than expected temperature for the gas in the photosphere. These molecules should be in the extended atmosphere, which is located above the photosphere, and cooler than the photosphere. The extended atmosphere is often found in AGB stars and is formed by pulsations. The detail of the extended atmosphere is discussed by Yamamura et al. and Cami et al. in these proceedings.

The dust mass-loss rate which we estimate from the 10 μm silicate emission is $\dot{M}_{\rm d} = 1.5 \times 10^{-11}\,{\rm M}_\odot\,{\rm yr}^{-1}$, which is consistent with previous estimates by Alcolea and Bujarrabal (1991). The CO-derived gas mass-loss rate from ^{12}CO J=1–0 and 2–1 lines is $\dot{M} = 2 \times 10^{-7}\,{\rm M}_\odot\,{\rm yr}^{-1}$ (Bujarrabal et al. 1988). CO emission represents the mass-loss rate 1000 years ago, while silicate emission represents current mass-loss rate. The implication is that the mass-loss rate of R Sct has reduced by about a factor of 100 (Alcolea & Bujarrabal 1991).

3. Period evolution and the evolutional stage

RV Tau stars are commonly assumed to be post-AGB stars with low initial mass. The reduction of the mass-loss rate is in favour for such an interpretation.

We have analysed published observations of R Sct to search for evidence of period decrease, which would be found if a star evolved from low temperature to high temperature in HR diagram. We used wavelet analysis (Bedding et al. 1998) to measure period evolution in both the 140-day period and its 70-day harmonic. Over the past two hundred years there is no evidence for a change of more than 2 days in the harmonic period, corresponding to 4 days in the main period (Fig. 2). This

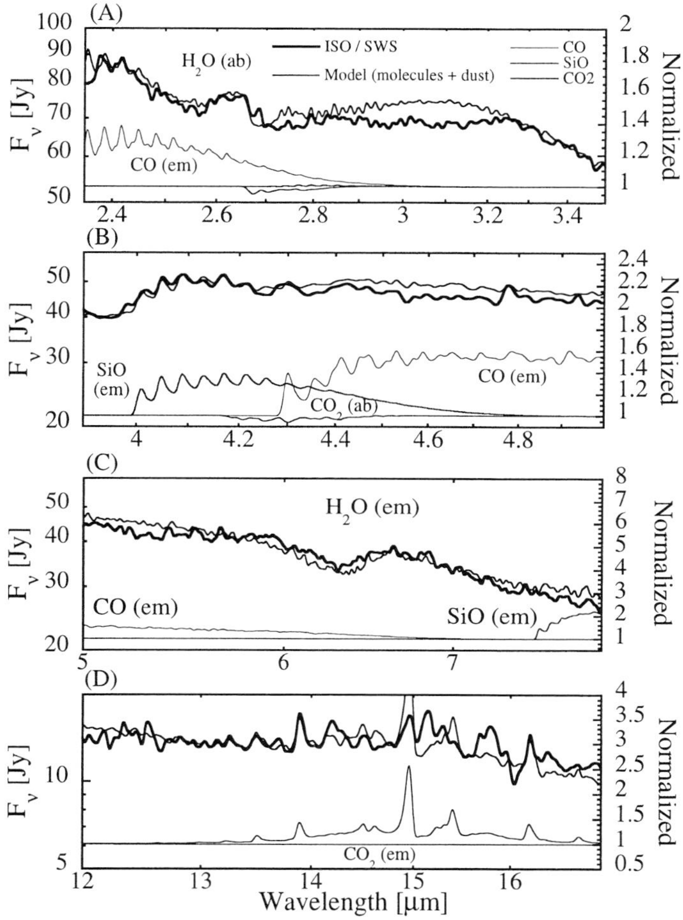

Figure 1. The ISO/SWS spectra of R Sct compared with the synthesized spectra (the left scale). Each molecular component is normalized to the background source and is indicated on the right scale. The quality of the ISO/SWS data is lower above 13 μm and the line intensity of each CO_2 line is not reliable.

limit is a factor of 5 below the predictions from effective temperature-luminosity relation.

It is therefore possible that R Sct is not a post-AGB star but still an AGB star. A sudden decrease of the mass-loss rate on the AGB is expected following the thermal pulse: the lower luminosity during the quiescent helium burning phase gives both a higher temperature and a lower mass-loss rate (e.g. Zijlstra et al. 1992, Vassiliadis & Wood 1993). The period evolution of R Sct slows down and reverses during the helium-burning phase. The phase lasts for ~ 5000 yr. This phase may fit the period evolution, and time scale for the detached dust shell, quite well.

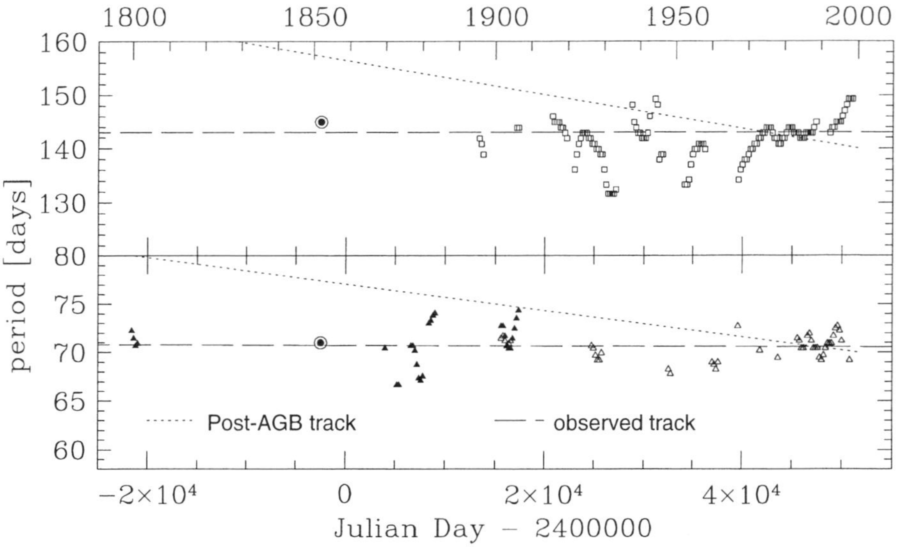

Figure 2. The period variation of R Sct over 200 years. The upper figure is for the primary period and the lower figure for the harmonic period. The predicted period variation from the post-AGB evolutionary track is shown by the dotted lines. The observed variation is shown by the dashed line, where the harmonic period is fitted with the linear function of $P = [70.71 \pm 0.33] - [0.03 \pm 0.10] \times 10^{-4} \times (\mathrm{JD} - 240\,000)$: i.e., the period is constant over 200 years (long dash line).

References

Alcolea, J., & Bujarrabal, V. 1991, A&A, 245, 499
Bujarrabal, V., Bachiller, R., Alcolea, J., & Martin-Pintado, J. 1988, A&A, 206, L17
Bedding, T. R., Zijlstra, A. A., Jones, A., & Foster, G. 1998, MNRAS, 301, 1073
Jura, M. 1986, ApJ, 309, 732
Kholopov, P. N., Samus, N. N., Frolov, M. S., et al. 1988, General Catalogue of Variable Stars. 4th Ed., Nauka Publishing House
Lèbre, A., & Gillet, D. 1991, A&A, 251, 549
Matsuura, M., Yamamura, I., Zijlstra, A.A., & Bedding, T.R. 2002, A&A, 387, 1022
Shenton, M., Monier, R., Evans, A., et al. 1994, A&A, 287, 866
Vassiliadis, E., & Wood, P. R. 1993, ApJ, 413, 641
Zijlstra, A. A., Loup, C., Waters, L. B. F. M., & de Jong, T. 1992, A&A, 265, L5
Yamamura, I., de Jong, T., Onaka, T., Cami, J., & Waters, L.B.F.M. 1999, A&A, 341, L9

MONOCHROMATIC RADII: A TOOL OF MIRA DIAGNOSTICS

M. Scholz

Institut für Theoretische Astrophysik der Universität Heidelberg, Tiergartenstrasse 15, 69121 Heidelberg, Germany, and School of Physics, University of Sydney, NSW 2006, Australia

scholz@ita.uni-heidelberg.de

Abstract The shape and the wavelength-dependence of the brightness distribution on the stellar disk at different variability phases and cycles provides a powerful tool of diagnostics of the physical and geometric structure of the atmosphere of M-type Mira variables. In particular, the stratification of outer atmospheric layers may strongly affect brightness distributions and derived monochromatic diameters in a characteristic way.

Keywords: monochromatic radii, Mira variables, Mira atmosphere

1. Introduction

The classical way of analyzing the physical structure of a stellar atmosphere is based on the appearance of spectral features (run of the continuum, equivalent widths and profiles of lines, strengths of molecular bands in cool stars). In the case of an M-type Mira variable, analysis is considerably complicated by the dynamic structure of the atmosphere involving a velocity stratification of outflowing and infalling matter as well as flat density gradients resulting in an extreme geometric extension of the atmosphere (cf., e.g., models of Bessell et al. 1989, 1996; Hofmann et al. 1998 (HSW98); Scholz & Wood 2000). On the other hand, models predict that the shape of monochromatic center-to-limb variations (CLVs) of intensity on the stellar disk may be highly sensitive to the physical and geometric structure of the Mira atmosphere. Thus, high-accuracy observation of diameters or even direct observation of CLV shapes at different wavelengths should be well suited for probing the atmospheric stratification.

Y. Nakada et al. (eds.), Mass-Losing Pulsating Stars and their Circumstellar Matter, 111–114.

2. Diameters

Conventionally, measurements of monochromatic diameters in the light of the continuum and of molecular bands of different strength are considered to be a first step towards determining the size of a Mira star as well as the geometry of its atmosphere. Since, however, diameters are no unique quantities but have to be defined in a sensible way in the case of a geometrically extended stellar atmosphere, and since the assignment of a diameter to the CLV (which is the only observable quantity) may be quite difficult, analysis procedures are not straight-forward (cf., e.g., Baschek et al. 1991; HSW98; Jacob & Scholz 2002 (JS02)).

(1) Monochromatic optical-depth radii, defined by the distance r from the star's centre where the monochromatic optical depth equals unity, $R_\lambda = r(\tau_\lambda{=}1)$, are usually well determined by the position of the steep flank of the CLV in case of a near-UD CLV shape. For instance, real-continuum CLVs in the optical to near-IR region are close to a UD in Miras. There is, however, no a-priori rule assigning R_λ to a specific point on the brightness distribution in case of a strongly darkened disk. In particular: (i) UD-fit radii may deviate substantially from R_λ and may be baseline-dependent in case of single-baseline observations. (ii) The FWHM of Gaussian-like CLVs predicted by models for strong molecular band absorption in the Wien parts of the Planck function cannot directly be related to a diameter-type quantity. (iii) Specific problems occur in cases of a two-component appearance of the CLV caused by a modestly absorbing outer "shell" in near-continuum bandpasses in the near-maximum to Rayleigh-Jeans parts of the Planck function.

(2) Observations in bandpasses of finite width and corresponding filter radii (Scholz & Takeda 1987) may be hard to interpret or even useless for analysis if bandpass integration includes different spectral features of different strength.

(3) The Rosseland radius of the star, $R = r(\tau_{\rm Ross}{=}1)$, which is used in model computations for describing global properties of the Mira variable, i.e. "the size" of the star, is no observable quantity but has to be correlated to measured monochromatic or filter radii via model considerations. Usually, real-continuum radii $R_{\rm cont}$ are close to R but special problems may occur for very cool and very extended Mira atmospheres in which molecular band absorption contributes significantly to the Rosseland opacity (HSW98). Also, the radius $R_{\rm p}$ of the hypothetical non-pulsating parent star of the Mira variable which is the relevant quantity in, e.g., the depiction of the evolutionary path or the period-radius-relation can only be found via model considerations.

3. Brightness distributions

Reconstruction of the monochromatic CLV from accurate multi- baseline interferometric observations allows direct comparison with CLVs predicted by different models. (If reconstruction of the full CLV is not possible, limited information about the CLV shape may be obtained from the single-baseline-dependence of UD-fit diameters (cf. JS02)). For instance, the width of the Gaussian-like CLV generated by strong TiO absorption at optical wavelengths depends on parameters, phases and cycles of Mira models and is extremely sensitive to changes of temperatures in the uppermost layers (Jacob et al. 2000). Observed Gaussian-like CLVs have, indeed, been reported (e.g. Tuthill et al. 1995; Haniff et al. 1995; Young et al. 2000; Hofmann et al. 2000, 2001), including changes with phase and cycle. Another example are two-component-type CLVs in near-continuum bandpasses in the near IR, generated by water absorption in the upper atmosphere, which also depend strongly on parameters, phases and cycles of Mira models, and which probe the position and thickness of the water "shell". Such a two-component-type CLV (protrusion-type water component, see JS02) was observed by Perrin et al. (1999) in R Leo in the K bandpass. According to Bedding et al. (2001), two-component-type CLVs with typical characteristics are also generated by dust formed in very high atmospheric layers. Thus, CLV shapes may provide detailed information about the physical and geometric structure of the Mira atmosphere (which in turn is related to the star's global parameters) as well as about flaws resulting from inadequate modelling assumptions.

4. UD-fit diameters in near-continuum bandpasses

The behaviour of UD-fit diameters inside near-continuum bandpasses in the near IR was recently studied by JS02. Mira models predict typical diameter differences within bandpasses which depend on stellar parameters, phases and cycles as well as on model assumptions. For instance, diameters are larger in K than in H and still larger in L and change with λ within the K bandpass, in qualitative agreement with observations (cf. JS02; see also contributions at this workshop by B. Mennesson, G. Perrin and H. Takami), and measured diameters may be compared to those predicted by models. As an example, observed UD-fit radii of S Lac in the λ-region of the K bandpass are entered in Figure 1 together with model predictions, showing a wide range from fair agreement to total disagreement with measurements and discriminating clearly between physically (or numerically) adequate and inadequate models.

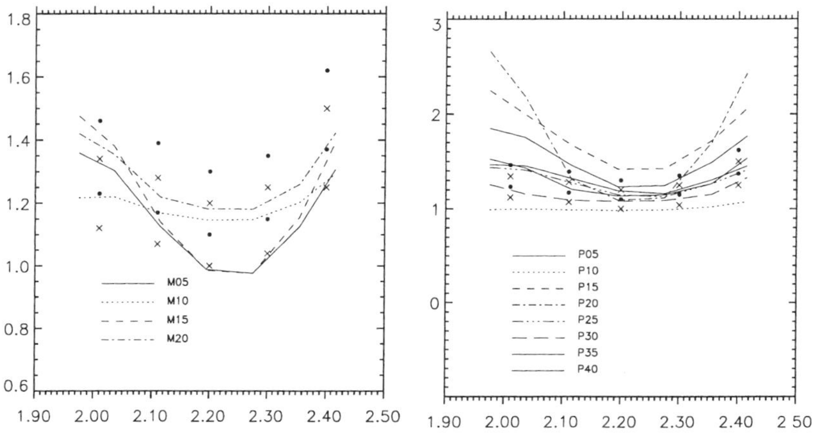

Figure 1. Predicted UD-fit radii (after JS02) in 5 narrow bands inside the K bandpass and 2 narrow bands just outside both edges of the K bandpass for the models of the M (left) and the P (right) series of HSW98 compared to observed UD-fit radii in S Lac (Thompson et al. 2002). The model series cover phases of several cycles (M05 = phase 0.5 model of the M series, etc.). Abscissa is central wavelength in μm of narrow bands. Ordinate is UD-fit radius in units of the parent star radius R_p. Observed values (slightly different bandpasses; not sorted for phases and cycles) are shown as crosses and dots where the value at 2.2μm was set to 1.0 (lower crosses), 1.1. (lower dots), 1.2 and 1.3 R_p.

References

Baschek, B., Scholz, & M., Wehrse, R. 1991, A&A, 246, 374
Bedding, T. R., Jacob, A. P., Scholz, M., & Wood, P. R. 2001, MNRAS, 325, 1487
Bessell, M. S., Brett, J. M., Scholz, M., & Wood, P. R. 1989, A&A, 213, 209
Bessell, M. S., Scholz, M., & Wood, P. R. 1996, A&A, 307, 481
Haniff, C. A., Scholz, M., & Tuthill, P. G. 1995, MNRAS, 276, 400
Hofmann, K.-H., Balega, Y., Scholz, M., & Weigelt, G. 2000, A&A, 353, 1016
Hofmann, K.-H., Balega, Y., Scholz, M., & Weigelt, G. 2001, A&A, 376, 518
Hofmann, K.-H., Scholz, M., & Wood, P. R. 1998, A&A, 339, 846 (HSW98)
Jacob, A. P., Bedding, T. R., Robertson, J. G., & Scholz, M. 2000, MNRAS, 312, 733
Jacob, A. P., & Scholz, M. 2002, to be published (JS02)
Perrin, G., Coude du Foresto, V., Ridgway, S. T. et al. 1999, A&A, 345, 221
Scholz, M., & Takeda, Y. 1987, A&A, 186, 200 (erratum: 196, 342)
Scholz, M., & Wood, P. R. 2000, A&A, 362, 1065
Thompson, R. R., Creech-Eakman, M. J., & van Belle, G. T. 2002, ApJ, in press
Tuthill, P. G., Haniff, C. A., & Baldwin, J. E. 1995, MNRAS, 277, 1541
Young, J. S., Baldwin, J. E., Boysen, R. C. et al. 2000, MNRAS, 318, 381

THE WIGNER DISTRIBUTION FOR PULSATION BEHAVIORS IN POST-AGB

Toshiki Aikawa
Tohoku Gakuin University
Sendai, 981-3193 Japan
aikawa@cs.tohoku-gakuin.ac.jp

Abstract Some of post-AGB stars show irregular pulsation with small amplitudes. Sometimes they appear as regular plusators but sometimes they show quite irregular behaviors. We believe that both these behaviors are realized in a steady state of nonlinear pulsation. The Wigner distribution have been used in many sound analyzes for a tool of the time-frequency distribution. So we apply the method to the time-frequency distribution of irregular pulsation of post-GAB stars to characterize the nonlinear pulsation. We first use this to the time-series data of model pulsation.

Keywords: nonlinear pulsation, post-AGB stars

1. post-AGB pulsation

The pulsation of typical post-AGB stars are characterized as follows:
(1) low-amplitude pulsation.
(2) irregular pulsation.
For example, we show the photometric variations of 89Her, and HD161796.

2. model pulsation

We use the following parameters for post-AGB pulsators as less-massive supergiants: mass and effective temperature are fixed as $M = 0.8M_\odot$ and $Te = 6300K$, Luminosity is treated as a free parameter with a range of $L = 1500L_\odot - 4000L_\odot$.

Linear analysis of radial pulsation is carried out with the following equation:

$$\begin{aligned} \omega^2 X &= G_1 X + \varepsilon G_2 Y \\ i\omega Y &= K_1 X + K_2 Y \end{aligned} \tag{1}$$

Y. Nakada et al. (eds.), Mass-Losing Pulsating Stars and their Circumstellar Matter, 115–116.

The parameter , ε controls the coupling between the acoustic wave and the thermal wave. We find that most of lower modes are pulsationally stable including the fundamental mode, but with the range of luminosity one strange mode which has strong negative dumping exists.
We performed simulations of nonlinear pulsation for those models. We found irregular pulsation with small amplitudes for luminous models, while regular pulsation for less luminous models. Thus, we conclude that the degree of complexity of pulsation increases with the values of luminosity in the models.

3. Wigner distribution

Time-series data of the light curves obtained from the nonlinear models are demonstrated with the Wigner distribution.
If the signal, $s(t)$, is sampled every Δt, then we obtain the discrete-time Wigner distribution as

$$W(m\Delta t, \omega) = 2\Delta t \sum_{n} s[(m+n)\Delta t]s^{*}[(m-n)\Delta t]exp(-j2\omega n\Delta t) \quad (2)$$

In practice, smoothing window functions are applied to the distribution (discrete-Pseudo Wigner Distribution). We applied the method to the photometric variations of the models.

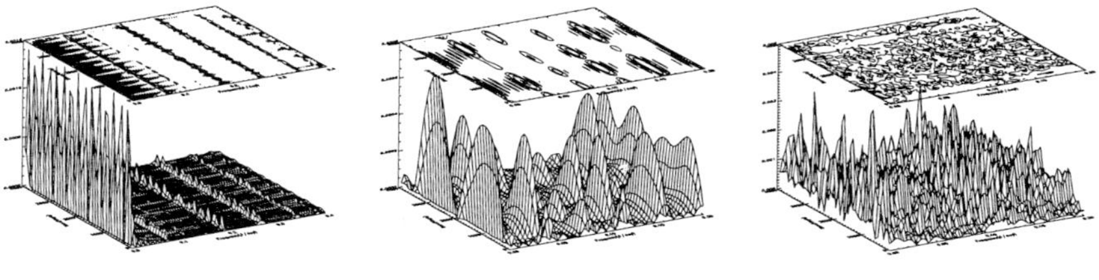

Figure 1. 3D maps of the Wigner distribution of light curve for models $L = 1500, 2800, 4000 L_{\odot}$. The frequency and the time in the distribution are drawn with x-axis, and y-axis.

4. Conclusions

We show the Wigner distribution on irregular pulsation of theoretical models of post-AGB stars. We find that the distribution can indicate the degree of the complexity contained in the time series data. As a next step, we are going to apply the method to the observational data of post-AGB pulsation like as 89 Her and HD167196.

AN IMPLICATION ON THE EXCITATION MECHANISM OF AGB VARIABLES

Mine Takeuti
Astronomical Institute, Tohoku University, Aoba, Sendai 980-8578 Japan
takeuti@astr.tohoku.ac.jp

Yasuo Tanaka
Faculty of Education, Ibaraki University, Mito 310-8212 Japan
tanaka@mito.ipc.ibaraki.ac.jp

Abstract Recent progress in observational and theoretical studies implies a scheme about the excitation mechanism of AGB variables. The coupling between the pulsation and the convective motion of the stellar envelope is to be studied as the cause of AGB variables.

Keywords: Variable stars, stellar pulsations, nonlinear dynamics, large-scale photometry, AGB variables

Recent Observational Results. Recent observational results of long-period variables suggesting the nature of their pulsations are **(a)** the multiplicity of the period-luminosity relation reported on the LMC (Wood, Alcock, Allsman, et al. 1999; Wood 2000; Cioni, Marquette, Loup, et al. 2001; Noda, Takeuti, Abe, et al. 2002); **(b)** the period transition of both semi-regular stars and Mira stars in the vicinity of the Sun (Kiss, Szatmáry, Cadmus, and Mattei 1999), and in the LMC (Noda, Takeuti, et al. 2002); and **(c)** the color-magnitude relation of red variables of the LMC studied in Noda, Takeuti, Abe, et al. (2002), and more completely in Noda, Takeuti, et al. (2002).

Low-dimensionality of Semi-regular Variability. The important progress of the theoretical study is the confirmation of low dimension nature of the semi-regular pulsation not only on RV Tauri stars but also on AGB variables. (Buchler and Kolláth 2001; Buchler, Kolláth, and Cadmus 2001; Ambika, Kembhavi, and Mattei n.d.). Such a semi-

Y. Nakada et al. (eds.), Mass-Losing Pulsating Stars and their Circumstellar Matter, 117–118.

regularity is evidence for highly adiabatic and/or the multimode pulsation.

Excitation Mechanism of Red Variable Instability Strip. Convection works to make the delay between the radial motion of gas and heat flow. This will make the pulsation enhanced (Xiong, Deng, and Cheng 1998). Such a mechanism is different to κ-mechanism.

Speculation on the AGB Variables. The most acceptable explanation of the multiplicity of the period-luminosity relation would be that each sequence corresponds to the different mode of stellar pulsation. This idea explains why the short-period sequence is bluer than the others on the color-magnitude diagram. It seems too early to discuss the modes of pulsation, if we have no plausible theoretical period-luminosity relation still now (Barthès and Luri 2001). The period transition will be evidence for the coupling of unsteady convection and pulsation. The occasional change of the period of stellar pulsation is hardly surprising.

Further Studies to Examine the Speculation. Monitoring of light curves of AGB variables both in the near-infrared and in the optical wavelengths, and spectroscopic studies to obtain more plausible data of the variation of the stellar radius are important. Modelling nonlinear oscillation including the effect of convection will be especially interesting.

References

Ambika, G., Kembhavi, A.K., & Mattei, J.A. : n.d., presented at the workshop on Celestial Mechanics and Dynamical Systems held at IUCAA, Pune, 2001.

Barthès, D., & Luri, X. 2001, A&A, 365, 519

Buchler, J.R., & Kolláth, Z. 2001, nlin.CD/0109028 (in French)

Buchler, J.R., Kolláth, Z., & Cadmus, R. 2001, astro-ph/0106329

Cioni, M.-R., Marquette, J.-B., Loup, C. et al. 2001, A&A, 377, 945

Kiss, L.L., K. Szatmáry, R.R. Cadmus, JR, & J.A. Mattei, 1999, A&A, 346, 542

Noda, S., Takeuti, M., Abe, F. et al. 2002, MNRAS, 330, 137

Noda, S., Takeuti, M. et al. 2002, in preparation

Wood, P.R. 2000, PASA, 17, 18

Wood, P. R., Alcock, C., Allsman, R.A. et al. 1999, in IAU Symp. 191, Asymptotic Giant Branch Stars, eds. T. le Bertre, A. Lèbre, & C. Waelkens (San Fransisco: ASP), 151

Xiong, D. R., L. Deng, L., & Cheng, Q.L. 1998, ApJ, 499, 355

EVOLUTION OF BINARIES CONTAINING AN AGB GIANT

Adam Frankowski
N. Copernicus Astronomical Center, ul. Rabiańska 8, 87-100 Toruń, Poland
frankowski@ncac.torun.pl

Abstract Results from numerical modelling of binaries containing an AGB giant and a companion accreting from its wind are presented.

Keywords: AGB, binary stars

1. Introduction to the model

This paper presents some results from numerical modelling of binaries containing an AGB giant and a companion accreting from its wind. Main assumption of the model are:

- The influence of the secondary on an AGB giant can be considered as a correction applied to a single star evolution.
- The secondary in a binary system affects directly only mass loss from the primary. The secondary accretes from the wind and can enhance it.
- Orbits are circular.
- The angular momentum loss and transfer considerations include orbital motions and rotation of the primary's envelope (rotation of the primary's core and of the secondary are neglected). Tidal interactions are taken into account.
- The rotating AGB giant can possess a magnetic field that causes the wind to corotate with the envelope up to Alfven radius.

Single star evolutionary tracks of Blöcker (1995a; 1995b) have been used. The formula for mass loss modification by the presence of a companion is similar to the one of Han *et al.* (1995) with the Reimers wind replaced by an appropriate Blöcker's formula. If the giant fills its Roche lobe, a check for stable RLOF versus CE is performed.

Y. Nakada et al. (eds.), Mass-Losing Pulsating Stars and their Circumstellar Matter, 119–120.

2. Results

Figures 1–2 present sample results for systems containing an AGB giant of ZAMS mass $3M_\odot$ (triangles) and $5M_\odot$ (squares). Logarithm of the mass ratio $q = M_1/M_2$ is plotted versus logarithm of the orbital period P (in days). Initial mass ratios vary from 0.1 to 1. Initial periods cover the range from 0.7 yr to 10 yr in the case of $3M_\odot$ primary and from 1.6 yr to 38 yr for $5M_\odot$ primary. Open symbols denote initial positions of the modelled binaries in this diagram, filled ones - their final loci. Skeletal symbols mark systems that have undergone stable RLOF. Fig. 1 shows the simplest case of no wind enhancement and no rotation of AGB envelope. Fig. 2 displays results from calculations with both rotation and magnetic field ($B = 10$G on the stellar surface, dipole geometry) included.

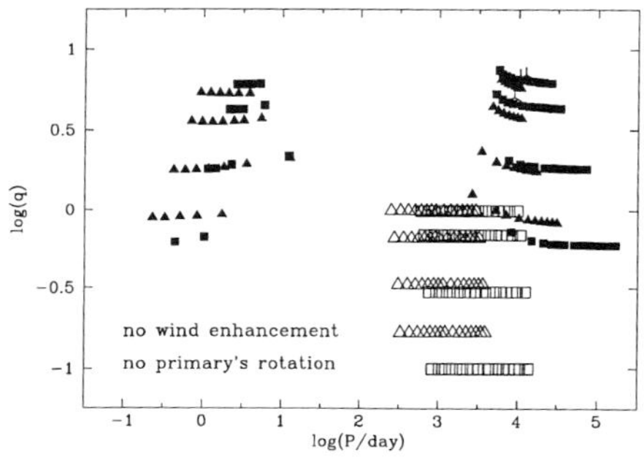

Figure 1. Mass ratio versus orbital period for no wind enhancement and no AGB star rotation case.

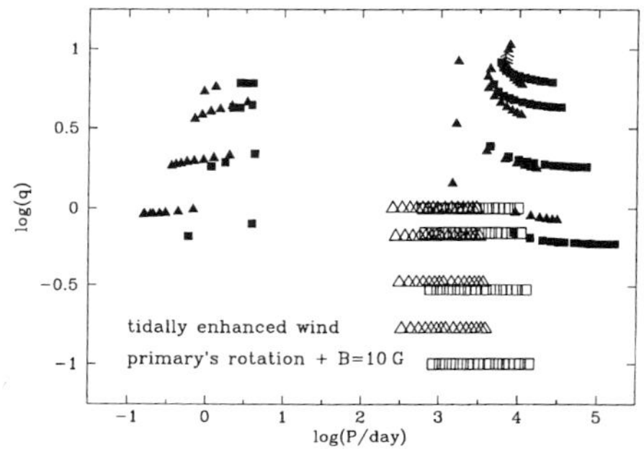

Figure 2. Same as Fig. 1, but with AGB star rotation and magnetic field.

Acknowledgments

This work was supported from the grant No. 2.P03D.020.17 of the Polish State Committee for Scientific Research.

References

Blöcker, T. 1995, A&A, 297, 727

Blöcker, T. 1995, A&A, 299, 755

Han, Z., Eggleton, P.P., Podsiadlowski, P., & Tout, Ch. A. 1995, MNRAS, 277, 1443

ON THE MOTION OF AN INFINITESIMAL MASS IN THE GRAVITATIONAL POTENTIAL OF A BINARY STAR SYSTEM

L.M. Saha
Department of Mathematics, Zakir Husain College, New Delhi-110002 India

M.K. Das
Department of Physics, Sri Venkateswara College, New Delhi, India

Yasuo Tanaka
Faculty of Education, Ibaraki University, Mito 310-8212 Japan
tanaka@mito.ipc.ibaraki.ac.jp

Abstract We have studied the motion of an infinitesimal mass in the gravitational potential of a binary star system. In this study the effect of finite amplitude motion, ϵ, on the stability of motion around the Lagrangian point L_4 has been investigated using a multiple time scale method. The binary orbit is taken to be of eccentricity e.

Keywords: Binaries, celestial mechanics, circumstellar matter, stability

In recent years several investigations have been made on the motion of an infinitesimal mass in the gravitational field of a binary star system (cf. Murray and Dermott 1999; Benest 1996). In this paper we confine ourselves to the planar motion only and further we take the orbit of the binary system to be elliptical i.e., the primary moves in an orbit of eccentricity, e. Also the effect of radiation, emitted from the primary, on the motion has been considered. This is mainly due to the fact that in interstellar space, the ratio of radiation to gravitational force, $\beta = F_r/F_g$, varies in the range of $3.4 \times 10^{-2} - 10^3$ (assuming the constituents to be water, quartz and graphite) for stars of spectral range $G_0 - O_5$. In binary star system orbital eccentricities could also be quite high e.g., for η CrB, $e = 0.28$; for α Cen, $e = 0.52$. This way both radiation and orbital eccentricity provide additional control parameters for the present

Y. Nakada et al. (eds.), Mass-Losing Pulsating Stars and their Circumstellar Matter, 121–122.

dynamical system. The mathematical formulation of the problem is given in the following section.

The equations governing the motion of an infinitesimal mass under the gravitational potential and radiation from the primaries rotating about its center of mass, taking into account the study of Ragos and Zagouras (1988). Following Bennet (1965), Szebehely (1967) and Simmons et al. (1985) we may write the nonlinear differential equations governing the present dynamical system as (cf. Saha et al. 2002) In order to study the stability of motion around L_4 using analytical methods, we must first consider the linearized system. We use the method of multiple time scale, and deduce the changes caused in the value of the parameter $\bar{\mu}$ due to finite amplitude motion, radiation effect of the primaries and finite eccentricity of the orbit. We observe that for the stability of motion, the finite amplitude motion, ϵ, around L_4 changes the limit $\bar{\mu}$ imposed by linear study, to $\mu = \bar{\mu} + \mu_p \epsilon^2$. Here for a given ϵ, μ_p represents the magnitude of the change caused by finite amplitude, radiation effect of the primaries and the eccentricity of the binary orbit.

We find that the finite amplitude motion around L_4 significantly affect the critical value $\bar{\mu} = 0.03859$ for stability of motion for both nonradiating and radiating binary system. Similar results are also obtained for different radiating binary system.

Acknowledgement. We thank the University Grants Commission, New Delhi for supporting the present work. LMS would like to thank the organizers of this meeting for the local hospitality.

References

Bennet, A. 1965, Icarus, 4, 177
Benest, D. 1996, A&A, 314, 977
Murray, C.D., & Dermott, M. 1999, Solar System Dynamics, CUP
Ragos, O., & Zagouras, C. 1988, Cel. Mech., 44, 135
Saha, L.M., Das, M.K., & Tanaka, Y. 2002, MNRAS, communicated
Simmons, J.F.L., Mcdonald, A.J.C. & Brown, J.C. 1985, Cel. Mech., 35, 145
Szebehely, V. 1967, Theory of Orbits, Academic Press

NUMERICAL INVESTIGATION OF THE STABILITY OF MOTION OF A MASS ELEMENT EJECTED FROM A COMPONENT OF A BINARY STAR SYSTEM

M.K. Das
Department of Physics, Sri Venkateswara College, New Delhi India

Manabu Yuasa
Research Institute for Science and Technology, Kinki University, Osaka 577-8502 Japan

L.M. Saha
Department of Mathematics, Zakir Husain College, New Delhi-110002 India

Reza K. Tavakol
School of Mathematical Sciences, Queen Mary and Westfield College, London U.K.

Abstract The numerical investigation of the stability of motion of a mass element ejected from a component of a binary star system has been made. The small mass element is initially taken as rest and the binary components are assumed to be revolving in a circular orbit around the center of mass. Only planar motion has been considered. The stability of motion has been investigated using the *Poincare surface of section method.*

Keywords: Celestial dynamics, binaries, circumstellar matter, numerical integration

The recent development in nonlinear dynamics and the new observations in the Solar system have resulted in the resurgence of interest in the restricted problem of three bodies where a small mass element of negligible mass, called the third body, moves in the gravitational potential field of two other massive bodies such that it does not affect their motion (cf. Murray and Dermott 1999). Further, the possibility of existence of planetary system in stars and also in binary or multiple system,

Y. Nakada et al. (eds.), Mass-Losing Pulsating Stars and their Circumstellar Matter, 123–124.

make such investigation more interesting from dynamical point of view (cf. Benest 1996).

In the present investigation, we have considered only the planar case and the third body is taken as moving in both the radiation and gravitation field of two stars moving in a circular orbit around their center of mass. The radiation from the binary component is characterized by the parameter β_i, the ratio of radiation to gravitational force. The range of variation of β_i is given in Saha et al. (2002). The significance of radiation effect was earlier studied by Mignard (1984). Scheurman (1980) studied the effect of radiation on the dynamics of the third body only in linear domain. In this paper, we obtain the trajectories of the third body directly by integrating the exact nonlinear differential equations of the system for different initial conditions. This enables us to investigate the nature of the various trajectories and their sensitivity to initial conditions. We resorted to the analysis of the phase space trajectories using *Poincare surface of section* method (cf Murray and Dumott 1999; Henon 1983).

We have computed the surface of section in the X-$\dot{X}$ plane for a given value of C (cf. Ragos and Zaouras 1988; Szebehely 1967). We have taken $C = 4.0$ and $\mu = 0.50$. Surfaces of sections were computed for the various initial conditions. Further we have considered both non-radiating binaries (Henon 1983) and radiating binaries with $q_1 = 0.90$, $q_2 = 1.0$. The result shows the transition from regular to chaotic motion as the initial condition changes, and how the nature of the phase space trajectories changes with the inclusion of radiation effect of one of the primaries. It is also important to note the significant changes caused by the radiation effect in the location of unstable points. Further results of the investigation will be reported in subsequent communication.

Acknowledgement. MKD and LMS would like to thank the University Grant Commitee, New Delhi for providing necessary support. LMS acknowledges the financial support provided by the organizers of the meeting.

References

Benest, D. 1996, A&A, 314, 983
Henon, M. 1983, Les Houches, Session XXXVI, 54
Mignard, F. 1984, Cel. Mech., 34, 275
Murray, C. D. & Dermott, S.F. 1999, Solar System Dynamics, CUP
Saha, L.M., Das, M.K. & Tanaka, Y. 2002, This proceeding
Schuerman, D.W. 1980, ApJ, 238, 337
Ragos, O., & Zagouras, C. 1988, Cel. Mech., 44, 135
Szebehely, V. 1967, Theory of Orbits, Academic Press

2:1 RESONANCE IN RV TAURI STARS

Toshihito Ishida
Nishi-Harima Astronomical Observatory, Sayo-cho, Hyogo 679-5313, Japan
ishida@nhao.go.jp

Abstract Behavior of two pulsationally unstable modes in 2:1 resonance is examined in relation to the alternating behavior of the deep and shallow minima of the RV Tauri stars. From results obtained from amplitude equations and numerical model calculations, 2:1 resonance seems to be one of the attractive origin of the alternation. Full results will be published elsewhere.

Keywords: RV Tauri stars, 2:1 resonance, Alternating minima

1. Introduction

RV Tauri stars are a group of yellow radial pulsating supergiants, and usually considered as post-AGB objects. Its light curves are characterized by alternative deep and shallow light minima with some irregularity. Takeuti and Petersen(1983) proposed 2:1 resonance between the fundamental(hereafter F) and the first overtone(hereafter 1O) radial modes as a candidate mechanism responsible for such alternation. More recently, Fokin(1994) found RV Tauri-like behavior in his hydrodynamic models and discussed that this behavior is caused from 2:1 resonance interaction between F and 1O. He also suggested that the RV Tauri-like behavior appears in lower Teff models, where F becomes unstable, namely in the region where both of F and 1O are pulsationally unstable.

There are some previous studies on the 2:1 resonance in pulsating stars, but because relation to the bump Cepheid phenomena is considered, only F is assumed to be unstable. In the present study, behavior of both of F and 1O are unstable is investigated at the first time.

2. Amplitude Equations

A set of amplitude equations for 2:1 resonance are derived by Aikawa (1984). Considering studies for nonresonant double-mode case into account, the following equations modified from Aikawa's equations to in-

Y. Nakada et al. (eds.), Mass-Losing Pulsating Stars and their Circumstellar Matter, 125–126.

clude "non-diagonal components" in the van der Pol's are considered,

$$\frac{dA_0}{dt} = \frac{1}{4}\sigma_0 C(0;0,2)A_0 A_2 \sin\theta + \frac{\sigma_0}{2}\mu_0 A_0(1 - \frac{\alpha_0^2}{4}A_0^2 - \frac{\alpha_{02}^2}{4}A_2^2), (1)$$

$$\frac{dA_2}{dt} = -\frac{1}{4}\sigma_2 C(2;0,0)A_0^2 \sin\theta + \frac{\sigma_2}{2}\mu_2 A_2(1 - \frac{\alpha_2^2}{4}A_2^2 - \frac{\alpha_{20}^2}{4}A_0^2), \quad (2)$$

$$\frac{d\theta}{dt} = \sigma_0 d_2 - \sigma_2 \frac{C(2;0,0)}{4}\frac{A_0^2}{A_2}\cos\theta + 2\sigma_0 \frac{C(0;0,2)}{4}A_2 \cos\theta. \quad (3)$$

Adopting coupling coefficients in Aikawa(1984), the behavior of the above amplitude equations is numerically examined. A complicated behavior is found when we adopt an appropriate set of parameters. Although the relation of this behavior to real RV Tauri stars is unclear, because the adopted coupling coefficients are for a classical Cepheid model, the very existence of such a complicated behavior in above set of amplitude equations is important. For further investigation, we need to estimate the coupling coefficients for RV Tauri star models.

3. Hydrodynamic Models

Linear nonadiabatic(hereafter LNA) and nonlinear models of the RV Tauri stars are investigated by Tuchman et al. (1993) and Fokin(1994), respectively, but both of them used old Los Alamos opacities. Results of LNA models and nonlinear models described in Ishida and Takeuti(1991) with OPAL opacity are presented here.

From the results of the LNA models, we found that the resonance condition is fulfilled in wide range in mass–Teff plane, however, when we limit to the region both of F and 1O are unstable, the room left for us is not so wide. Moreover, some representative cases are picked up and studied with time-evolving models. Although affected by strong shocks, analysis of the resultant velocity curves indicate that complicated time-evolving behavior of two finite amplitudes, implicating RV Tauri-like behavior, appears only when both of F and 1O are unstable,

Although Buchler(in these proceedings) found 2:1 resonance with one unstable mode from analysis of observational data of R Scuti , the results presented here indicate that complicated behavior near 2:1 resonance is an another attractive explanation of the behavior like RV Tauri stars.

References

Aikawa, T. 1984, MNRAS, 206, 833
Fokin, A.B. 1994, A&A, 292, 133
Ishida, T., & Takeuti, M. 1991, PASJ, 43, 795
Takeuti, M., & Petersen, J.O. 1983, A&A, 117, 352
Tuchman, Y., Lebre, A., Mennesier, M.O., Yaari, A. 1993, A&A, 271, 501

THE DETERMINATION OF INTRINSIC POLARIZATION FOR RV TAURI STARS

Kazuo Yoshioka
The University of the Air, Maebashi, 371-0032, Japan
yoshioka@u-air.ac.jp

Keiichi Saijo
National Science Museum, Tokyo, 169-0073, Japan
saijo@kahaku.go.jp

Hideo Sato
National Astronomical Observatory, Mitaka, 181-8588, Japan
satohd@cc.nao.ac.jp

Abstract We made the multicolor polarimetric observations of 17 RV Tauri stars. We report the characteristics of intrinsic polarization of these stars. The interstellar polarizations were determined by the modified near-neighbor method. Generally speaking, the degrees of interstellar polarization are not large, because they do not increase monotonously with distance. Thus, the characteristics of intrinsic polarization do not differ markedly from those of observed polarization.

Keywords: RV Tauri stars, intrinsic polarization

Introduction

The RV Tauri stars are semiregular variables which lie between the Cepheids and the Mira-type variables in the HR diagram. Their light curves are characterized by alternate deep and shallow minima. The RV Tauri stars are generally thought as post-AGB stars, and their circumstellar dust envelopes(CDE)'s are thought to be formed as a result of mass loss at the final stage of the AGB phase.

Y. Nakada et al. (eds.), Mass-Losing Pulsating Stars and their Circumstellar Matter, 127–129.

On the basis of spectroscopic characteristics in an optical region Preston et al.(1963) divided the RV Tauri stars into 3 subgroups, group A,group B, and group C. The group A generally shows anomalously strong TiO bands at light minima whose strength corresponds to early M-type supergiants, while intensities of metallic lines indicate G or K-type. The group B shows spectra to which a definite spectral type cannot be assigned. The most distinctive characteristics is that near light minima CH and CN bands appear with considerable strength indicative of an enhanced carbon abundance. The group C shows all the characteristics of the group B except that the carbon features are weak or not present.

1. The Observations with MCP

We made the multicolor polarimetric observations of 17 RV Tauri stars between 1993 October 23 and 1998 October 29, using the multi-channel polarimeter(MCP) attached to the 91 cm reflector at the Dodaira Station of the National Astronomical Observatory. MPC measures linear polarization simultaneously at 8 colors(0.36, 0.42,0.455, 0.53, 0.64, 0.69, 0.76, and 0.88 μ m), whose colors are indicated with the number of the channel in order of wavelength.

2. The Determination of Interstellar Polarization

We determined the interstellar polarization by the modified near-neighbor method by Bastien(1985). The main modification point is that a distance is used as the parameter for obtaining the degree of interstellar polarization, instead of $E(B-V)$. The reason for this modification is that for many RV Tauri stars the observed $E(B-V)$ includes not only the interstellar components but also the CDE component. We adopted the empirical formula given by Whittet et al.(1992) for a wavelength dependence of the degree of interstellar polarization. We used the interstellar polarization database compiled by Hirata(1999), which contains 13969 data for several wavelength collected from 45 literatures.

We have determined the interstellar polarizations for 11 stars, TW Cam, EQ Cas, V360 Cyg, SS Gem, SU Gem, AC Her, EP Lyr, U Mon, TT Oph, R Sct, and RV Tau. Among the above stars, the interstellar polarization for 4 stars, TW Cam, SS Gem, U Mon, and R Sct have been already determined by other observers. Our values for the position angle of interstellar polarization $\theta_{\rm IS}$ are close to those determined by other observers. On the other hand, in some cases our values for the maximum degree of polarization $p_{\rm max}$ differ markedly from those by other observers. Generally speaking, our $p_{\max}$ values are smaller than other values. This difference seems to be mainly due to the parameter used to estimate $p_{\max}$ values. We used the distance as the parameter, and the degree of interstellar polarization p_{IS} do not always increase monotonously with distance.

3. The Characteristics of Intrinsic Polarization

Generally speaking the characteristics of intrinsic polarization do not differ markedly from those of observed polarization, because the $p_{\max}$ values do not large. For example, the stars belonging to the A group show the tendency for p values to take a maximum at an intermediate wavelength. On the other hand, the stars belonging to the B group show the tendency for p values to take a minimum at an intermediate wavelength or their p values increase or decrease with wavelength.

The differences in the wavelength dependence of p value seem to reflect the differences in the structure of CDE among subgroups. For example, the B group stars may have more than two CDE's and each of CDE's has a different grain size distribution. In fact, on the basis of multiwavelength observations, Shenton et al.(1992) suggest the presence of at least two distinct CDE's for the B group star, AC Her. Furthermore, Jura et al.(2000) found a dust ring around AC Her by mid-infrared observations with 10m Keck reflector and suggest that there are two types of grain with the sizes of less than 0.1 μ m and of more than 200 μ m from the analysis of the ring.

Lastly, the p values for AC Her show periodic changes not only with the pulsational period of 75.5days but also with the revolutional period of 1194days which was determined by Winckelet al.(1998).

MULTI-COLOR INTERFEROMETRIC OBSERVATIONS OF MIRA STARS

Probing the Extended Atmosphere Structure and Chemistry

B. Mennesson, M.J. Creech-Eakman, R.R. Thompson
Jet Propulsion Laboratory, California Institute of Technology, 4800 Oak Grove Drive Pasadena CA 91106, USA. e-mail: bmenness@huey.jpl.nasa.gov

G. Perrin, V. Coudé du Foresto, A. Merand
Laboratoire d'Etudes Spatiales et d'Instrumentation en Astrophysique, 5 Place Jules Janssen F-92195 Meudon, France. e-mail: guy.perrin@obspm.fr

S. Ridgway
National Optical Astronomy Observatories, Tucson, AZ 85719, USA. e-mail: ridgway@noao.edu

Abstract Interferometric observations in the atmospheric windows of the near infrared (1 to 4 μm) can be efficiently used to probe the chemical composition of Miras atmosphere and provide direct measurements of extended gas layers around these stars. This is illustrated by recent Miras observations carried out with the FLUOR instrument of the IOTA interferometer (Mennesson et al. 2002, and Perrin et al. proceedings of this conference) and with the PTI test-bed (Thompson 2002, Thompson et al. 2002). These visibility measurements show evidence for continuum emission from very extended (2-3 stellar radii) semi-transparent gaseous atmospheric layers, and large apparent diameter changes with pulsation phase. Interestingly these observations are consistent with the extended molecular gas layers (H_2O, CO ...) already inferred around some of these objects from ISO and high resolution ground-based FTS infrared spectra.

Keywords: Mira atmosphere, infrared interferometry

Y. Nakada et al. (eds.), Mass-Losing Pulsating Stars and their Circumstellar Matter, 131–135.

1. IOTA/FLUOR observations

IOTA (Infrared and Optical Telescope Array) is a 3 telescope interferometer (Traub 1998) located on Mount Hopkins, Arizona. It is operated by the Harvard-Smithsonian Institution at the CfA. Its three 45 cm telescopes can be relocated over an L shaped set of stations providing baselines ranging in length from 5 to 38 m. All observations reported here were obtained with the FLUOR (Fiber Linkee Unit for Optical Recombination) instrument (Perrin 1996, Coudé du Foresto 1998 et al. 1998) operating in the K' band (2.0 to 2.35 μm) and with TISIS (Thermal Infrared Stellar Interferometric Set-up, Mennesson et al. 1999) FLUOR's extension to the L' band (3.4 to 4.1 μm).

Nine bright O-rich Miras have been observed with IOTA in both K' and L' broad-band filters, in most cases at variability phases differing by less than 0.1. These results synthesize two observing campaigns carried out in February/March 2000, and then in October/November 2000. Surprisingly all Miras showed strong increases (from 20% to 100%) in measured uniform disk diameters between the K' and L' bands (Mennesson et al. 2002, Changon et al. 2002). This apparent enlargement with wavelength is far more important than what can be accounted for by conventional differential limb darkening between the 2 bands.

Substantial absorption lines of molecular H_2O, CO, SiO and CO_2 are visible in some O-rich Miras near infrared spectra (Yamamura et al. 1999, Tsuji et al. 1997), but they appear outside the K' and L' filters. Even though some wing absorption features may still be present, the overall stellar flux remains largely overwhelmed by continuum emission for broad-band observations in any of the H, K' and L' filters (Scholz 2001). There is then no spectral evidence for molecular bands strong enough to produce a large opacity difference between the K' and L' broad-band regions. This leads us to an interpretation based on a continuum effect. We propose a model consisting of a cool, semi-transparent gaseous shell extending far above the classical photosphere, typically 3 stellar radii away (figure 1). We suggest that the Planck weighting of the emission from the two layers will suffice to make the L' UD diameter appear larger than the K' UD diameter. Owing to the wavelength dependence of the Planck function, the extended cooler ($\simeq$ 1500 K to 2000 K) gas layer contributes a larger fraction to the overall stellar flux at 3.8 μm than at 2.2 μm for instance. More generally when the star is observed in infrared regions that are beyond the peak emission of the "classical" photosphere ($\lambda > 1 - 1.5\,\mu$m), emission from this extended region can become important.

Figure 2 shows as an example the best fit obtained to R Leonis K' and L' visibility measurements using this simple 2-layer model, and a 1-D radiative transfer equation, solved for each line of sight. Central photosphere is considered as a blackbody with temperature T^*, radius R^*. To keep the model simple, temperature inside the envelope is assumed to decrease as $r^{-0.5}$ (radiative equilibrium), and density goes as r^{-2} (mass flux conservation).

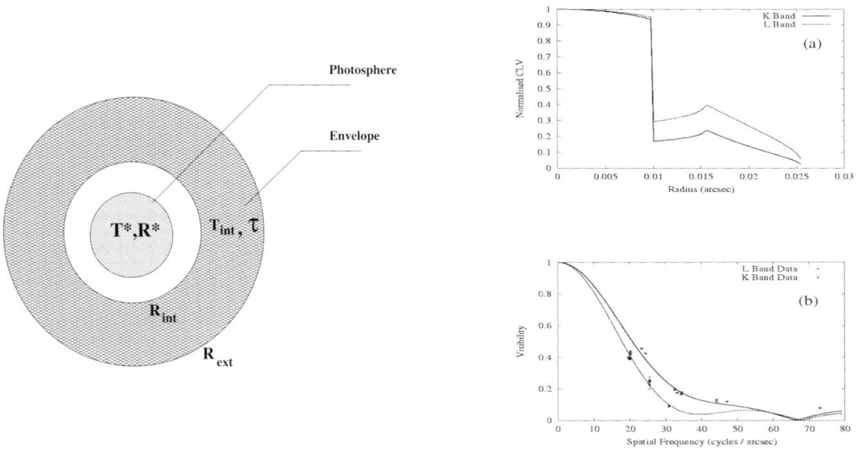

Figure 1. Left: adopted spherically symmetric model (see text). Envelope has inner radius R_{int}, outer radius R_{ext}, inner temperature T_{int} and optical depth τ common to both K' and L' bands. Right: results of R Leonis observations modeling using a 2-layer model (see text). Predicted Center to Limb Variations (CLV) and visibility curves are given in full line in K and dashed line in L'. (a) CLV. (b) Model visibility and observations: March 1997 (K') and November 2000 (L').

2. PTI observations

The Palomar Testbed Interferometer (PTI) Mira program has been following a group of about 70 M, C and S-type miras since June, 1999. A description of PTI can be found in Colavita et al. (1999) and the standard reduction method for PTI data is given in Boden et al. (1998). Observations were generally taken every 2-3 weeks during the observing season, with a system visibility estimated instantaneously for a given night's data based on the observations of mostly unresolved (0.8<mas) calibration standards, interleaved with the target observations. All diameters assume a uniform disk(UD) model.

A wealth of data is present in the PTI sample, and only part of it has been analyzed yet (Thompson 2002, Thompson, Creech-Eakman, &

van Belle, 2002). We only report here on the results obtained on one O-rich Mira (S Lac) and one C-rich (RZ Peg) Mira. The UD diameter vs wavelength dependency observed for these 2 stars is quite characteristic for the O-rich and C-rich Miras of the sample. S Lac exhibits a minimum apparent size in the center of the K-band, typical of oxygen-rich Miras (Thompson 2002), which is not evident in the non-Mira counterpart (Fig 2). Sources of opacity, such as H_2O probably account for the larger sizes in the 2.0 and 2.4 μm channels. The carbon-rich Mira RZ Peg depicts a maximum apparent size in the 2.3 and 2.4 μm channels, owing to C_2H_2, HCN and CO as sources of opacity. The shape in RZ Peg is also typical of carbon-rich Miras (Thompson 2002).

Multi-epoch data on S Lac (see Thompson et al. 2002, figure 2) show a phase lag between the 2.2 and 2.4 μm apparent sizes, suggesting the source of opacity (H_2O) lies significantly above the $\tau(2.2\,\mu m)=1$ photosphere. This agrees nicely with the 2-layer model invoked to fit the FLUOR K'/L' data. Conversely, no phase lags are seen between the three K-band channels for the C-rich RZ Peg, suggesting the sources of opacity lie at the $\tau(2.2\,\mu m)=1$ photosphere or very close.

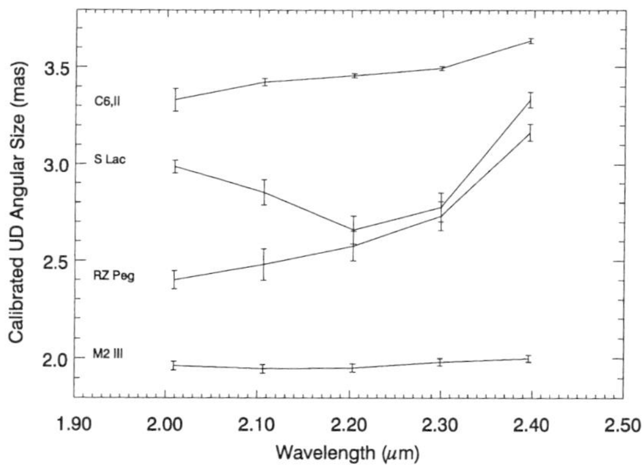

Figure 2. Calibrated UD spectral angular sizes for Miras and non-Miras. Both Mira stars exhibit greater degrees of size departures from their continuum than do the non-Mira chemical counterparts. These representative data were taken on the same night (JD 2451715). Top to Bottom: HIP 92194 (C6,II), S Lac (M4-M8e), RZ Peg (C5-6e,I), HD 193347 (M2 III)

3. Summary

We have reported on a systematic and rather surprising result obtained with the IOTA interferometer on 9 O-rich Miras: a strong increase in the apparent broad-band UD diameters from K' (around 2.15

μm) and L' (around 3.8 μm) bands. We interpreted the observed diameter shift as a continuum or pseudo-continuum effect, and suggested that a very extended ($\simeq$ 3 stellar photospheric radii) semi-transparent gas layer was responsible for it, as its contribution to the overall flux increases with IR wavelength.

ISO detection of hot water in several Miras atmospheres, and recent narrow band measurements obtained by PTI on a large number of O-rich Miras indicate that the source of opacity in the outer layer could be mainly molecular H_2O. Further analysis of the PTI sample is necessary to see if S Lac's observed behaviour is common among O-rich Miras. Yet, the detection of a phase lag between S Lac's UD diameters observed in the continuum and in H_2O absorption bands appears also compatible with a 2-layer extended gaseous atmosphere around this type of stars.

Acknowledgments

Work related to the PTI observations was performed at JPL, CIT under contract from NASA. Data were obtained using PTI, which is supported by NASA contracts to JPL. Science operations with PTI are possible through the efforts of the PTI collaboration. MCE and BBT would like to thank especially thank J. Mueller and K. Rykovski. This research has made use of the SIMBAD and AFOEV databases.

References

Boden, A.F., Colavita, M. M., van Belle, G.T., & Shao, M. 1998, Proc. SPIE, 3350, 872

Chagnon, G., Mennesson, B., Perrin, G., et al. 2001, AJ, submitted

Colavita, M. M., et al. 1999, ApJ, 510, 505

Coudé du Foresto, V., Perrin, G., Ruilier, C., Mennesson, B., Traub, W. & Lacasse, M. 1998, in SPIE Conf. Ser. 3350, Astronomical Interferometry, ed. R. Reasenberg (Bellingham: SPIE), 854

Mennesson, B., Perrin, G., Chagnon, G., Coudé du Foresto, V., Ridgway S., et al. 2002, ApJ, in press

Perrin, G. 1996, PhD dissertation, Université Paris VII

Scholz, M. 2001, MNRAS, 321, 347

Thompson, R. R. 2002, PhD Thesis, Univ. of Wyoming

Thompson, R. R., Creech-Eakman, M. J., & van Belle, G. T. 2002, ApJ, in press

Traub, W.A. 1998, in SPIE Conf. Ser. 3350, Astronomical Interferometry, ed. R. Reasenberg (Bellingham: SPIE), 848

Tsuji, T., Ohnaka, K., Aoki, W., & Yamamura, I. 1997, A&A, 320, L1

Yamamura, I., de Jong, T., & Cami, J. 1999, A&A, 348, L55

ORIGIN AND PROPAGATION OF SHOCKS IN THE ATMOSPHERES OF MIRA-TYPE STARS

Georgij M. Rudnitskij
Sternberg Astronomical Institute, Moscow State University
13 Universitetskij prospekt, Moscow, 119899 Russia
gmr@sai.msu.ru

Abstract In Mira-type stars the optical emission lines and molecular maser lines (SiO, H_2O, OH) probe different layers of the stellar atmospheres and circumstellar envelopes, affected by shock waves. Balmer emission appears slightly above the stellar photosphere. As the shock reaches the masering region in the circumstellar envelope, the SiO and H_2O maser intensity grows. The lack of the radio continuum, expected to accompany the Hα emission in Miras, probably indicates that the source of the Balmer emission lines is small, e.g., connected with an active region on the star or with a shock generated by a close low-mass companion.

Keywords: Mira variables, shocks, extrasolar planets

Emission lines appearing in the optical spectra of Mira-type variables are considered as signatures of shock waves propagating in the atmospheres of the stars. To produce, e.g., emission lines of the hydrogen Balmer series, the shock should be strong enough, with a velocity $v_{\rm sh}$ of at least 15–20 km s^{-1}. If the shocks are generated by stellar pulsations, are spherical and have a size covering the entire stellar disc, the propagation of such a shock should also yield radio continuum emission at centimeter wavelengths: it can be shown that the postshock ionised layer of gas is optically thick at these frequencies (Rudnitskij 1990). However, the centimetre radio continuum in Miras is quite weak (Knapp et al. 1995; Reid and Menten 1997; Chapman and Rudnitskij 2002), suggesting shock velocities well below 15 km s^{-1}.

This difficulty can be resolved if the source of hot Balmer emission is small and covers only a small fraction of the stellar disc. It can emit enough Balmer photons to account for the observed hydrogen emission. At the same time its angular size is small, and it is below the detection limit of the radio continuum observations. There are at least two possibilities of such a small source.

Y. Nakada et al. (eds.), Mass-Losing Pulsating Stars and their Circumstellar Matter, 137–138.

1. Active region on the red giant with large-scale magnetic loop, similar to solar active regions (Bychkov and Panchuk 1977).

2. Hot spot connected with the supersonic motion of a close low-mass companion (Rudnitskij 2000; Struck et al. 2002). Such a migrating hot spot may have been observed in the infrared on *o* Cet (Lopez et al. 1997).

We will concentrate on the latter possibility. At present more than eighty planetary systems have been detected around main-sequence stars (Schneider 2002). These planets probably can survive as their central star enters the red giant stage of the evolution (Struck-Marcell 1988; Rybicki and Denis 2001). Then a planet revolving at a distance $r \sim$ 1 A.U. from the centre of its star will move at a velocity of $\sim$30 km s^{-1}. This strongly supersonic motion will produce a strong shock in front of the planet and trailing behind it. This shock is a source of the Balmer emission. In this case the observed doubling of the Balmer lines (Udry et al. 1998) can be naturally explained as the planet is crossing the stellar disc and we see the approaching and receding walls of the shocked cone. Propagating farther out to $r \sim 10^{14}$ cm, the shock can excite enhanced SiO and H_2O maser emission. The stars with rare flares of the Hα and H_2O emission (Rudnitskij et al., this volume) may have planets in eccentric orbits, producing events only at the periastron passage once per 10–15 years. Eccentric orbits may resist the circularising effect of the motion in a viscous medium if the periastron passages are accompanied by enhanced mass-loss rate episodes (Soker 2000).

References

Bychkov, K.V., & Panchuk, V.E. 1977, SvA, 21, 189

Chapman, J.C., & Rudnitskij, G.M. 2002, PASA, submitted

Knapp, G.R., Bowers, P.F., Young, K., & Phillips, T.G. 1995, ApJ, 455, 293

Lopez, B., et al. 1997, ApJ, 488, 807

Reid, M.J., & Menten, K.M. 1997, ApJ, 476, 327

Rudnitskij, G.M. 1990, in From Miras to Planetary Nebulae: Which Path for stellar Evolution? ed. by Mennessier, M.O. & Omont, A. Gif-sur-Yvette, Editions Frontières, 268

Rudnitskij, G.M. 2000, in ASP Conf, Ser. 203, The Impact of Large-Scale Surveys on Pulsating Star Research, (San Francisco: ASP), 384

Rybicki, K.R., & Denis, C. 2001, Icarus, 151, 130

Schneider, E. 2002, http://www.obspm.fr/planets/

Soker, N. 2000, A&A, 357, 557

Struck, C., Cohanim, B.E., & Willson, L.A. 2002, ApJ, 572, L83

Struck-Marcell, C. 1988, ApJ, 330, 986

Udry, S., Jorissen, A., Mayor, M., & Van Eck, S. 1998, A&AS, 131, 25

THEORETICAL COLOR MASS-LOSS RELATIONS FOR O-RICH MIRA STARS

K.S. Jeong[1,2], J.M. Winters[3], T. Le Bertre[1], E. Sedlmayr[2]
[1] *LERMA, Observatoire de Paris, 61 av. de l'Observatoire, 75014 Paris, France*
[2] *TU Berlin, ZAA, PN 8-1, Hardenbergstr. 36, 10623 Berlin, Germany*
[3] *MPI für Radioastronomie, Auf dem Hügel 69, 53121 Bonn, Germany*
Kyung.Jeong@obspm.fr

Mass-loss plays a key rôle for the stars on the Asymptotic Giant Branch (AGB), by dominating the subsequent evolution of the star (Olofsson 1999) and also, it is in this phase that low and intermediate mass stars contribute the most to the replenishment of the interstellar medium in the solar neighborhood (Sedlmayr 1994). It is therefore essential to estimate quantitatively the mass loss rates of AGB stars in the solar neighborhood and ideally, in different chemical environments such as the Bulge. Empirical correlations between IR color indices and mass loss rates have been found for Mira stars (Le Bertre & Winters 1998, hereafter LW98) which can provide an easy tool to measure directly their mass loss rates simply via infrared color indices, independent of the distance to the star. In this work, we present synthetic relations between infrared color indices and mass loss rates derived from time-dependent model calculations for circumstellar dust shells (CDS) around oxygen-rich AGB stars (Jeong et al. 2002, in preparation) and compare these relations to the empirical relations in LW98.

The modeling method applied here to the CDS of AGB stars is a consistent time dependent treatment of hydrodynamics, thermodynamics, equilibrium chemistry, dust formation including growth and evaporation, and radiative transfer (Fleischer et al. 1992, Winters et al. 1997, Jeong et al. 2002). The models are defined by the prescription of the 4 fundamental stellar parameters $(M_*, L_*, T_*, \varepsilon_i)$ and of the 2 pulsation parameters (P,Δu). We consider a spherically symmetric atmosphere and heterogeneous dust formation with TiO_2 nuclei (Gail & Sedlmayr 1998, Jeong et al. 1999). The parameter range for the model calculations is $M_* = [1.0, 1.6]\,M_\odot$, $L_* = [1.0, 2.0]\,10^4\,L_\odot$, $T_* = [2000, 2400]\,K$, $\varepsilon_C/\varepsilon_O = 0.5$, $\Delta u = [5, 10]\,kms^{-1}$, and $P = [300, 650]\,d$.

Y. Nakada et al. (eds.), Mass-Losing Pulsating Stars and their Circumstellar Matter, 139–141.

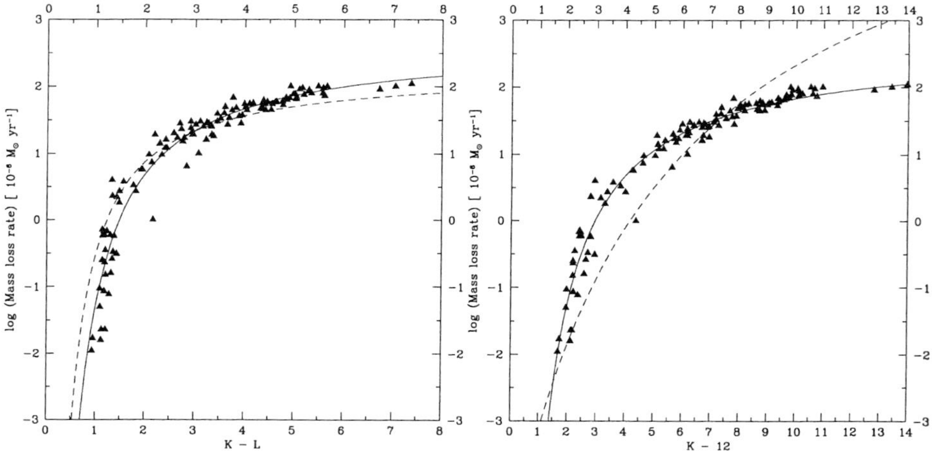

Figure 1. The relations K-L vs. $\dot{M}$ (left) and K-[12] vs. $\dot{M}$ (right). Symbols represent the model calculations, solid lines are fits to the model calculations and the dashed lines give the empirical relations in LW98.

Based on the model calculations for O-rich Miras, we have derived relations between IR colors and $\dot{M}$, which provide clear correlations:

$\log \dot{M} = -4.27/(J - K + 0.65) + 2.20,$

$\log \dot{M} = -3.91/(K - L) + 2.65,$
$\log \dot{M} = -7.69/(K - [12]) + 2.60,$
$\log \dot{M} = -5.26/(K - [7]) + 2.26,$
$\log \dot{M} = -6.83/(K - [15]) + 2.22 \quad$ with $\dot{M}$ in $[10^{-6}\,M_\odot\,yr^{-1}]$.

The relations between J-K and $\dot{M}$ and between K-L and $\dot{M}$ (Fig. 1) show quite good agreement with the empirical relations derived from observations (LW98). The relation between K-[12] and $\dot{M}$ (Fig. 1) gives higher mass loss rates than the empirical relation by LW98. Instead, the agreement is better with the empirical values by Whitelock et al. (1994). For [7]-[15] and IRAS colors, the correlations between colors and $\dot{M}$ are rather coarse. The IR color–mass loss rate relations can be applied to the observations with ISO and with ASTRO-F in order to quantitatively estimate the mass loss rates of AGB stars.

Acknowledgments

The calculations were performed on the Cray vector and parallel computers accessible through ZIB Berlin. This work was partly supported by the *FAZIT-Stiftung*, the *BMBF* (grant 05 3BT13A 6), the PROCOPE program (D/9822849 (DAAD) and F/99001 (MAE)), and through a *Marie Curie Fellowship* (HPMF-CT-2000-01041).

References

Fleischer, A.J., Gauger, A., & Sedlmayr, E. 1992, A&A, 266, 321
Gail, H.-P., & Sedlmayr, E. 1998, Faraday Discussion no. 109, 303
Jeong, K.S., Winters, J.M., & Sedlmayr, E. 1999, IAU Symp. 191, 233
Le Bertre, T., & Winters, J.M. 1998, A&A, 334, 173 (LW98)
Olofsson, H. 1999, IAU Symp. 191, 3
Sedlmayr, E. 1994, IAU Colloquium 146, 163
Whitelock, P.A., Menzies, J., Feas,t M.W., et al.., 1994, MNRAS, 267, 711
Winters, J.M., Fleischer, A.J., Le Bertre, T., & Sedlmayr, E., 1997, A&A, 326, 305

HIGH-PRECISION MEASUREMENTS OF COOL GIANT STARS WITH THE VLTI

Andrea Richichi
European Southern Observatory
Karl-Schwarzschildstr. 2 - 85748 Garching - Germany
arichich@eso.org

Markus Wittkowski, Markus Schöller
European Southern Observatory
Casilla 19001 - Santiago 19 - Chile
mwittkow@eso.org, mschoell@eso.org

Abstract We present first observations of cool giant stars obtained at the ESO Very Large Telescope Interferometer (VLTI), both with two small test telescopes on a 16 m baseline, and by combining two 8m telescopes on a 100 m baseline. In particular, we devote special attention to a study of Mira (*o* Cet), with measurements clearly showing deviations from spherical symmetry and time variations. The VLTI will constitute an ideal instrument for such studies in the future. In particular it will permit detailed and accurate investigations of southern AGB stars, such as accurate measurements of surface structure parameters (diameters, diameter variations, asymmetries, center-to-limb variations, special features like hot spots) and of circumstellar envelopes.

Introduction

The ESO Very Large Telescope Interferometer (VLTI) is located on Cerro Paranal, in northern Chile. This facility consists of the four fixed 8.2m Unit Telescopes (UT), and of a number of 1.8m Auxiliary Telescopes (AT) which can be moved over an array of 30 stations.

At present, small test telescopes (40cm siderostats) are being used routinely, mainly for commissioning purposes. Also, some UTs telescopes have been used successfully in several runs. Both the MIDI (operating in the mid-infrared) and the AMBER (near-IR) instruments are in an advanced stage of construction, and are scheduled to begin operations

Y. Nakada et al. (eds.), Mass-Losing Pulsating Stars and their Circumstellar Matter, 143–144.

between the end of 2002 and the beginning of 2003. The AT telescopes will be progressively added starting from 2003.

1. Results

A list of cool giant stars observed by the VLTI in its first year of commissioning is reported in Table 1. In a large number of cases, these represent first-time observations with an interferometer, also thanks to the southern location of the VLTI.

Table 1. List of cool giant stars observed by the VLTI in its first year of commissioning

1 Oph	KQ Pup	chi Aqr	T Eri	AK Hya	omi Cet	U Ant
gam Sge	lam Aqr	Delta-Vir	alf Her	RX Lep	R Cnc	V Hya
HR 8685	29 Leo	eta Gem	AR Cet	R Peg	R Hya	TX Psc
alf Ori	eta Sgr	NZ Gem	bet Gru	S ORI	S Scl	HR 977
alf Sco A	gam Cru A	2 Cen	S Lep	X Hya	R Aqr	R Scl
alf Cet	sig Lib	del02 Gru	V744 Cen	SW Vir	W Hya	U Hya
pi. Eri	tau04 Eri	62 Sgr	L02 Pup	U Her	U Ori	W Ori
gam Eri	13 Gem	psi Phe	T Ari	R Crt	R Leo	X Cnc
119 Tau	30 Psc	T Cet	T Lep	Y Cen	UU For	R Lep

One of the best studied cases is that of Mira (o Cet). The main conclusions of our observations of Mira are related to the detection of a significant variation of the visibility with azimuth of the baseline, i.e. of the position angle on the sky.

While the hypothesis of a simple stellar disk is certainly inadequate in the case of Mira, the simple interpretation presented in our poster shows very dramatically that the shape of the object is far from being circularly symmetric. Similar conclusions on the asymmetry of Mira were reached also by a few previous observers. It is interesting however to note that the direction of the elongation seems quite different in the optical and in the near-IR. It can be speculated that emission is arising from different stellar and circumstellar components in the visible and near-IR, and/or from scattering.

Due to limitations of space, it is impossible to report in these proceedings the information presented in the poster, the figures and the references. Work is in progress to publish the VLTI results on Mira and other stars. The main contribution of the present work is to strengthen and expand the set of such observations, with new accurate results to measure the asymmetry of Mira over a larger range of position angles in the near-IR, and to monitor its pulsation over at least one cycle.

THE STABILITY OF MAGNETO-HYDRODYNAMIC STELLAR WINDS

Yasumichi Saito
Physics Department, Faculty of Education, Iwate University

Abstract Magnetohydrodynamic winds flowing out from fast-rotating stars are analyzed. Acceleration of the winds is briefly discussed with relation to the location of the singular points of the basic equations. The local stabilities of the winds are examined.

Weber and Davis (1987) proposed a steady-state wind model with rotation and magnetic field. They illustrated its structure for the location and types of singular points of the basic equations and discussed an influence of the magnetic torque on the angular momentum loss. The applications of this model for stellar winds and the expanding phenomena of stellar atmospheres were made by some authors (e.g., Saitō 1974; Limber 1974; Yeh 1976: Belcher, MacGregor 1976). Furthermore, this model is extended for high-speed winds including relativistic or radiative effects (e.g., Michel 1967; Goldreich, Julian 1970; Okamoto 1978; Kennel et al 1983; Friend, MacGregor 1984; Nerney, Suess 1987; Daigne, Drenkhan 2002).

The original basic equations are analyzed in part by Weber and Davis and done successively (Yeh 1976; Saito, Saitō 1986, Chakrabarti 1990) and the wind-type solutions are classified into two modes. All the authors but Saitō and Limber obtained super-fast mode wind solutions and Saitō and Limber did sub-fast mode ones.

The former solutions are dominantly accelerated by the magnetic torque and pass through the fast magnetosonic point. The solutions are selected favorably from a viewpoint of stability. The latter ones are accelerated by the centrifugal force or/and the magnetic torque and applied only for the slow-expanding envelopes of Be stars. Either of the solutions is well accelerated inside distances a few times as large as the the Alfvénic radius.

Every wind flow is confined to the equatorial plane, but it is pointed out that the flow diverge away from this plane for distances much larger

Y. Nakada et al. (eds.), Mass-Losing Pulsating Stars and their Circumstellar Matter, 145–147.

than the Alfvénic radius (Suess, Nerney 1973; Nerney, Suess 1975,1985). Nevertheless, Nerney and Suess (1987) obtained modified super-fast mode wind solutions confined to the equatorial plane for fast-rotating, hot stars. They mainly examined angular momentum loss, but the wind solutions of original or variously modified Weber-Davis equations are generally believed to be confined near the equatorial plane.

Modified equations including a viscous stress are expressed according to Saitō's (1974) formulation. Shakra-Sunyaev's viscosity is assumed and then a force originated from the viscosity is derived as $(-\alpha/r^3)\mathrm{d}(r^3P)/\mathrm{d}r$ where r is the radial distance from the center of the star, P is the gaseous pressure and α is a dimensionless parameter. A sub-fast mode wind solution is expressed in terms of four physical variables, i.e., the radial and azimuthal speeds and the density of gas and the azimuthal Alfvén speed.

In order to examine the stability of the sub-fast mode wind solution, the modified equations are converted into time-dependent ones and the four stationary variables are replaced straightforward by four time-dependent ones. We assume that each of the replaced variables is made up of a fluctuation part plus the stationary, sub-fast mode part and confined to the equatorial plane and the fluctuation part is proportional to $\exp[\mathrm{i}(\mathbf{k}\cdot\mathbf{r}-\omega\mathbf{t})]$. Here, t is time. The parameters $\mathbf{k}$ and ω, respectively, are a wave-number vector which is suitably given at any radial distance and a angular frequency which should be determined. When the four variables are substituted into converted equations, these equations are reduced to four linear combinations of the fluctuation terms and a dispersion relation which is a quartic equation of ω is derived.

Values of the parameters but the polytropic index γ are the same as adopted in Saitō (1974). If $\gamma = 1$, the terminal speed becomes zero and a value of $\gamma = 1.2$ is adopted. Values of $\alpha = 0.1, 0.2$ are adopted. The vector $\mathbf{k}$ is parallel to $\mathbf{r}$ and values of k $(= |\mathbf{k}|)$ are suitably given at any distance.

Four solutions of ω have been determined and their imaginary parts have been positive or negative. If the part is negative, the wind solution is stable and if it is positive, vice versa. The imaginary part is positive for distances smaller than several tens of the Alfvénic radius where is much smaller than the fast magnetosonic one. When k decreses, it can be negative near the distances. The positive values of the imaginary part are large or small, according as $\alpha = 0.1, \mathrm{or} 0.2$. For distances larger than hundred of the Alfvénic radius the part is negative and nearly equal to zero.

If the sub-fast mode wind flow is not unstable near the photosphere, it can be stable over the Alfvénic radius.

References

Belcher, J.W., & MacGregor, K.B. 1976, ApJ, 210, 498
Chakrabarti, S.K. 1990, in Theory of Transonic Astrophysical Flows (World Scientific, Singapore), chapt 8
Dargne, F., & Drenkhan, G. 2002, A&A, 381, 1066
Kennel, C.F., Fujimura. F.S., & Okamoto, I. 1983, Geophys. Astrophys. Fluid Dynamics. 26, 147
Limber, D.N. 1974, ApJ, 192, 429
Michel, F.C. 1969, ApJ, 156, 727
Nerney, S.F. & Suess, S.T. 1975, ApJ, 196, 837
Nerney, S.F. & Suess, S.T. 1985, ApJ, 296, 259
Nerney, S.F. & Suess, S.T. 1987, ApJ, 321, 355
Okamoto, I. 1974, MNRAS, 185, 69
Saitō, M. 1974, PASJ, 26, 103
Saito, Y. & Saitō , M. 1986, Ap & SS, 119, 33
Suess, S.T. & Nerney, S.F. 1973, ApJ, 184, 17.
Weber, E.J. & Davis, L. 1967, ApJ, 148, 217
Yeh, T. 1976, ApJ, 206, 768

III

OPTICAL AND IR OBSERVATIONS OF CIRCUMSTELLAR MATTER

CONFRONTING CONSISTENT DUST SHELL MODELS WITH OBSERVATIONS OF LONG-PERIOD VARIABLE STARS

Jan Martin Winters

Max-Planck-Institut für Radioastronomie, Auf dem Hügel 69, D-53121 Bonn, Germany

jwinters@mpifr-bonn.mpg.de

Abstract The ingredients necessary for a consistent description of dust forming atmospheres around pulsating stars are addressed and the results of corresponding hydrodynamical model calculations are compared with respective observations.

Keywords: hydrodynamics, mass-loss, outflows, dust

Introduction

The thermodynamical conditions encountered in the extended atmospheres of pulsating Asymptotic Giant Branch (AGB) stars allow for the efficient formation of small solid particles ('dust grains') out of the gas phase. Due to their huge photon cross sections, the grains efficiently transmit energy and momentum from the stellar radiation field to the atmospheric gas. Thereby, the emerging dust component changes the local conditions of its own formation. Moreover, the highly non-linear self-regulating dust formation process induces important non-local feedbacks on the hydrodynamical and thermodynamical structure of the dust forming system, leading to several dynamical phenomena (such as dust induced shock waves and the existence of an eigen-timescale of the circumstellar shell) with specific observable consequences.

1. Modeling approach

Due to the coupling of the dust component to its surroundings, a realistic treatment of the grain formation process requires a self-consistent treatment of the dust forming system, taking into account the mutual interrelations among the involved physical processes. Only by such an approach we can hope to finally arrive at a physical interpretation of the

Y. Nakada et al. (eds.), Mass-Losing Pulsating Stars and their Circumstellar Matter, 151–158.

observational data being acquired by instruments of ever increasing resolution and sensitivity. Moreover, only on the basis of a self-consistent treatment is it possible to make predictions from the modeling which can be tested by future observations.

The problem to be solved includes as a central part the classical stellar atmosphere problem, consisting of hydrodynamics, thermodynamics, radiative transfer, and of the chemistry of the gas, augmented by an equation of state. Already this system is nonlinearly and non-locally coupled. The solution to the atmosphere problem is completely described by the four stellar parameters: stellar mass, stellar luminosity, stellar temperature and (photospheric) element abundances.

In order to treat a dust-forming stellar atmosphere, the chemistry needs to comprise the dust forming species and a physical description of the dust nucleation and growth process has to be provided.

Concerning the first point, suitable high-temperature condensates have to be identified according to the considered type of chemical composition (i.e. oxygen-rich or carbon-rich) and the availability of suitable growth species in the gas phase has to be investigated. This generally constitutes a difficult problem (e.g., Gail & Sedlmayr 1986, Gail & Sedlmayr 1999, Jeong et al. 1999, Woitke 1999) which is mainly related to the nucleation process. Nucleation of seed particles (critical clusters) is treated here in the framework of modified classical nucleation theory (see Patzer et al. 1998 for a detailed discussion) and requires the knowledge of the change of the Gibbs free energy $\Delta G(N)$ associated with the formation of a cluster of N monomers from the vapor. These data may be obtained from laboratory experiments or by quantum-mechanical calculations (e.g., Patzer et al. 1999, Chang et al. 2000, Jeong et al. 2000). Once seed particles have formed, addition of molecules from the gas phase and possibly subsequent reactions on the surface of the growing grain leads to the formation of (heterogeneous) macroscopic particles. This growth process is treated here by means of a moment method developed by Gail & Sedlmayr (1988) including the process of grain evaporation (Gauger et al. 1990) and heterogeneous growth processes (Dominik et al. 1993, Jeong et al. 1999).

Grain formation is coupled to the gas phase chemistry by depletion of the condensing elements and by changing the gas phase abundances by grain evaporation processes and possible grain surface reactions. From a proper treatment of the dust formation process, the grain-size distribution function can be obtained (Dominik et al. 1989, Gauger et al. 1990, Sedlmayr & Winters 1997, Winters et al. 1999), from which the transport coefficients of the dust component can be derived. Due to its huge cross section with respect to the stellar radiation field and by its frictional coupling to the gas, the dust strongly couples to the hydrodynamic structure by introducing the dominating acceleration term in the equation of motion and to the thermodynamic structure by providing the dominating opacity in the energy equation. By a physical treatment of the dust formation problem as outlined above, no additional parameters (like, e.g., a "condensation temperature") are introduced.

In order to account for a pulsating star, we apply a sinusoidally moving inner boundary condition below the photosphere which is characterized by two additional quantities, the velocity amplitude at the inner boundary and the pulsation period. This somewhat artificial ingredient is introduced due to a presently lacking consistent physical description of the pulsation driving zone of AGB stars.

2. Model structures

2.1 Massive AGB winds

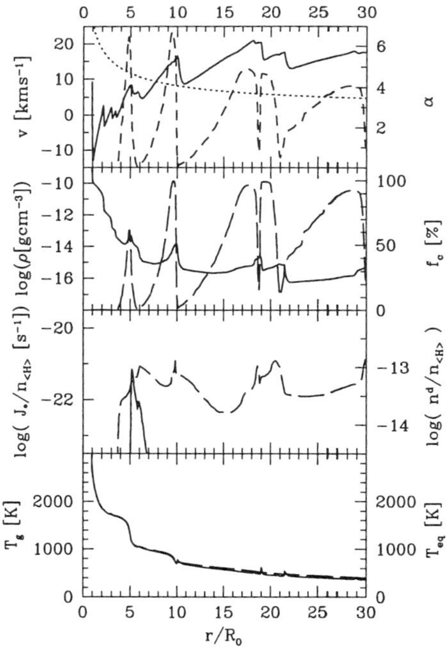

Figure 1. Radial structure of a high mass-loss rate model (w35 in Winters et al. 2000b). Solid lines: l.h.s. ordinate

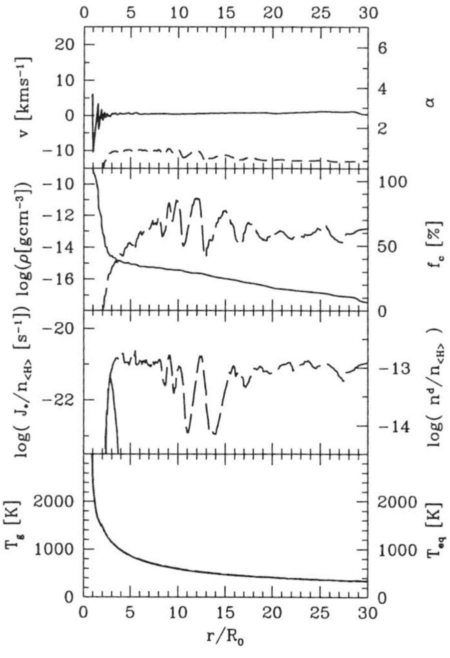

Figure 2. Radial structure of a low mass-loss rate model (w127 in Winters et al. 2000b)

Figure 1 displays, for a fixed instant of time, the result of one such model calculation, which produces a massive wind, typical of mass-losing dusty AGB stars, and shows several distinct features which are only revealed by the consistent treatment:

– Dust is concentrated in discrete layers (see the degree of condensation f_c), caused by the changing temperature, density, and chemical composition in the dust formation zone.

– Strong shocks are propagating through the circumstellar shell. These shocks are induced by radiation pressure on dust ($\alpha = g^{\rm rad}/g^{\rm grav}$).

– Steps in the temperature stratification indicate a pronounced back-warming, which cuts off dust nucleation (J_*) at the inner edge of the dust nucleation zone.

The combined effects of time-dependent dust formation, radiative acceleration, back-warming, and radial dilution of the dust layers introduces a new intrinsic time scale to the system (Fleischer et al. 1995, Höfner et al. 1995) which is longer than the typical pulsation periods of AGB stars and which leads to observable modulations of the light curves (e.g., Winters et al. 1994,2000b), long term variations in the spatial intensity profiles (e.g., Winters et al. 1995,1999) and shape variations in the synthetic CO infrared line profiles (Winters et al. 2000a).

These features are in good agreement with corresponding observations of 'high mass-loss rate' AGB stars, e.g., R For (Winters et al. 1994), AFGL 3068 (Winters et al. 1997), and IRC +10216 (Winters et al. 2000a)

2.1.1 Dust induced shocks. Several instabilities occur in this kind of models (see Woitke 2001), for example a dynamical instability leading to the dust induced shock waves: An initial compression of the gas enhances the collision rates between dust grains and molecules, leading to a more efficient growth of the grains. This, in turn, increases the dust opacity and thereby the hydrodynamic acceleration of the dust forming zones by radiation pressure on the dust grains and frictional momentum transfer to the gas. The larger velocities of the dust forming zones lead to a further compression of the medium. This positive feedback constitutes a control loop which amplifies initial perturbations in the system.

2.2 Low mass-loss rate winds

Figure 2 displays the radial structure of a model, in which a much smaller amount of dust forms. Consequently, the dust opacity remains small (e.g., the ratio of radiative to gravitational acceleration α stays below 1) and the dust has only a minor influence on the model structure. As a result, the model produces only a low mass-loss rate and a low outflow velocity and its structure (beyond $\approx 5R_*$) resembles that of a stationary wind. The characteristics of this type of models (mass-loss rate $\dot{M} < 3 \cdot 10^{-7} M_{\odot}\mathrm{yr}^{-1}$, outflow velocity $v_{\infty} < 5\,\mathrm{km\,s^{-1}}$) are in good agreement with, e.g., the properties of the central stars of detached shell objects (e.g., Olofsson et al. 2000, Winters & Le Bertre 2001) and also offers an explanation for the peculiar hydrodynamic structure detected in the semi-regular variable L^2 Pup (Jura et al. 2002, Winters et al. 2002).

3. Topology of solution space

The limit $\alpha = 1$ clearly separates the models into two classes: Class A models produce high mass-loss rates in excess of $3 \cdot 10^{-7} M_{\odot}\mathrm{yr}^{-1}$ and wind outflow velocities larger than $5\,\mathrm{km\,s^{-1}}$, while Class B models stay below these values (Winters et al. 2000b). Since the value of α is determined by the stellar parameters, a star evolving along the AGB could change between both states, e.g., by increasing its luminosity and decreasing its temperature it could move from region B to A, i.e. change from a low mass-loss rate state to a high mass-loss rate state. As was demonstrated by combining these wind models with stellar evolution calculations (Schröder et al. 1998,1999), low initial mass stars could even 'loop' from B to A and back to B during a thermal pulse, thereby producing detached gas and dust shells around the star. The characteristics expected for the high mass-loss rate episode (duration about 800 yr, mass-loss rate of the order of $10^{-5} M_{\odot}\mathrm{yr}^{-1}$, outflow velocity of about $15\,\mathrm{km\,s^{-1}}$) match very well the properties of the well observed detached shell around the carbon star TT Cyg (Olofsson et al. 2000).

4. IRC +10216

The proto-typical dust enshrouded carbon star IRC +10216 shows different features which seem to be at least in qualitative agreement with our model predictions:

a) The long-term infrared light curves of IRC +10216 obtained over the past 35 years show a modulation similar to the variations seen in our multi-periodic models.

b) The CO (2-0) first overtone line profiles observed between 1979 and 1990 show monotonic shape variations within 10 years. This can be interpreted in terms of the formation of a new dust layer (Winters et al. 2000a).

c) The dust shell of IRC +10216 appears spherically symmetric on large spatial scales showing a distinct multi-layered structure which indicates mass-loss variations on timescales between about 40 and 800 years (Mauron & Huggins 2000; see also the contributions by P. de Laverny and by Y. Simis, this volume). However, on small spatial scales ($\approx 200\,\mathrm{mas}$) the circumstellar environment clearly displays a fragmented structure consisting of four main components ('A' - 'D'), which show a dramatic evolution in separation and relative brightness over the past ≈ 6 years (Weigelt et al. 1998,2002; Osterbart et al. 2000, Tuthill et al. 2000; see also the contribution by J. Monnier, this volume). These observed changes occur on a time scale longer than the stellar pulsation period. However, the interpretation of the observed structures ('holes' in the circumstellar shell?, dust clouds?) is yet unclear. In particular, there is a debate even about the star's location in the high resolution images. Essentially two scenarios have been proposed in the literature: i) the star is located at the brightest (2 μm) component ('A') (see Weigelt et al. 1998, Haniff & Buscher 1998, Tuthill et al. 2000) and ii) the star is located at the initially second brightest component ('B') (Osterbart et al. 2000, Men'shchikov et al. 2001, Weigelt et al. 2002).

In order to investigate this question in terms of our dust formation models, the consequences of both scenarios for the increasing separation between A and B and for the fading of component B w.r.t. A are considered in the framework of consistent 1D models.

4.1 Scenario 1: The star is located at A

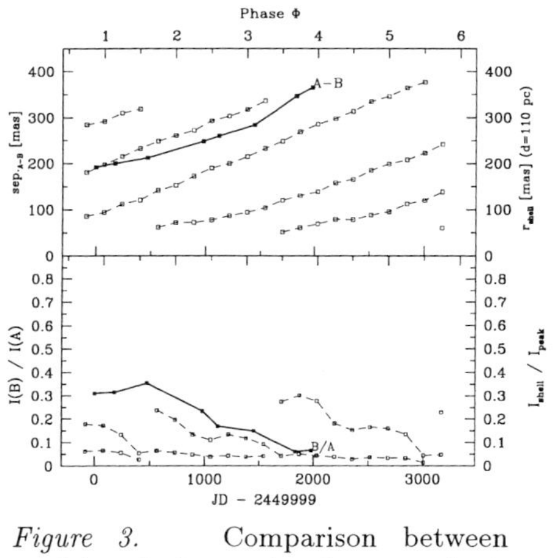

Figure 3. Comparison between model and observation for scenario 1. Upper panel: observed separation between A and B (solid line) and location of successive dust layers in the model (dashed lines); lower panel: observed (solid) and synthetic (dashed) intensity ratios

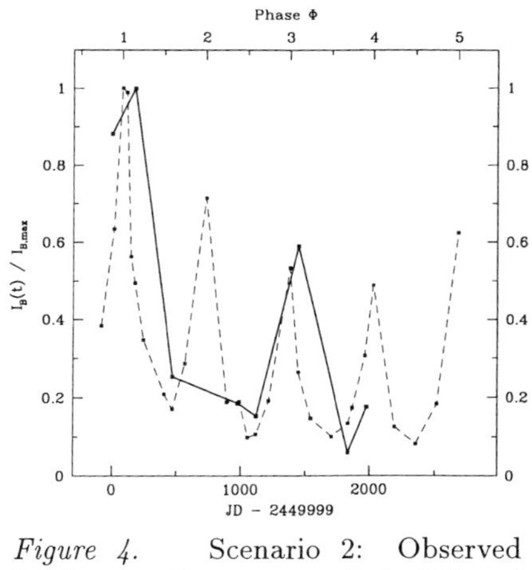

Figure 4. Scenario 2: Observed brightness of component B (solid) and synthetic intensity towards the model star (dashed), both normalized to their respective maximum values.

In this scenario, component B is assumed to be produced by thermal emission from a dust cloud, blown away from the star. In terms of the models we consider component B as being a fragment of a spherical shell.

Fig. 3 depicts a direct comparison between observation and model results over almost 6 pulsation periods. From the upper panel of Fig. 3 we conclude, that the observed component B would correspond (at least) to the second layer in the model and that B is moving almost in the plane of the sky, since its relative velocity w.r.t. component A equals the outflow velocity of $\approx 15\,\mathrm{kms}^{-1}$ (for a distance d = 110 pc). Between 1996 and 1997 (around day 500), a new layer (cloud) should have formed according to the model. This is exactly the time when component A starts noticeably changing its shape in the images (see Weigelt et al. 2002).

The observed variation of the intensity ratio between B and A is consistent with the model:

After an initial phase of almost constant relative intensity B/A the observed intensity ratio rises to its maximum value around day 500, then it decreases till day 1200 where it settles in a plateau till day 1500, from which on another steep decrease follows. Note, that a new dust layer is forming in the model during the time interval when the B intensity starts decreasing. The behavior of the synthetic intensity ratios is very similar (see lower panel of Fig. 3): The initial intensity decrease is interrupted by an intermediate phase, when the intensity ratio settles in a temporary plateau before it decreases further. Also, the time scale of the observed variation is in agreement with the time scale predicted by the model. However, there is a difference of about a factor 3 between the model intensity ratio and the observed ratio for the time interval used for direct comparison in Fig. 3. From this discrepancy, it seems unlikely that B is produced by thermal emission from a dust cloud.

4.2 Scenario 2: Component B is the star

In this case, the observed fading of B might be caused by increasing obscuration of the star due to the formation of new dust along the line of sight (see Weigelt et al. 2002) and/or by a steadily increasing mass-loss rate as proposed in Men'shchikov et al. (2001).

In order to allow for a quantitative comparison of the results of the 1D model calculation and the observation, the brightness of component B itself has to be determined (instead of the ratio B/A). For this purpose, the integrated flux in the full field of view of the images (2".05) was compared to coeval K-band photometry of IRC +10 216.

The resulting (absolute) brightness of component B, normalized to its maximum value (which occurred in April 1996) is plotted in Fig. 4, connected by the solid line. The dashed line connects the synthetic K-band intensities towards the model star. The general behavior of the synthetic intensity variation, such as the rate of fading and the occurrence of a plateau around day 1000 are in rather good agreement with the observation, lending independent support to the interpretation that the star may indeed be located at position B (Men'shchikov et al. 2001) and that components A, C, and D are rather produced by 'holes' in the circumstellar dust shell than by dust clouds (Niccolini et al. 2002).

Acknowledgments: I thank the organizers for providing a generous travel support and for a very enjoyable conference. The calculations were performed on the Cray vector and parallel computers accessible through the ZIB Berlin. This work was supported by the *BMBF*, the PROCOPE program, and by the CNRS.

References

Chang, C., Patzer, A.B.C., Sedlmayr, E., et al. 2000, Chem. Phys. Lett., 324, 108
Dominik, C., Gail, H.-P., & Sedlmayr, E., 1989, A&A, 223, 227
Dominik, C., Sedlmayr, E., & Gail, H.-P. 1993, A&A, 277, 578
Fleischer, A.J., Gauger, A., & Sedlmayr, E. 1995, A&A, 297, 543
Gail, H.-P., & Sedlmayr, E. 1986, A&A, 166, 225
Gail, H.-P., & Sedlmayr, E. 1988, A&A, 206, 153
Gail, H.-P., & Sedlmayr, E. 1999, A&A, 347, 594
Gauger, A., Gail, H.-P., & Sedlmayr, E. 1990, A&A, 235, 345
Haniff, C.A., & Buscher, D.F. 1998, A&A, 334, L5
Höfner, S., Feuchtinger, M.U., & Dorfi, E.A. 1995, A&A, 297, 815
Jeong, K.S., Chang, C., Sedlmayr, E., & Sülzle, D. 2000, J. Phys. B, 33, 3417
Jeong, K.S., Winters, J.M., & Sedlmayr, E. 1999, IAU Symp. 191, 233. ASP.
Jura, M., Chen, C., & Plavchan, P. 2002, ApJ, 569, 964
Mauron, N., & Huggins, P.J. 2000, A&A, 359, 707
Men'shchikov, A.B., et al. 2001, A&A, 368, 497
Niccolini, G., Woitke, P., & Lopez, B. 2002, A&A, submitted.
Olofsson, H., Bergman, P., Lucas, R., et al. 2000, A&A, 353, 583
Osterbart, R., et al. 2000, A&A, 357, 169
Patzer, A.B.C., Chang, C., Sedlmayr, E., & Sülzle, D. 1999, Eur. Phys. J. D, 6, 57
Patzer, A.B.C., Gauger, A., & Sedlmayr, E. 1998, A&A, 337, 847
Schröder, K.-P., Winters, J.M., Arndt, T.U., & Sedlmayr, E. 1998, A&A, 335, L9
Schröder, K.-P., Winters, J.M., & Sedlmayr, E. 1999, A&A, 349, 898
Sedlmayr, & E., Winters, J.M. 1997, Lecture Notes in Physics 497, 89
Tuthill, P., Monnier, J., Danchi, W., & Lopez, B. 2000, ApJ, 543, 284
Weigelt, G., Balega, Y., Blöcker, T., et al. 1998, A&A, 333, L51
Weigelt, G., Balega, Y., Blöcker, T., et al. 2002, A&A, in press
Winters, J.M., Fleischer, A.J., Gauger, A., & Sedlmayr, E. 1994, A&A, 290, 623
Winters, J.M., Fleischer, A.J., Gauger, A., & Sedlmayr, E. 1995, A&A, 302, 483
Winters, J.M., Fleischer, A.J., Le Bertre, T., & Sedlmayr, E. 1997, A&A, 326, 305
Winters, J.M., Keady, J.J., Gauger, A., & Sada, P.V. 2000a, A&A, 359, 651
Winters, J.M., & Le Bertre, T. 2001, in Post-AGB objects as a phase of stellar evolution, 93
Winters, J.M., Le Bertre, T., Jeong, K.S., Helling, C., & Sedlmayr, E. 2000b, A&A, 361, 641
Winters, J.M., Le Bertre, T., & Keady, J.J. 1999, IAU Symp. 191, 261
Winters, J.M., Le Bertre, T., Nyman, L.-A., Omont, A., & Jeong, K.S. 2002, A&A, 388, 609
Woitke, P. 1999, in Astronomy with Radioactivities, 163, MPE Report 274
Woitke P. 2001, Rev. Mod. Astron., 14, 185

MID-INFRARED OBSERVATIONS OF MASS-LOSING RED GIANTS

M. Jura
Department of Physics and Astronomy
UCLA, Los Angeles CA 90095-1562, USA
jura@astro.ucla.edu

Abstract Within the past 3 years, mid-ir imagers have become available at 10-meter class telescopes. As a result, it is now possible to spatially resolve many sources which previously appeared to be point-like, and here we report mid-infrared observations at Keck of two stars, L_2 Pup and OH 231.8+4.2. For L_2 Pup, one of the nearest mass-losing red giants, we find that the wind may be driven by the stellar pulsations, with radiation pressure on dust being relatively unimportant. The stellar pulsations that drive the wind may be nonradial. For OH 231.8+4.2, we find both an extended bipolar flow and an unresolved central source. We suggest that the unresolved central source can be explained by an opaque, flared disk with an outer radius of 5×10^{15} cm and an outer temperature of ~130 K. One possible model to account for this flaring is that the material in the disk is orbiting the central star and not simply undergoing a radial expansion.

Introduction

The evolution and death of stars is a central topic in modern astrophysics. In their advanced evolutionary phases, stars can lose more than half of their initial mass; this mass loss plays a critical role in determining the final fate of the star.

When stars become red giants with luminosities greater than 10^3 $L_{\odot}$, dust grains form in the mass outflow. These dust grains reprocess the photospheric light emitted by the star into an infrared excess typically at $\lambda > 3$ μm. The infrared excess can be used to study the mass loss from the star.

In the past, it has been possible to measure the integrated infrared excess of a mass-losing star, but we have had little, if any, spatial information on the dust distribution. However, there have been two techno-

Y. Nakada et al. (eds.), Mass-Losing Pulsating Stars and their Circumstellar Matter, 159–164.

logical advances in the past several years which now enable us to image many of these circumstellar envelopes. First, 10m-class telescopes have been built at excellent infrared sites such as Mauna Kea. Second, mid-infrared imagers have been built and are now routinely used at these telescopes. At 10 μm and 20 μm, the ideal resolutions, λ/D, of a 10m telescope are 0.2" and 0.4", respectively. It is reasonably common for the observing conditions to be such as to achieve 0.4" image quality at both wavelengths at Keck.

Consider a source with flux, F_ν, If the source is an opaque black body of temperature, T, and it subtends solid angle, Ω, then

$$F_\nu \;=\; B_\nu(T)\,\Omega \tag{1}$$

where B_ν denotes the Planck function. Assuming that we can resolve a solid angle of 3.0×10^{-12} ster, the solid angle subtended by a source with radius 0.2", then opaque sources with temperatures of 100K and 200 K can be resolved at 17.9 μm if their fluxes are 6.7 Jy and 380 Jy, respectively. The most obvious candidates to resolve are sources that are nearby, sources that are cold and/or sources that are optically thin.

In this paper, we discuss two stars, L_2 Pup and OH 231.8+4.2, whose circumstellar shells we have recently studied at Keck. These data illustrate the power of the infrared imagers and may also illustrate pathways for future work.

1. L_2 Pup

Our work on this star has been discussed in Jura, Chen & Plavchan (2002a); here we summarize the results.

L_2 Pup (= HD 56096 = HR 2748) has a distance measured by the *Hipparcos* satellite of 61 $\pm$ 5 pc, thus making it, along with R Dor, one of the pulsating, mass-losing red giants that is nearest to the Sun. In the Yale Bright Star Catalog it is classified as M5IIIe and assigned a visual magnitude of 5.10. It is a semiregular variable with a period of 140.6 days according to the General Catalog of Variable Stars, but with a 137 day period according to Whitelock et al. (2000).

With a time-averaged measure that m_K = - 2.24 mag and thus m_{bol} = 0.73 mag (Whitelock et al. 2000), the luminosity of L_2 Pup is 1500 $L_\odot$. The time-averaged value of (J-K) is 1.31 mag (Whitelock et al. 2000), and therefore, from the models presented by Bessell et al. (1989) and Houdashelt et al. (2000), the time-averaged effective temperature of L_2 Pup is about 3400 K. Alternatively, for Mira stars, according to Van Belle et al. (1996), the average effective temperature of a Mira is

$$T_{eff} \;=\; (Msubclass) * (-273) \;+\; 4690 \tag{2}$$

This equation was derived for a fit to the data for Miras of spectral type M6 and later. Extrapolating this expression to L_2 Pup whose spectral type is M5 implies that this star has an effective temperature of 3300 K, We infer a time-averaged radius of L_2 Pup of 7.8 $\times$ 10^{12} cm. With this luminosity and temperature, it is possible that L_2 Pup is either a first ascent red giant or an Asymptotic Giant Branch Star (Lattanzio 1991, Girardi et al. 2000). Because of its inferred space motion near 90 km s^{-1} (Feast & Whitelock 2000), this star probably belongs to the Thick Disk and the mass of its main sequence progenitor was probably near 1 $M_{\odot}$.

Our 11.7 μm and 17.9 μm images of the dust shell around L_2 Pup are shown in Jura et al. (2002a). The shell is resolved and azimuthal variations of the mass loss by as much as a factor of 2 are detected. The inferred mass loss rate is about 3 $\times$ 10^{-7} $M_{\odot}$ yr^{-1} if the grain speed is 10 km s^{-1}. [Since the particles are thought to be driven supersonically through the gas by radiation pressure, the grain speed is inferred to be larger than the gas outflow speed of about 3 km s^{-1} derived by Kerschbaum & Olofsson (1999) from measurements of the emission from circumstellar CO.] From the stellar models, it seems that the star will spend about 5 $\times$ 10^5 years in its current phase of evolution, regardless of whether it is on the first or second ascent up the red giant branch (Lattanzio 1991; Girardi et al. 2000). With this estimate for the duration of its current state, the star might lose 15% of its initial mass during this phase of its evolution.

The mass loss from L_2 Pup can be explained by the "Case B" models of Winters et al. (2000). In these models, the mass loss is driven by pulsations and radiation pressure on grains plays a minor role in the dynamics. In view of the anisotropic mass loss in our Keck images, the time-variable optical polarization (Magalhaes et al. 1986), and the difference in the amplitude and timing of the near infrared vs. mid-infrared (see Jura et al. 2002a), it is plausible that the star is undergoing non-radial pulsations.

Finally, similar studies of other mass-losing red giants within ~100 pc of the Sun such as R Cas, R Leo and χ Cyg might serve as a powerful tool for studying the winds from these systems.

2. OH 231.8+4.2

OH 231.8+4.2 appears to be an evolved member of the open cluster M 46 (Jura & Morris 1985), and on the basis of the main sequence turnoff from this system, this mass-losing star probably had an initial main sequence mass of about 3 $M_{\odot}$. Although the star is buried in an optically opaque circumstellar dust cloud, scattering in bipolar lobes perpendicular to the nominal disk indicates the presence of a luminous M9 star (Cohen & Frogel 1977). The star is variable with a period which might be as short as 648 days (Feast et al. 1983) or as long as 708 days (Kastner et al. 1992), and its luminosity ranges between 10^4 $L_\odot$ and 2×10^4 $L_\odot$ (Kastner et al. 1998). At this conference, Whitelock has noted that the star may not have a single period. The luminosity and effective temperature can be explained if the star lies on the AGB.

The optical and near-infrared morphology of OH 231.8+4.2 indicates that there is a massive disk surrounding the central star (see, for example, Kastner al. 1992). Additionally, OH 231.8+4.2 displays a bipolar outflow at radio and optical wavelengths which is inferred to carry at least 0.3 $M_\odot$ (Reipurth 1987, Morris et al. 1987, Sanchez-Contreras et al. 2000, Alcolea et al. 2001). The bipolar morphology exhibits discrete clumps (Alcolea et al. 2001).

Jura, Chen & Plavchan (2002b) have reported 11.7 μm and 17.9 μm images of this system with the Keck I telescope. We detected both a central unresolved point source with a diameter less than 0.5" producing $F_\nu(17.9\ \mu\mathrm{m}) = 60$ Jy, and emission extended more than 1" away from the star which is aligned with the bipolar outflow seen on larger scales. We find that the unresolved central source can be explained by emission from a black body with a radius of $\sim 5 \times 10^{15}$ cm and a temperature of $\sim$130 K.

A natural way to explain this unresolved black-body central source is to use models of flared disks computed by Chiang & Goldreich (1997, 1999). Recasting those models, Jura et al. (2002b) show that the expected temperature of the disk, T, at distance R from the star is given by the expression:

$$T_{disk} = \left(\frac{1}{7}\right)^{2/7} \left(\frac{R_*}{R}\right)^{3/7} \left(\frac{2k_B T_* R_*}{G M_* \mu}\right)^{1/7} T_* \qquad (3)$$

where k_B is Boltzmann's constant, T_*, R_*, and M_* are the temperature, radius and mass of the orbited star, G is the gravitational constant, and μ is the mean molecular weight of the material. At $R = 5 \times 10^{15}$ cm, this expression gives $T = 130$ K, in agreement with the observations.

In this model, we assume that we are witnessing the outer, flared portion of the disk which we suggest lies at 5×10^{15} cm from the central star. We suggest that the outer radius of the disk is determined by the amount of angular momentum available to "spin-up" the disk. Although there is no direct spectroscopic evidence to support this view, we suspect that OH 231.8+4.2 is a binary system. We hypothesize that some of the mass loss from the AGB star gets entrapped in an orbiting ring; this ring than expands under the action of tidal torques (Pringle 1991). The largest size that the ring may acquire is determined by the amount of angular momentum, J, that can be extracted from the central binary. Jura et al. (2002b) estimate that

$$J = M_{disk}\sqrt{\frac{GM_* R_{out}}{\pi}} \tag{4}$$

Plausible values of the angular momentum in the binary ($\sim 10^{53}$ g cm^2 s^{-1}) and mass in the disk (0.1 $M_\odot$) allow for $R_{out} = 5 \times 10^{15}$ cm, as suggested by the data.

Jura et al. (2002b) argue that if there is a central opaque disk, then it is quite natural that the material is orbiting the star rather than expanding as is usually assumed. Jura & Kahane (1999) have discussed the possibility that post main sequence red giants possess orbiting material; OH 231.8+4.2 appears to be another member of this class with a disk with a total mass near 0.1 $M_\odot$. Eventually, large macroscopic solids and even planets might form from this disk.

As discussed by Waters et al. (1993) and Jura & Kahane (1999), there are other objects besides OH 231.8+4.2 of long-lived orbiting material around mass-losing Asymptotic Giant Branch stars. These models may pertain to the recent result by Melnick et al. (2001) who detected H_2O emission in the outflow of IRC+10216, a carbon-rich star. It is very difficult to explain the observation that $[H_2O]/[H_2] \sim 10^{-6}$ with any model of gas-phase chemistry. Instead, Melnick et al. propose that a cloud with $\sim 10^{29}$ g of comets is in orbit around IRC+2016, and these comets are evaporating and producing the observed gas-phase water. It should be noted, however, that other carbon-rich stars such as BM Gem exhibit oxygen-rich species in their circumstellar envelopes (see Kahane et al. 1998). The most likely model to explain these data is that there is some kind of orbiting reservoir of oxygen-rich material. Therefore, an alternative explanation for the presence of H_2O in the outflow around IRC+10216 is that there is a long-lived disk, somewhat similar to what is found around OH 231.8+4.6, rather than a cloud of comets. The heating and evaporation of this disk material might produce the observed gas-phase H_2O. This picture is similar to that proposed by Melnick et al. (2001) except that the reservoir of orbiting material was created analogously to the orbiting disk around OH 231.8+4.2 and does not require a massive cloud of long-lived comets. We have proposed that there is currently $\sim 10^{30}$ g of solids in orbit around OH 231.8+4.2, more than enough oxygen-rich material to account for the H_2O outflow from IRC+10216.

3. CONCLUSIONS

The use of mid-ir imagers on 10m class telescopes such as Keck provides a powerful tool for probing circumstellar environments. Above, we describe two recent applications of this technique. In the future, we expect that many more sources will be observed with the consequence that we will learn much more about how red giant stars evolve and lose mass.

Acknowledgments

This work has been partly supported by NASA.

References

Alcolea, J., Bujurrabal, V., Sanchez Contreras, C., Neri, R. & Zweigle, J. 2001, A&A, 373, 932
Bessell, M. S., Brett, J. M., Scholz, M. & Wood, P. R. 1989, A&Ap, 213, 209
Chiang, E. I. & Goldreich, P. 1997, ApJ, 490, 368
Chiang, E. I. & Goldreich, P. 1999, ApJ, 519, 279
Cohen, M. & Frogel, J. A. 1977, ApJ, 211, 178
Feast, M. W., Catchpole, R. M., Whitelock, P. A., Roberts, G., Jones, J. S. & Carter, B. S. 1983, MNRAS, 203, 1207
Feast, M. W. & Whitelock, P. A. 2000, MNRAS, 317, 460
Girardi, L., Bressan, A., Bertelli, G. & Chiosi, C. 2000, A&AS, 141, 371
Houdashelt, M. L., Bell, R. A., Sweigart, A. V. & Wing, R. F. 2000, AJ, 119, 1424
Jura, M., Chen, C. & Plavchan, P., 2002a, ApJ, 569, 964
Jura, M., Chen, C. & Plavchan, P. 2002b, ApJ, 574, in press
Jura, M. & Kahane, C. 1999, ApJ, 521, 302
Jura, M. & Morris, M. 1985, ApJ, 292, 487
Kahane, C., Barnbaum, C., Uchida, K., Balm, S. P. & Jura, M. 1998, ApJ, 500, 466
Kastner, J. H., Weintraub, D. A., Merrill, K. M., & Gatley, I. 1998, AJ, 116, 1412
Kastner, J. H., Weintraub, D. A., Zuckerman, B., Becklin, E. E., McLean, I. & Gatley, I. 1992, ApJ, 398, 552
Kerschbaum, F. & Olofsson, H. 1999, AÂS, 138, 299
Lattanzio, J. C. 1991, ApJS, 76, 215
Malgahaes, A. M., Codina-Landaberry, S. J., Gneiding, C. & Coyne, G. V. 1986, A&A, 154, 1
Melnick, G. J., Neufeld, D. A., Ford, K. E. S., Hollenbach, D. J. & Ashby, M. L. N. 2001, Nature, 412, 160
Morris, M., Guilloteau, S., Lucas, R. & Omont, A. 1987, ApJ, 321,888
Pringle, J. E. 1991, MNRAS, 248, 754
Reipurth, B. 1987, Nature, 325, 787
Sanchez Contreras, C., Bujurrabal, V., Neri, R. & Alcolea, J. 2000, A&A, 357, 651
van Belle, G. T., Dyck, H. M., Benson, J. A. & Lacasse, M. G. 1996, AJ, 112, 2147
Waters, L. B. F. M., Waelkens, C., Mayor, M. & Trams, N. R. 1993, A&A, 269, 242
Whitelock, P., Marang, F. & Feast, M. 2000, MNRAS, 319, 728
Winters, J. M., Le Bertre, T., Jeong, K. W., Helling, Ch., & Sedlmayr, E. 2000, A&A, 361, 641

HIGH-RESOLUTION IMAGING OF DUST SHELLS

Asymmetries and Temporal Evolution

John D. Monnier
Harvard-Smithsonian Center for Astrophysics
60 Garden Street, Cambridge, MA 02138 USA
jmonnier@cfa.harvard.edu

Peter G. Tuthill
University of Sydney

William C. Danchi
NASA - Goddard Space Flight Center

Anita Richards
Jodrell Bank Radio Observatory

Abstract Advances in high-resolution imaging at visible, infrared, and radio wavelengths have allowed the circumstellar envelopes of evolved stars to be probed in unprecedented detail. While adequate to explain broadband spectral energy distributions, simplistic mass-loss scenarios (assuming spherical symmetry and continuous dust formation) have been shown to be grossly inadequate to understand the clumpy and often asymmetric dust shells recently revealed.

In this paper, we will highlight a few sources that illustrate qualitatively new mass-loss phenomena, including the presence of circumstellar "disks" and evidence for sporadic, asymmetric mass ejections. Perhaps important physics is still missing from our picture of mass-loss on the AGB, for instance the effects of magnetic fields or large-scale convective motions.

Y. Nakada et al. (eds.), Mass-Losing Pulsating Stars and their Circumstellar Matter, 165–172.

Keywords: Mass-loss; interferometry; dust; pulsation; masers

1. Why is Imaging Important Now?

A number of new facilities have recently begun delivering high quality, high spatial resolution images. For instance, harnessing the diffraction-limit of the world's largest optical telescope (Keck Observatory) for imaging complex objects has only recently been made possible in the infrared using aperture masking techniques (Tuthill et al., 2000b; Monnier, 1999). Other workers have achieved similar results using speckle interferometry on slightly smaller telescopes (e.g., Weigelt et al., 1998). The diffraction-limit of a 10-m aperture telescope is approximately 50 milliarcsecond (mas) at 2.2μm, which is quite comparable to that of the Hubble Space Telescope in the visible regime. Further, MERLIN and the VLBA can achieve even higher resolution in the radio using very long baseline interferometry, especially for maser studies.

Not only are multiwavelength studies now possible with the new suite of modern observatories, but the achievable spatial resolution is fine enough to fully resolve the largest and closest circumstellar dust shells. That is, the 50 milli-arcseconds resolution mentioned above corresponds to a physical scale of just 10 AU for nearby carbon stars and miras (distance ~200 pc). Given typical 20 km/s motions, we then expect about 20 mas per year proper motions, easily detectable in only a single year. For the nearest red supergiants (~1.5 kpc), physical scales smaller than 75 AU can be directly probed. With the previous (smaller) generation of telescopes, there were just no sources close enough to Earth to allow the detailed mapping now possible.

1.1 Mysteries to be Addressed

With sufficient high resolution imaging, there are number of important topics we can address:

- Mass-loss Mechanisms. Current thinking (e.g., Lafon and Berruyer, 1991) is that radiation pressure on dust grains drive the high mass-loss rates in the winds of evolved stars. The ultimate test of this theory would be to watch dust form *in situ* and to measure the acceleration. Perhaps magnetic fields play a larger role than previously assumed (e.g., Kemball and Diamond, 1997)?

- Dust Formation Theory. Does dust form continuously and smoothly throughout a stellar pulsational cycle? Do we see evidence for dust formation instabilities (e.g., Winters et al., 1995) seen in spherically-symmetric dynamical simulations? Does dust form at the densities, pressures, and temperatures we expect?

- Asymmetric Mass-loss. Planetary nebulae typically show asymmetries and it is not known if these asymmetries originate before or after the end of the AGB phase. What could cause strong deviations from spherical symmetry?

As you will see, the dust shells around evolved stars are complex and dynamic. It is impossible to answer the above questions by analyzing spectral energy distributions alone, and high resolution imaging is essential for making new progress in this field. We argue that the best approach for understanding such complex phenomena is through multi-wavelength, multi-epoch, and multi-resolution case studies. In this short review, we will briefly introduce some targets which have recently been extensively studied and refer the interested reader to the references included in each section for detail.

2. Case # 1: Red Supergiant VY CMa

The evolved red supergiant VY CMa has been the subject of recent attention by a number workers in high-resolution imaging (Smith et al., 2001; Monnier et al., 1999; Wittkowski et al., 1998; Kastner and Weintraub, 1998; Richards et al., 1998); please see these papers for detailed analysis.

In Figures 1 & 2, we have shown different views of the VY CMa envelope and each panel is described in the captions. Important conclusions from these images include:

- Mass-loss can be wildly asymmetric and inhomogeneous. Clearly there are important mechanisms operating beyond spherically- symmetric pulsation and dust-driven winds. Smith et al. (2001) interpret recent HST images to suggest "multiple, asymmetric ejection episodes due to localized events on VY CMa's surface." Could such events (whose physical origins are unexplained) be happening on AGB stars as well or only for Red Supergiants?

- Even the highest resolution imaging in the infrared can be severely affected by optical depth effects. The circumstellar "disk" is so embedded that it can not be directly imaged in the thermal infrared, although it is revealed by the distinctive one-sided scattering nebulosity.

- Interpreting maser distributions is not straightforward. The water masers do not show a clear correlation with the infrared dust

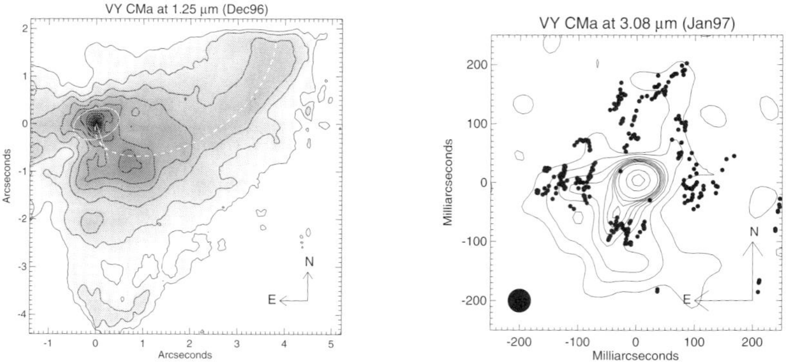

Figure 1. This figure shows three different views of the VY CMa dust envelope. The left panel shows an adaptive optics image at 1.25μm (Monnier et al., 1999) revealing light scattered off of dust in the asymmetric and clumpy envelope. The one-sided scattered light nebula can be produced by a thick "disk" of dust oriented roughly NW-SE in the inner envelope. The right panel shows a zoom-up (contours) of the inner dust shell at 3.1μm (Monnier et al., 1999). Even at this wavelength, where the dust is optically thinner and the star itself is clearly seen, the putative inner disk is not visible and we see only emission from the South "pole" due to optical depth effects. Overlaid are water maser spots (Richards et al., 1998) which ring the star, and but do not show any obvious disk structure or correlation with the IR emission (note there is some uncertainty on how to register these masers and IR image since the star is not directly detected in the radio).

emission (possibly corrupted by optical depth effects), while the OH masers in the outer envelope do show correspondence to the large dust ejections.

3. Case # 2: Carbon Star CIT 6

The dust shell around CIT 6 was recently studied in detail using archival HST imagery and Keck aperture masking (Monnier et al., 2000). Figure 3 shows a side-by-side comparison of the visible-light and infrared image of this nebula; see figure caption for more details.

Even before high-resolution imaging was possible, this source was thought to be a bipolar nebula due to its strong visible and infrared linear polarization (e.g., Trammell et al., 1994). A large 90° change in polarization angle was observed by spectropolarimetry between the blue and red wavelengths and was interpreted in terms of a torus of dust surrounding the carbon star. The HST imaging using a blue filter presented an unexpected morphology (not shown here), suggesting that in fact light from a close (hot) companion was responsible for the blue polarization change.

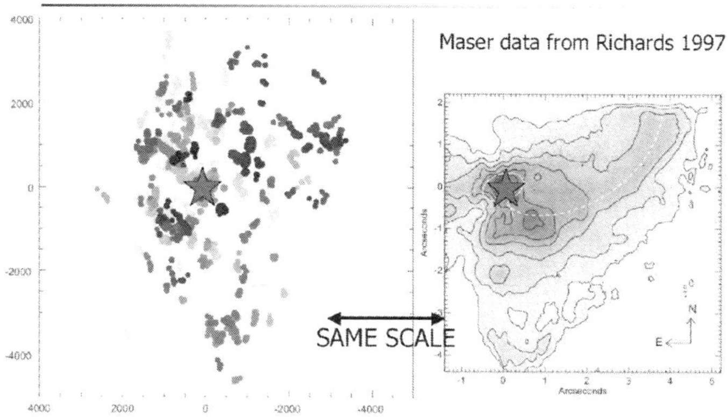

Figure 2. In this figure, we show that the large scale dust inhomogeneities do show some correspondence with the OH mainline maser components at 1667 MHz (Richards, 1997). Combining this data, especially for the south arc of dust, can lead to three-dimensional velocities of the material and mass estimates, crucial pieces of information for determining the ejection mechanism.

CIT 6 is another example of the importance of imaging and illustrates that polarization measurements can also provide crucial information necessary to accurately interpret images.

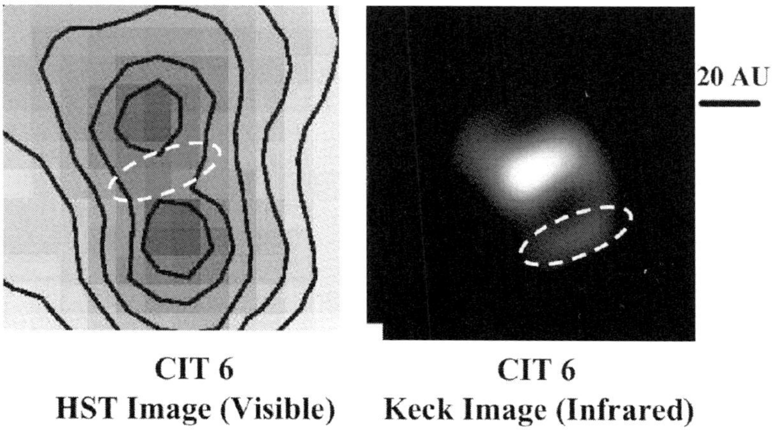

Figure 3. The Hubble Space Telescope visible-light image and the Keck aperture masking IR image of the CIT 6 nebula are shown here (Monnier et al., 2000) at roughly the same spatial resolution. The location of the hypothesized circumstellar "disk" is located by the dashed ellipse in each image. The bright IR source (right panel) is likely scattered light in the northern lobe of this bipolar nebula, and relative astrometry indicates it is associated with the bright northern lobe in the visible. The faint IR source indicated by the ellipse is likely thermal emission from the "disk" and appears in the visible image as a dark lane separating the two scattered-light lobes.

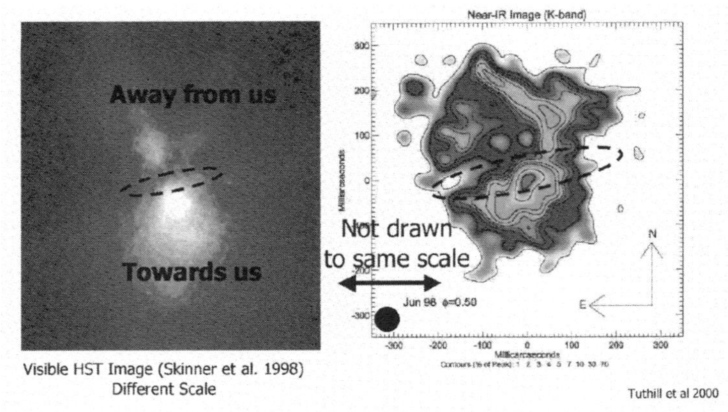

Figure 4. This figure shows the visible-light image of the IRC+10216 nebula from HST (Skinner et al., 1998) and an infrared image from the Keck (Tuthill et al., 2000a). Clearly this nebula is showing strong departures from spherical symmetry and evidence for clumpy and localized mass-loss.

4. Case # 3: Carbon Star IRC+10216

The last case study to be mentioned is IRC+10216, the closest carbon star. We can only briefly mention this source, and we refer the reader to a number of recent papers (Men'shchikov et al., 2001; Tuthill et al., 2000a; Osterbart et al., 2000; Weigelt et al., 1998; Skinner et al., 1998; Haniff and Buscher, 1998).

In figure 4, we present visible-light and infrared images of the IRC +10216 nebula. We have marked the hypothesized location of the circumstellar "disk" according to the analysis of Tuthill et al. (2000); other workers prefer a different interpretation (e.g., Men'shchikov et al., 2001). Even with this superb spatial resolution, the interpretation of this clumpy and inhomogeneous dust shell is controversial (see references above for some attempts).

Figure 5 shows some recent data from our Keck aperture masking experiment, which might be revealing new dust forming and being ejected from the system. By monitoring the time evolution of this dust clump, we hope to definitively confirm (or contradict) the dust-driven wind model for mass-loss.

5. Summary

High-resolution imaging has revealed a number of interesting and unexpected aspects of mass-loss around evolved stars. We see clear evidence for very clumpy dust shells and large-scale asymmetries (we note that the circumstellar "disks" observed are likely due to enhanced equa-

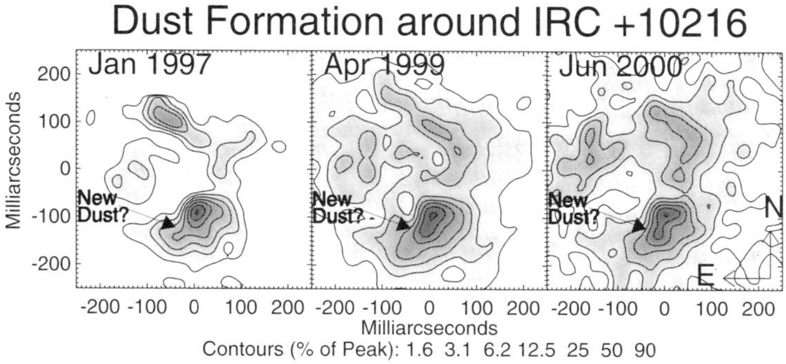

Figure 5. The IRC+10216 nebula at 3 epochs using Keck aperture masking. We see evidence of new dust being formed and ejected from the system. If the bright compact emission arises from the star itself (see Tuthill et al., 2000a), then it is difficult to explain (under the current mass-loss paradigm) how this new dust can form so close to the star and be moving with the proper motion implied by these data.

torial mass-loss and not similar to accretion disks around young stellar objects.) We even see some signs for localized mass ejections from the surface of the stars themselves, suggesting a new mass-loss mechanism. These observations should motivate a careful re-analysis of mass-loss mechanisms for evolved stars.

Interpreting the complex environments is highly uncertain without multi-wavelength data, since optical depth effects are often important. Even with high-quality multi-wavelength data as for IRC+10216, VY CMa, and CIT 6, multiple interpretations are still possible. By monitoring these prototype sources for a decade, proper motions will allow the dust shell dynamics to be revealed and hopefully clarify the situation.

We want to close with a warning that the case studies introduced here are by no means "representative" all of evolved stars, but are probably extreme examples. Dust shells around less evolved stars are usually optically thinner and not able to be imaged as easily. We note that our preliminary results on more "typical" sources (e.g., Monnier et al., 2001) do indicate very clumpy inner dust shells around most systems, but usually without the dramatic global asymmetries shown in the examples discussed here.

References

Haniff, C. A., & Buscher, D. F. 1998, A&A, 334, L5

Kastner, J. H., & Weintraub, D. A. 1998, AJ, 115, 1592

Kemball, A. J., & Diamond, P. J. 1997, ApJ, 481, L111

Lafon, J. P. J., & Berruyer, N. 1991, A&A Rev., 2, 249

Men'shchikov, A. B., Balega, Y., Blöcker, T., Osterbart, R., & Weigelt, G. 2001, A&A, 368, 497

Monnier, J. D. 1999, PhD thesis, University of California at Berkeley

Monnier, J. D., Millan-Gabet, R., Tuthill, P. G., Traub, W., Carleton, N., Coude du Foresto, V., Danchi, W. C., Lacasse, M., Morel, S., Perrin, G., & Porro, I. 2001, BAAS, 198

Monnier, J. D., Tuthill, P. G., & Danchi, W. C. 2000, ApJ, 545, 957

Monnier, J. D., Tuthill, P. G., Lopez, B., Cruzalebes, P., Danchi, W. C., & Haniff, C. A. 1999, ApJ, 512, 351

Osterbart, R., Balega, Y. Y., Blöcker, T., Men'shchikov, A. B., & Weigelt, G. 2000, A&A, 357, 169

Richards, A. 1997, PhD thesis, University of Manchester

Richards, A. M. S., Yates, J. A., & Cohen, R. J. 1998, MNRAS, 299, 319

Skinner, C. J., Meixner, M., and Bobrowsky, M. 1998, MNRAS, 300, L29

Smith, N., Humphreys, R. M., Davidson, K., Gehrz, R. D., Schuster, M. T., & Krautter, J. 2001, AJ, 121, 1111

Trammell, S. R., Dinerstein, H. L., & Goodrich, R. W. 1994, AJ, 108, 984

Tuthill, P. G., Monnier, J. D., Danchi, W. C., & Lopez, B. 2000a, ApJ, 543, 284

Tuthill, P. G., Monnier, J. D., Danchi, W. C., Wishnow, E. H., & Haniff, C. A. 2000b, PASP, 112, 555

Weigelt, G., Balega, Y., Bloecker, T., Fleischer, A. J., Osterbart, R., & Winters, J. M. 1998, A&A, 333, L51

Winters, J. M., Fleischer, A. J., Gauger, A., & Sedlmayr, E. 1995, A&A, 302, 483

Wittkowski, M., Langer, N., & Weigelt, G. 1998, A&A, 340, L39

ISO/SWS OBSERVATIONS OF THE EXTENDED ATMOSPHERES AND THEIR TIME VARIATIONS*

Issei Yamamura
The Institute of Space and Astronautical Science
Yoshino-dai 3-1-1, Sagamihara, Kanagawa, 229-8510, Japan
yamamura@ir.isas.ac.jp

Abstract The Infrared Space Observatory (ISO) has greatly extended our scope of evolved stars. Data in the wavelength region inaccessible from the ground unveil new aspects of the objects. Here we review the studies of time variations of molecular bands in the near- and mid-infrared spectra of oxygen-rich stars observed by the Short-Wavelength Spectrometer (SWS). These molecular bands are thought to arise in the extended region of the atmosphere and the inner part of the circumstellar envelope. It is seen that there are two regions in the extended atmosphere; one follows the optical variability and the other does not. We argue that pulsation shocks running through the atmosphere play an essential role on the variations.

Keywords: Stars: AGB and post-AGB – Stars: atmospheres – Stars: late-type – Infrared: stars

1. Introduction

Stars in the late stage of their evolution are often observed as pulsating variables of Mira or Semi-Regular (SR) types. It is commonly known that the stellar pulsation pushes up the matters from the surface of the star and extends the atmosphere. At the outer part of this extended atmosphere, temperature is cool enough for dust formation. As soon as dust is formed, radiation from the star blows out the dust together with surrounding gas, and causes mass loss. As the most essential processes

*Based on observations with ISO, an ESA project with instruments funded by ESA Member States (especially the PI countries: France, Germany, the Netherlands and the United Kingdom) with the participation of ISAS and NASA. The SWS is a joint project of SRON and MPE.

Y. Nakada et al. (eds.), Mass-Losing Pulsating Stars and their Circumstellar Matter, 173–180.

of mass loss occurs in the extended atmosphere, it is important to know how the matters move out and what kind of processes form molecules and dusts at which part in the extended atmosphere, as the star pulsates. Monitoring the extended atmosphere along the pulsation cycle is therefore crucial for the understanding of mass loss. Since the temperature of this interesting region is approximately 500–2000 K, infrared observation is a suitable tool.

During its 29 months mission, the Infrared Space Observatory (ISO; Kessler et al. 1996) observed hundreds of evolved stars in various observing modes. Thanks to its broad and continuous wavelength coverage (2.35–45.2 μm) where is mostly inaccessible from the ground and its resolution power (200 up to 2500), the Short-Wavelength Spectrometer (SWS; de Graauw et al. 1996) has innovated studies of evolved stars.

In this paper we discuss time variations of molecular bands in the spectra of oxygen-rich Mira variables. There are a few related papers in this volume. Cami et al. give general overview of the molecular bands in the ISO/SWS spectra of the oxygen-rich red-giants. Matsuura et al. analyze the SWS spectrum of the RV Tau star, R Sct. Onaka et al. and Sung et al. discuss the time variation of dust bands in the oxygen-rich and carbon-rich red-giants, respectively. Time variations of infrared molecular bands in carbon-rich stars are studied by Hron et al. (1998) and Loidl et al (1999).

2. Time Variation of Infrared Spectra

There are a few ISO proposals which intended to monitor the variable stars (e.g., Onaka et al. 2002; Hron et al. 1998). More observations by different PIs are now available in the ISO data archive (`http://www.iso.vilspa.esa.es/`). In total there are more than 10 variable stars with useful multiple observations.

Figure 1 shows the ISO/SWS spectra of the Mira variable, T Cep from 2.4 to 10 μm. The star was observed seven times between August 1996 to December 1997 with approximately an equal interval. The resultant dataset covers the variability phase from 0.4 to 1.6 with an interval of $\sim$0.2. The infrared spectra of this star are dominated by the H_2O bands. Especially the band around 2.5 μm is prominent. CO bands at 4.3 and 2.3 μm and the SiO first overtone at 4.0 μm are less clear, but do present. Noticeable features in the spectra are those of two dioxides, CO_2 at 4.2 μm and SO_2 at 7.3 μm. They are detected for the first time by the ISO.

It is seen that not only the flux level but also shape of the spectral features changes with the time. The most prominent variations are seen in 3.5–4.0 μm (H_2O), at 4.2 μm (CO_2), and at 7.3 μm (SO_2). They are discussed in section 4.

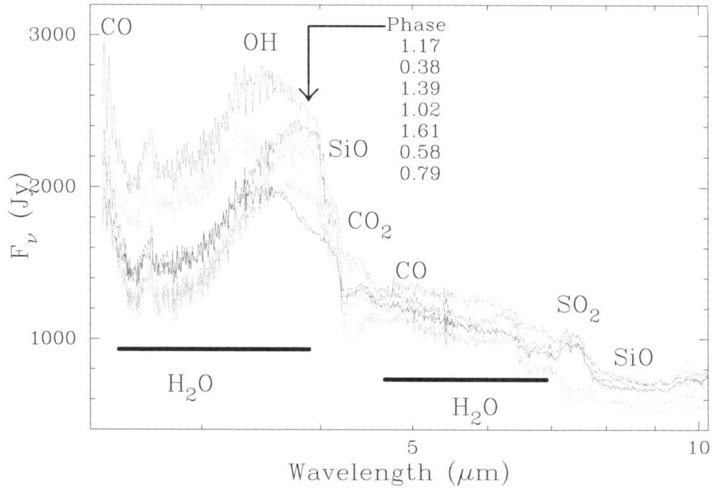

Figure 1. ISO/SWS spectra of the Mira variable, T Cep, at seven different variability phases. The observed phases are indicated in the same order of the spectra at 3.8 μm. Major spectral features are indicated.

3. The Slab-Model Approach

Physically self-consistent models of stellar atmospheres have been constructed and successfully used for various spectral type stars (see, Decin et al. 2000). Inclusion of spherical effects has improved the reality of the models for late-type giants.

What ISO discovered is that the observed spectra quite often deviate from the theoretical predictions. Tsuji et al. (1997) reported an unexpected detection of H_2O absorption bands in an M2 star β Peg. The band is detected in even earlier type stars (Tsuji 2001; Matsuura et al. 1999). Obviously there are something that have not been considered in the conventional atmosphere models, and it is generally thought to be stellar pulsation. The efforts to include the dynamical effects into the model atmospheres have been continued (e.g., Höfner et al. 1998; 1999), and the models have succeeded in some extent to explain the observations at least qualitatively. However, the limitation of the computation facility prevents to establish the models ready for quantitative comparison with the SWS spectra of stars with large variability.

The approach used throughout in this paper is different. In order to grab the most essential nature of the stars, we apply a very simple, empirical model called 'slab-model' (Yamamura et al. 1999a; Matsuura et al. 2002). Figure 2 (right) illustrates the model. We assume a plane-parallel geometry. Each molecular layer has three parameters; an excitation temperature, a column density, and a size of the layer with respect to the background continuum source; the star. Light from the star passes through the molecular layers one after the other and forms a spectrum. If a layer is larger than one behind, it forms additional emission to the spectrum (this represents the extended atmosphere), which fills in the absorption caused at the center. If the layer is large enough, the net spectrum shows emission. We assume a Local-Thermodynamical Equilibrium (LTE) for the excitation of the molecules.

We would like to stress that the slab-model approach does not intend to describe the every processes taking place in the atmosphere and is not competitive with the sophisticated atmosphere models. What the slab-model tells us is that we need such molecular components in order to reproduce the observed spectra. With these results, we have to proceed our study to the next step to consider what are the physical and chemical processes behind. We believe that such information is useful for the construction of dynamical model atmospheres. In addition, the model provides satisfactorily consistent results with a spherical geometry model (see, Matsuura et al. 2002).

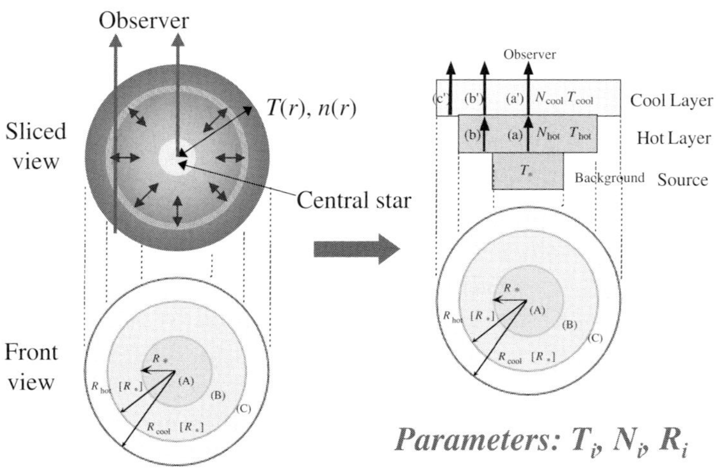

Figure 2. The *slab-model* simplifies some aspects of the stellar atmosphere *(left)*, such as sphericity, variation of the density and the excitation temperature, and velocity field. It describes the atmosphere by overlaying independent, uniform layers with three most important parameters; an excitation temperature, a column density, and a radius *(right)*.

4. Time Variation of Molecular Bands

4.1 H_2O

Yamamura et al. (1999b) analyzed spectra of two Mira variables, *o* Cet and Z Cas, with the slab-model and showed that the spectra of 2.4–4.0 μm region were predominantly represented by a combination of a *hot* (~2000 K) and a *cool* (~1200 K) H_2O layers. *o* Cet was observed by ISO at optical maximum and it showed H_2O emission features in 3.5–4.0 μm. They argued that it is because the hot H_2O layer expands as large as 2 $R_\star$ at maximum. Matsuura et al. (2002) adopted the analysis to 20 spectra of four Mira variables to see time variations of the H_2O layers. Figure 3 shows the derived size of the hot H_2O layer ($R_{\rm hot}$) for four sample stars. In any stars $R_{\rm hot}$ at maximum is as twice large as that at minimum. Note that the *real* size of the layer should be corrected by the absolute size of the continuum source. The stellar radius at the continuum wavelength shrinks by less than 20 % (Bessell et al. 1996) at maximum. This gives the net variation of the layer size of 50–100 %. Matsuura et al. (2002) argue that the change of the layer size is caused by pulsation shocks propagating through the extended atmosphere.

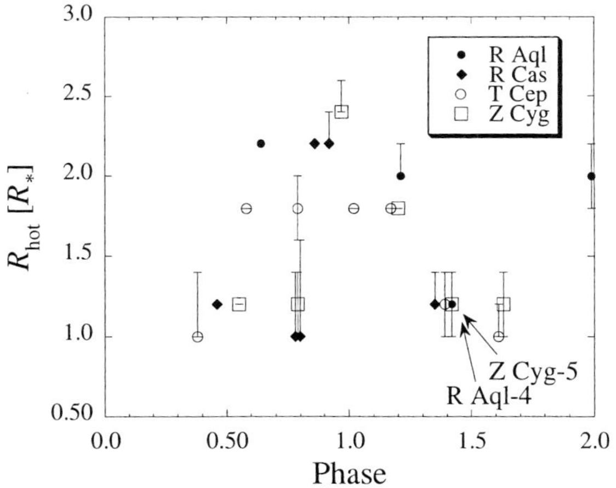

Figure 3 The radius of the *hot* H_2O layer derived by the slab-model analysis for four Mira variables. The radius is in unit of the background continuum source, i.e., the central star. This figure is taken from Matsuura et al. (2002).

4.2 SO_2

Detection of SO_2 ν_2 band at 7.3 μm is first reported by Yamamura et al. (1999a) in three Mira variables, UX Cyg, *o* Cet, and T Cep. Search for the band in the archive data concludes that the molecule is quite common in the atmosphere of red giants (Cami et al. this volume). It is noted that the time variation of the band in T Cep does not coincide with the optical variability. In Figure 1 the band is clearly seen in emission at the first three phases (ϕ =0.38, 0.58, 0.79; around the first minimum) while it is in absorption at the last two observations (ϕ =1.39, 1.61; around the next minimum).

Analysis with the slab-model tells us that the molecule has an excitation temperature of ~600 K and is located at 2–6 $R_\star$. The density of the molecule is estimated in the order of 10^4 cm^{-3}. This implies that the relative abundance of the molecule is as much as or more than 10^{-6}; a large fraction of sulfur atoms should be in the form of SO_2. The result is in contrast to the prediction from the thermal-equilibrium chemistry, which expects a negligible amount of SO_2. Probably, non-equilibrium chemistry is taking place in the region (see Duari et al. 1999; Beck et al. 1992).

4.3 CO_2

It is somewhat surprising that not only the 15 μm ν_2 fundamental band but also several high-excitation bands are prominent in the spectra of M-giants (Justtanont et al. 1998). These *satellite bands* are usually seen in low mass-loss rate stars and are always in emission, while the fundamental band is sometimes seen in absorption. The model analysis shows that the excitation temperatures of these bands are 700–300 K or even lower (Ryde et al. 1999; Cami et al. 2000), indicating that the molecules distribute in wide range in the extended atmosphere and the circumstellar envelope.

No systematic study of time variations of the CO_2 bands has been carried out. The right panel of Figure 4 shows the seven spectra of T Cep in the 15 μm region. It is seen that the high-excitation bands at 13.8 and 16.2 μm are strongest at maximum and weaker in minimum. On the other hand, 15 μm band is seen in absorption at the beginning but gradually turn to emission toward the last observation, just in the opposite way of the SO_2 band. Time variation of the 4.2 μm band is even more complicated. The left panel of Figure 4 shows again the T Cep spectra. Because of contamination of CO and H_2O, it is not straightforward to see the change of the CO_2 band. But it is already obvious that the change does not follow the optical variability. This remind us that the molecules producing the band are located in the similar region with those for the 15 μm band and the SO_2 band.

5. Summary

Time variations of the molecular bands in T Cep are summarized in Table 1. What the present analyses tell us are: (1) Time variations of molecular bands can be classified into two groups; one follows the optical

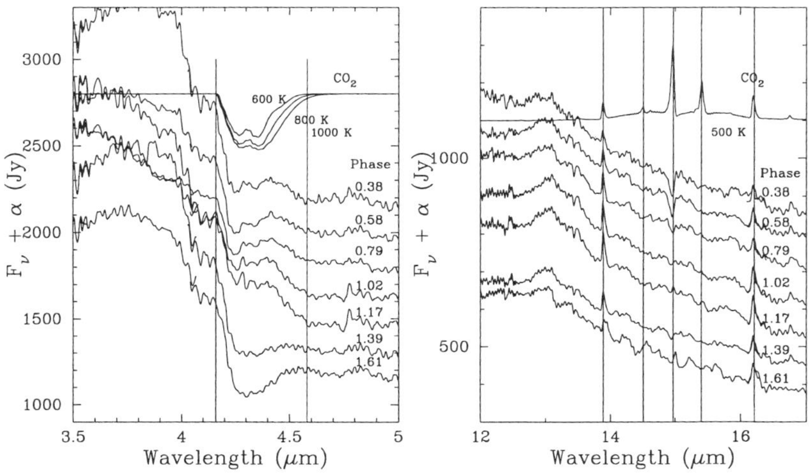

Figure 4. Time variation of CO_2 ν_3 band *(left)* and ν_2 band *(right)* in T Cep. Spectra are scaled arbitrary to see the change of the features clearly.

variability and the other does not. (2) The former group arises in the region closer to the photosphere and hotter than the latter group. The boundary between two regions is at ~3 $R_\star$. This view is consistent with a theoretical calculation by Höfner et al (1998). (3) The hot (~2000 K) molecular layer is extended as large as 2 $R_\star$. (4) Some contribution from the molecules in the circumstellar envelope is expected. For such molecules the time variation is expected to follow the optical variability again, through the radiative excitation. (5) A pulsation shock plays an important role in the extended atmosphere. It pushes up the matters, changes the density and temperature structure, and also triggers non-equilibrium chemical processes.

There are plenty of unsolved problems. Formation of the extended atmosphere of Mira variables has not been fully understood. It is not clear that the same mechanism works for weak pulsators such as SRb and Lb variables. The density of the SO_2 layer at 4–6 $R_\star$ seems to be extremely high. Furthermore, CO_2 and SO_2 are more abundant than in the thermal-equilibrium condition. Time-dependent chemical calculation including shock propagation and probably dust formation process is indispensable. It is possible that non-LTE excitation is taking place in the extended atmosphere (González-Alfonso & Cernicharo 1999).

Improvement of dynamical model atmospheres is in progress. In parallel to such activity, further analysis of observed spectra and investigation of individual physical and chemical processes have to be carried out.

Table 1. Time variations of molecular bands in T Cep.

	Phase	0.38	0.58	0.79	1.02	1.17	1.39	1.61
H_2O	R_{hot}	1.0	1.8	1.8	1.8	1.8	1.2	1.0
	3.8 μm	a	~0	e	e	~a	a	a
SO_2	7.3 μm	e	e	e	e	~0	a	a
				⟶weaker⟶			⟶deeper⟶	
	14 & 16 μm	e	e	e	e	e	e	e
		weak		⟶	strong	⟶		weak
CO_2	15 μm	a	a	a	0	0	0	e
		⟶weaker⟶						
	4.2 μm	———	———	———	very complicated	———	———	———

'a': seen in absorption. 'e': seen in emission. '0': no clear feature recognized.

Acknowledgments

The author acknowledges M. Matsuura for careful reading of the manuscript and the comments. I.Y is supported by Grant-in-Aid for Encouragement of Young Scientists (No. 13740131) from Japan Society for the Promotion of Science.

References

Beck, H.K.B., Gail, H.-P., Henkel, R., & Sedlmayr, E. 1992, A&A,265, 626
Bessell, M.S., Scholz, M., & Wood, P.R. 1996, A&A, 307, 481
Cami, J., Yamamura, I., de Jong, T., et al. 2000, A&A, 360, 562
Decin, L., Waelkens, C., Eriksson, K., et al. 2000, ESA-SP, 456, 289
de Graauw, Th., Haser, L.N., & Beintema, D.A., et al. 1996, A&A, 315, L49
Duari, D., Cherchneff, I., & Willacy, K. 1999, A&A, 341, L47
González-Alfonso, E., & Cernicharo, J. 1999, in ESA SP-427, The Universe as seen by ISO, eds.P. Cox, M.F. Kessler, (Noordwijk), 325
Höfner, S., Jørgensen, U.G., Loidl, R., & Aringer, B. 1998, A&A, 340, 497
Höfner, S. 1999, A&A, 346, L9
Hron, J., Loidl, R., Höfner, S., et al. 1998, A&A, 335, L69
Justtanont, K., Feuchtgruber, H., de Jong, T., et al. 1998, A&A, 330, L17
Kessler, M., Steinz, J.A., Anderegg, M.E., et al. 1996, A&A, 315, L27
Loidl, R., Höfner, S., Jørgensen, U.G., & Aringer, B. 1999, A&A, 342, 531
Matsuura, M., Yamamura, I., Murakami, H., et al. 1999, A&A, 348, 579
Matsuura, M., Yamamura, I., Cami, J., et al. 2000, A&A, 383, 972
Onaka, T., de Jong, T., & Yamamura, I. 2002, A&A, 388, 573
Ryde, N., Eriksson, K., & Gustafsson, B. 1999, A&A, 341, 579
Tsuji, T. 2001, A&A, 376, L1
Tsuji, T., Ohnaka, K., Aoki, W., & Yamamura, I. 1997, A&A, 320, L1
Yamamura, I., de Jong, T., Onaka, T., et al. 1999a, A&A, 341, L9
Yamamura, I., de Jong, T., & Cami, J. 1999b, A&A, 348, L55

THE CIRCUMSTELLAR ENVIRONMENT OF AGB STARS STUDIED BY POLARIMETRY

Pierre Bastien
Département de physique, and Observatoire du mont Mégantic,
Université de Montréal, B.P. 6128, Succ. Centre-ville
Montréal, QC H3C 3J7 Canada
bastien@astro.umontreal.ca

Abstract Polarimetry of AGB stars is a very useful tool for learning about the distribution of their circumstellar material. After a brief overview of mechanisms and techniques for dealing with polarimetry of AGB stars, specific polarization observations that illustrate each technique are presented. The data show that most stars do not have a spherical envelope, but rather that their circumstellar material is found to be mostly axially symmetric. A bipolar geometry is quite common for these stars.

Keywords: Circumstellar material, dust grains, polarization, scattering

1. Introduction

This review of the polarization properties of AGB stars is meant to be illustrative rather than exhaustive. Some topics, for example, the polarization at radio wavelengths, are not be covered here. Previous reviews dealing with polarimetry of AGB stars are by Magalhaes (1988) and by Coyne (1988). Other papers in the same book might also be of interest.

2. Polarizing Mechanisms and Techniques

2.1 Polarizing Mechanisms

In this review, we will be concerned with three mechanisms that give rise to polarization. A mechanism which is particularly effective in the interstellar medium is dichroic extinction by non-spherical aligned grains. There are many alignment mechanisms; many of them include

Y. Nakada et al. (eds.), Mass-Losing Pulsating Stars and their Circumstellar Matter, 181–188.

a magnetic field. Grains can have various shapes but for our purpose they can be modelled as oblate or prolate. Collisions in the presence of a magnetic field act such that the short axis of the grains will precess around the direction of the field. The components of the electric field of the radiation will be more absorbed along the longest direction of the grain. The net result is a radiation beam polarized parallel to the magnetic field. The interstellar polarization has a smooth bell-shaped wavelength dependence which is described by the Serkowski law in the visible and the near-infrared.

When grains are aligned, their thermal emission is not symmetric along the line of sight. The geometry is such that light is polarized perpendicular to the field, as shown in Figure 1. This is mostly important in the K band and longward, including the submm where the emission comes from cold grains in the cores of molecular cores.

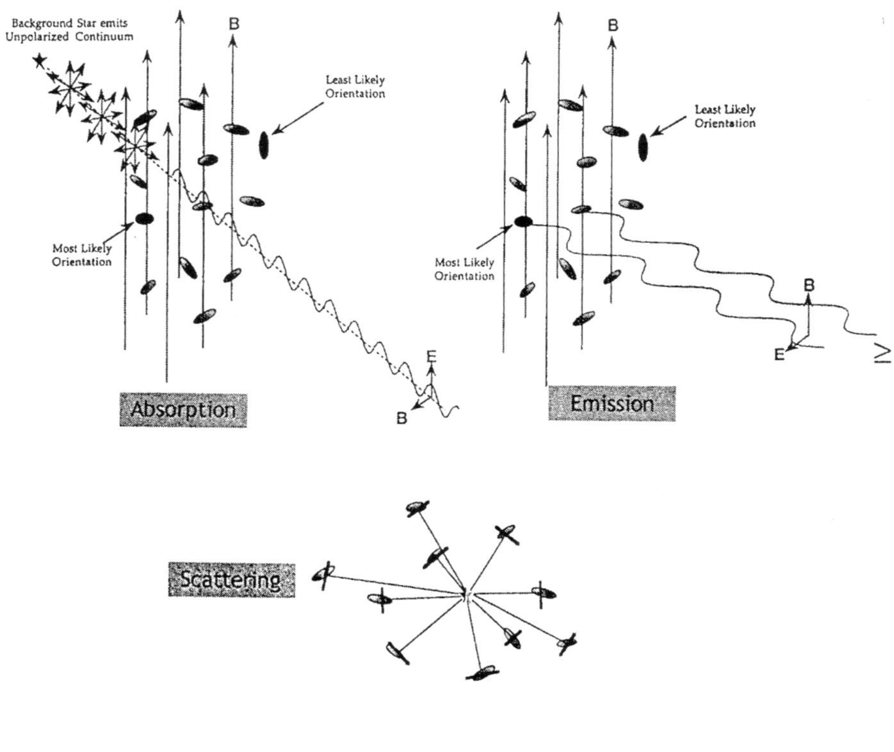

Figure 1. Illustration of how the absorption, the emission, and the scattering of light produces polarization. From Weintraub et al. (2000a).

Light can also be scattered by dust grains, by molecules and by electrons. The scattering plane is defined by the light source (i.e., the star), the scatterer and the observer. Usually, the beam is polarized perpendicular to the scattering plane. This fact can be used to find the position of the illuminating source (see below). For spherical grains, the scattering process is described by the Mie theory. Scattering can also be computed for ellipsoidal grains, and even for irregular grains with the T-matrix formulation. Multiple scattering can be important for optical depths as small as 0.1.

In some sources, many mechanisms can contribute to the observed polarization. In the visible and near-infrared, the stellar light can be scattered close to the star and then polarized by the interstellar medium. By measuring the polarization as a function of wavelength and time, one can usually decipher the various contributions.

2.2 Techniques

One can divide the different techniques used into three different categories: aperture polarimetry, also called broad-band polarimetry, where the polarization of the light passing through a diaphragm is measured. When a bidimensional detector such as a CCD or an infrared array is used, one gets a polarization map. Spectropolarimetry is obtained by using a spectrograph with a polarizing device. Good books dealing with polarization in Astronomy are by Tinbergen(1996), and by Leroy (1998).

3. Observations

3.1 Aperture Polarimetry

CO line profiles provide a diagnostic for spherically symmetric winds by probing the velocity field of AGB stars. Stars with "normal" symmetric CO lines with sharp edges have a spherical envelope while those with "abnormal" profiles have non-spherical winds. Kahane et al (1997) observed the linear polarization of a sample of 68 stars at the Mont-Mégantic Observatory between 1994 and 1996. The sample included 39 "abnormal" and 18 "normal" profile candidates. One would expect a correlation since polarization is also a diagnostic of symmetry in the distribution of the circumstellar (CS) material. The histogram of the polarization for the two samples are shown in Figure 2. A Kolmogorov-Smirnov test confirms what we see: there is no significant difference between the two samples. The interpretation by Kahane et al is that the difference is due to projection effects: the most favourable orientations for high polarization and anomalous CO profiles are in fact perpendicular to one another.

However, the data revealed an unexpected correlation, shown in Figure 3: Carbon-rich envelopes are less polarized than Oxygen-rich en-

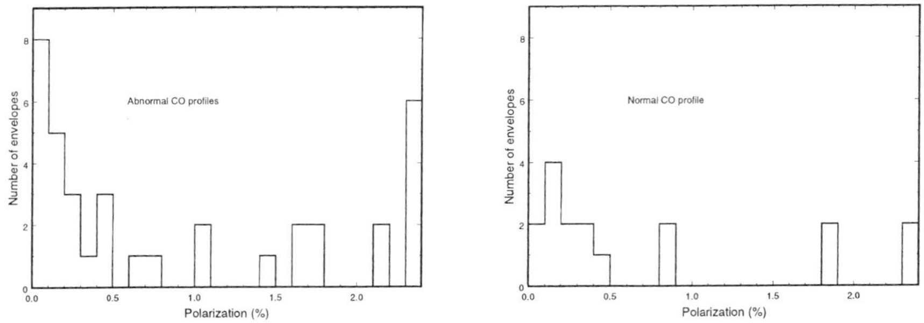

Figure 2. Histograms of the polarization distribution for AGB stars with abnormal and normal CO line profiles. Stars with a polarization larger than 2.5% are all included in the last bin.

velopes. The most likely explanation is that the silicates in the envelopes of O-rich stars are less efficient for scattering and polarizing light than carbon grains around in C-rich stars.

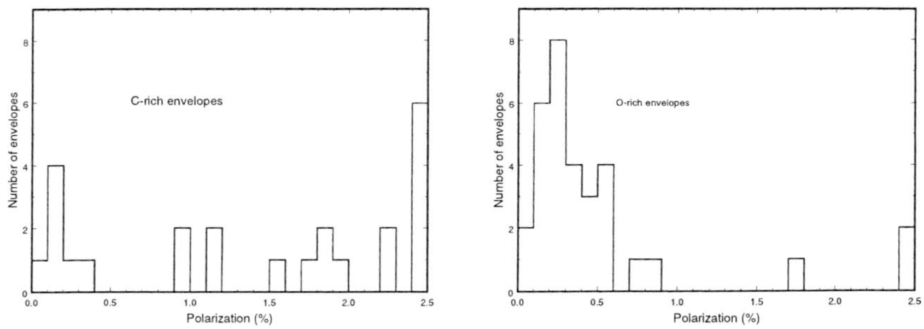

Figure 3. Histograms of the polarization distribution for AGB stars with Carbon-rich and Oxygen-rich envelopes. Stars with a polarization larger than 2.5% are all included in the last bin. The two histograms are clearly different from each other.

The CO data can be used to estimate the mass loss rate of the stars (Loup et al 1993). Stars with a high polarization are found to have on average a higher mass loss rate, which could mean that the scattering efficiency increases with envelope density.

The wavelength of polarization was also measured for 18 AGB stars and only one star was found to display the Serkowski interstellar polarization law, a clear indication that the polarization in these stars in mostly intrinsic (although in many stars there is also an IS contribution).

Finally, the polarization was found to be time-variable in 17 out of 31 stars with enough measurements. The position angle is also variable in many cases. However, no correlation was found with the light curve.

Another example is the carbon star IRAS 06088+1909 observed by Richichi et al (1998). The $P(\lambda)$ data can not be fitted by the Serkowski law; the polarization at K is perpendicular to that at R. The model, based also on lunar occultation data, include a small disk; the optical depth is large at R so that P results from single scattering by material in the lobes. At K the optical depth is small and hence the polarization in perpendicular to the disk because it arises from single scattering in the disk.

3.2 Imaging Polarimetry

Imaging polarimetry can be used to find the position of the illuminating source, even if it is not visible directly because the optical depth to the source is too large. This has been done with HST by Sahai et al (1999) for the proto-planetary nebula (PPN) Roberts 22 and by Weintraub et al (2000b) for the well-known Egg nebula (Figure 4). In the latter case, single scattering produces a centro-symmetric pattern of polarization vectors around the illuminating source. However, they found that the illuminating source is not the (obvious) candidate source A at the tip of the northern lobe. Source A is found to be extended and if it is a companion to the post-AGB at position B, it is too far ($>$ 550 AU) to have had any influence in producing the bipolar geometry.

For Roberts 22, the PPN shows bipolar lobes with a centro-symmetric pattern. The polarization reaches levels of 40-50%, evidence for single scattering in these regions. The central region shows a complex polarization pattern and multiple scattering must be involved there.

Gledhill et al (2001) observed and classified 16 PPN into bipolar, shell and core types. There is clear overall evidence for axisymmetric distribution in most of the sources. Delgado (this conference) presented polarization maps of some AGB stars.

3.3 Spectropolarimetry

Landstreet & Angel (1977) were the first ones to draw attention to the fact the polarization in Mira variables changes across atomic or molecular absorption features. This shows that there are at least two intrinsic components to the polarization: one in the continuum arises from the CS dust shell, and another one which must arise in the stellar atmosphere.

Further spectropolarimetry of AGB stars was presented by Trammell et al (1994) and showed clearly changes in polarization across the Hα

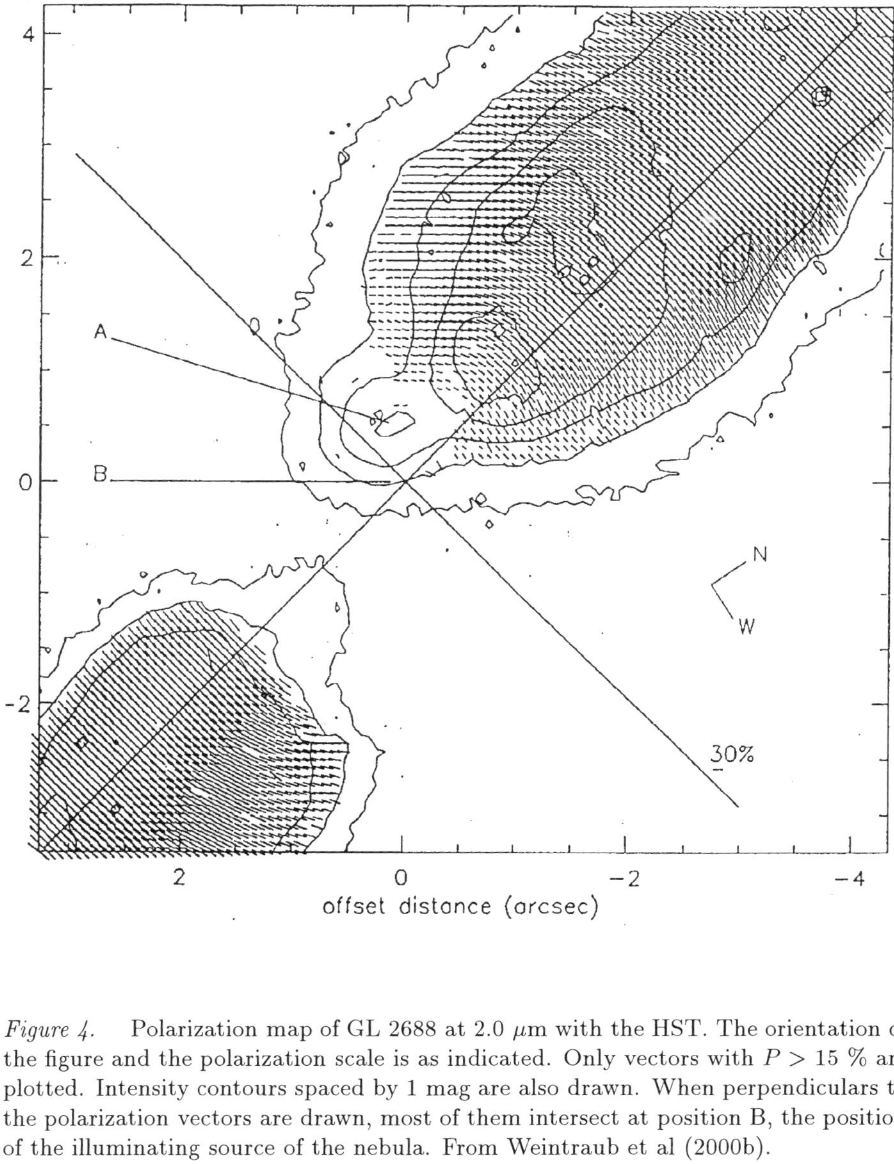

Figure 4. Polarization map of GL 2688 at 2.0 μm with the HST. The orientation of the figure and the polarization scale is as indicated. Only vectors with $P > 15$ % are plotted. Intensity contours spaced by 1 mag are also drawn. When perpendiculars to the polarization vectors are drawn, most of them intersect at position B, the position of the illuminating source of the nebula. From Weintraub et al (2000b).

and Na I emission lines, again a proof for two mechanisms for producing the polarization (see Figure 5).

4. Interpretation

The polarization vector of most AGB stars is variable in both its magnitude and its direction. This polarization arises from scattering by

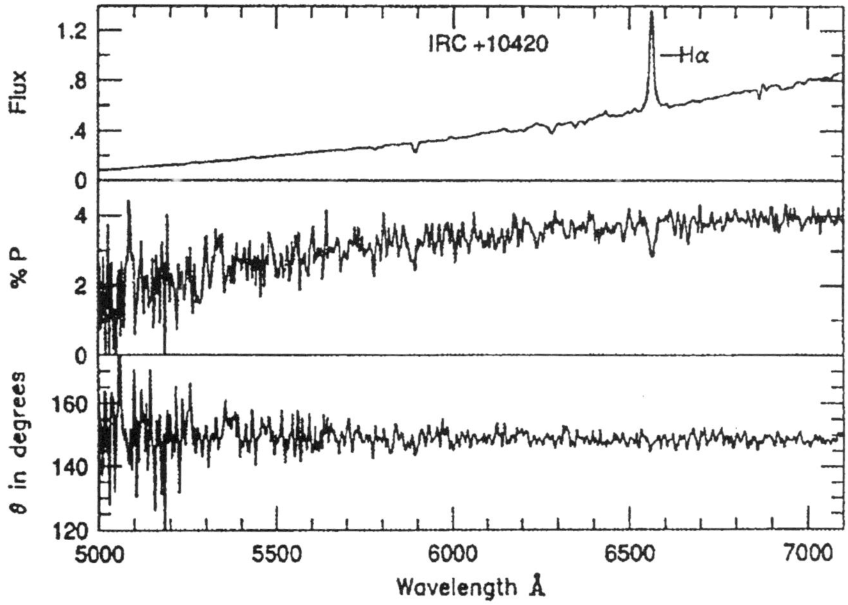

Figure 5. The total flux, the polarization, and its position angle as a function of wavelength for IRC+10420. The flux spectrum is normalized to $6x10^{-13}$ erg $\AA^{-1}$ cm^{-2} s^{-1}. P changes across Hα, but the position angle stays constant at this wavelength. From Trammell et al (1994).

circumstellar material. The distribution of the material can not be spherically symmetric, otherwise there would be no polarization. The position angle variations in many cases are not random but wander around a particular value; this is indicative of an axis of symmetry. A bipolar geometry can be seen directly on many images, which supports this interpretation.

The polarization can be quite large in the bipolar lobes, as a result of single scattering in these regions. Closer to the central source, the polarization is typically 10-20%, due to multiple scattering in regions of large optical depths. This global picture of the CS material of AGB stars is similar to pre-main sequence stars (Bastien 1996).

In addition to these contributions from the CS material, there is also, at least in some stars, some polarization arising in the atmosphere and the wind very close to the star, as shown by spectropolarimetry. Hence, there is already as the material leaves the star, a significant asymmetry in many AGB stars.

Acknowledgments

The author wishes to thank Drs. M. Seki, and M. Matsumura for their kind invitation and the SOC for financial support. This research is supported by NSERC.

Appendix: Questions

P. Whitelock. Are there more stars that show a change in polarization across the Na I line?

In addition to IRAS 20056+1834 which I showed, there is also GL 2699 in the same paper by Trammell et al. Landstreet & Angel also have a few stars that show this effect in Na I. There may be others in other papers.

M. Seki. You mentioned the importance of multiple scattering in AGB star environment. Can we expect the case where the degree of polarization may be increased due to multiple scattering effect as compared with the case of single scattering?

Very good point. If there is an optically thick disk around the central star, it is impossible to view the star when it is edge-on. Instead, we see light that has made its way around the disk, by being scattered above or below the disk, and then on the near side of the disk. The final polarization is thus perpendicular to the last scattering plane, or parallel to the disk. The polarization can reach up to 20% in this region. If the optical depth is smaller, for example in the K band, then we may see the star directly, and the light may be scattered mostly once by dust grains in the plane of the disk. The polarization is perpendicular to the disk.

References

Bastien, P. 1996. Polarimetry of the Interstellar Medium. W. Roberge, D. C. B. Whittet, eds., San Francisco: Astronomical Society of the Pacific, p. 297

Coyne, G. V. 1988. Polarized Radiation of Circumstellar Origin. Coyne et al. eds., Vatican City: Vatican Observatory, p. 511

Gledhill, T. M., Chrysostomou, J. H., Hough, J. H., & Yates, J. A. 2001, MNRAS, 322, 321

Kahane, C., Viard, E., Ménard, F., Bastien, P., & Manset, N. 1997, ApSS, 251, 223

Landstreet, J. D., & Angel, J. R. P. 1977, ApJ, 211, 825

Leroy, J.-L. 1998. La polarisation de la lumière et l'observation astronomique. Amsterdam: Gordon and Breach.

Magalhaes, A. M. 1988. Polarized Radiation of Circumstellar Origin. Coyne et al. eds., Vatican City: Vatican Observatory, p. 461

Richichi, A., Stecklum, B., Herbst, T. M., Lagage, P.-O., & Thamm, E. 1998, A&A, 334, 585

Sahai, R., Zilstra, A., Bujarrabal, V., & Hekkert, P. te L. 1999, ApJ, 117, 1408

Tinbergen, J. 1996. Astronomical Polarimetry. Cambridge: Cambridge University Press.

Trammell, S. R., Dinerstein, H. L., & Goodrich, R. W. 1994, AJ, 108, 904

Weintraub, D. A., Goodman, A. A., Akeson, R. L. 2000a. Protostars and Planets. IV., V. Manning, A. P. Boss, S. Russell, eds., Tucson: Univ. of Arizona Press, 247

Weintraub, D. A., Kastner, J. H., Hines, D. C., Sahai, S. 2000b, ApJ, 531, 401

VIOLET SPECTRA OF CARBON STARS ASSOCIATED WITH SILICATE FEATURES

Hideyuki Izumiura
Okayama Astrophysical Observatory
National Astronomical Observatory of Japan
izumiura@oao.nao.ac.jp

Abstract We have observed the violet spectra of three carbon stars, BM Gem, V778 Cyg, and EU And, that are associated with silicate dust emission features as well as that of a proto-typical J-type carbon star, Y CVn, with the high dispersion spectrograph (HDS) on SUBARU telescope. The spectrum of BM Gem has revealed the presence of Balmer lines up to H_{23}, all of which are found to have P-Cyg type line profile, as well as the Paschen and Balmer continuum emission. The blue-shifted absorption gives the gas outflow velocity of about 400 km s^{-1}, that is very unusual for a cool carbon star. Our results strongly suggest the existence of a low luminosity companion surrounded by an accretion disk in BM Gem system, which have long been postulated but has never been confirmed. We have not detected either continuum or line emission in the violet region shortward of 4000Å in the other two silicate carbon stars.

Keywords: Stellar evolution, mass-loss, thermal pulses, carbon stars, silicates, Balmer lines, BM Gem

Introduction

There is a small group of carbon stars that show silicate dust emission features in the mid-infrared at 10 and 18 μm, so called "silicate carbon" stars (Little-Marenin 1986, Willems and de Jong 1986). Silicates are the signature of oxygen-rich chemistry in the circumstellar dust shell, while the designation of a carbon star means the presence of C_2 and CN band spectra due to carbon-rich chemistry in the photosphere. The oxygen-rich chemistry in their circumstellar envelopes was immediately confirmed by the detection of 22 GHz water vapor masers (Benson and Little-Marenin 1987, Nakada, Izumiura, Onaka et al. 1987). Later, carbon stars showing silicate absorption features (Jiang et al. 2000) and

Y. Nakada et al. (eds.), Mass-Losing Pulsating Stars and their Circumstellar Matter, 189–196.

a crystalline silicate dust shell (Molster et al. 2001) were also discovered with the Infrared Space Observatory (Kessler et al. 1996).

The co-existence of oxygen-rich and carbon-rich chemistries in a single system is controversial. Two explanations were immediately proposed. One was the binary system consisting of a carbon star and a dust enshrouded M-type giant (OH/IR) star (Benson and Little-Marenin 1987). The other explanation invoked a transition from an oxygen-rich star to a carbon star by the third dredge-up of carbon produced by the thermal pulses in the stellar interior, later mixed to the surface (Willems and de Jong 1986, 1988). These earlier scenarios were, however, rejected on the basis of the steadinesses of the color-indices and silicate features (Lloyd-Evans 1990) as well as of the absence of the spectroscopic signatures of a luminous M-type (OH/IR) companion (Lambert et al. 1990).

Now there are two revised pictures for the peculiar silicate carbon stars. Kahane, Barnbaum, Uchida et al. (1998) and Jura and Kahane (1999) detected CO emission lines of unusually narrow widths in BM Gem and EU And, respectively. They proposed the existence of a circumbinary reservoir of oxygen-rich gas (and dust) around the stars, which is presumably given rise to by the interaction with an unseen companion to explain the unusually narrow CO widths. Yamamura, Dominik, de Jong et al. (2000) analyzed the ISO spectrum of V778 Cyg to find that the silicate emission features did not change in the last 14 years, and proposed a circumstellar disk of oxygen-rich material formed around an invisible companion when the primary was an oxygen-rich mass-losing star. They concluded that the silicate features emanated from the oxygen-rich material blown out from the disk by the primary.

It is now very important to examine if they indeed have a companion star, in order to understand the origin of silicate carbon stars. It would be very difficult, however, to detect a postulated low luminosity companion in the optical or in the longer wavelength bands because carbon stars are very bright intrinsically there. It is necessary to examine other unexplored wavelength regions. In this respect, it has long been known that cool carbon stars exhibit extreme violet flux deficiency (Shane 1928), the agent of which has not been identified yet explicitly ("violet opacity problem"). Indeed cool carbon stars are very dim in the violet (eg. U-V of Y CVn is 8.9, while those of a M5 giant and a dwarf are 4.2 and 2.8, respectively). This problem can be turned into an advantage if we try to detect an assumed companion in the violet and ultraviolet. Some certain signatures of the companion star may be revealed in the violet spectral region. We thus performed a high sensitivity spectroscopic observations of three silicate carbon stars in the violet spectral region with HDS on SUBARU.

1. Observations

Observations were made with the high dispersion spectrograph (HDS, for details readers should consult Noguchi et al. 2002) installed at the Nasmyth focus of SUBARU telescope (Kaifu et al. 2000). We used an atmospheric dispersion corrector. A log of observation is shown in Table 1. We observed the spectral range between 3550 – 5200Å simultaneously with the two mosaiced CCDs of 2048×4100 pixels from EEV. On-chip binning by 2×2 pixels was done for V778 Cyg and EU And to improve the S/N ratios in the violet region. A bright J-type carbon star, Y CVn was also observed for comparison, because all silicate carbon stars examined spectroscopically were found to be J-type. J-type stars are those showing low $^{12}C/^{13}C$ abundance ratios in the atmosphere. A flux calibration source, Feige34 was also observed but with a slightly different setting which gives a coverage of 3100 – 4750Å.

Table 1. Log of observations

Star	Slit Width	Resolution	Exposure	Date(UT)	Seeing	secZ
BM Gem	0.72"	50,000	1,800s×2	2001-01-29	~0.80"	1.01
	0.72"	50,000	1,800s×1	2001-04-14	~0.85"	1.20
V778 Cyg	1.0"	38,000	2,400s×1	2001-07-30	~1.2"	1.42
EU And	1.0"	38,000	2,400s×1	2001-07-30	~1.2"	1.13
Y CVn	0.72"	50,000	900s×1	2001-02-01	~0.80"	1.18
Feige 34	4.0"	9,500	300s×1	2001-01-28	~0.60"	1.10

Data reductions were performed using the echelle package on NOAO IRAF. Standard procedures were taken. Cosmic ray events were removed manually. Flux calibration has not been applied to the spectra in the following figures.

2. Results and Discussion

The resulting spectra reduced to a resolution of 1Å of BM Gem, V778 Cyg, and EU And are shown in Figure 1 together with that of Y CVn. We scaled them by adjusting the slope of the spectrum between 4050Å and 4200Å for comparison. The spectrum of BM Gem was found to be very intriguing. It has a significant level of continuum emission in the violet, which is clear when it is compared with Y CVn which does not show such continuum emission. Both Paschen and Balmer continuum emission is observed clearly. It also shows a series of emission lines, which have been identified with Hydrogen Balmer lines from H_γ to H_{23}. These spectral features resemble to those of a certain kind of planetary

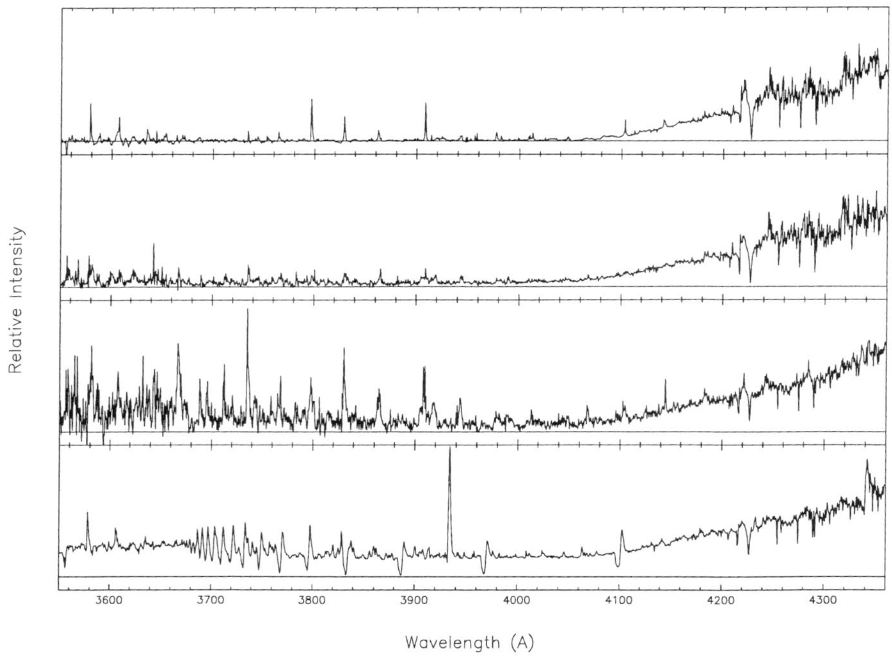

Figure 1. Violet spectra of target stars. From top to bottom, Y CVn, EU And, V778 Cyg, and BM Gem (Jan'01). The horizontal lines shows the zero level.

nebulae and those of YY Ori type young stellar objects. What is more interesting is that all the lines show P-Cyg type line profile. Strong CaII K emission was also detected but CaII H emission was absent. We should note here that the emission-like features seen below 4000Å in Y CVn, EU And, and V778 Cyg are artifacts due to ghost spectra of a much longer wavelength region given rise to by the cross-disperser grating. The ghost spectra run on the CCD with inclinations different from the true spectra, and cross them periodically, which results in a series of emission-like features in the extracted one-dimensional spectrum. They also show up in BM Gem, but do not stand out due to the considerable continuum emission.

In Figure 2, we show close-up's of the vicinity of H_δ, H_ϵ, and H_8 lines together with CaII K line, observed on two nights separated by 75 days. In the figure, we employ the systemic velocity of 73.2 km s^{-1} (V_{LSR}) of BM Gem (Kahane, Barnbaum, Uchida et al. 1998, Jura and Kahane 1999) obtained from the mm-wave CO emission lines. The line of sight velocity of BM Gem determined by optical spectroscopy coincides with that of CO lines (Barnbaum et al. 1992). The line profiles show that the emitting gas is expanding at a velocity of 400 km s^{-1}. Besides, variability of the Balmer lines, in particular, in the line of sight velocity of the absorption component is evident in the figure. On the other hand, spectral features in the region dominated by the photospheric CN and C_2 band absorption lines reveal little variation in radial velocity as well as in profile.

We estimate the continuum level of BM Gem by comparing it with a standard star, Feige 34. At around 3646Å (Balmer limit), we have obtained a flux density ratio (BM Gem/Feige 34) of 6.9×10^{-3} (uncertainty should be less than a factor of 1.5), assuming the slit transmission efficiencies of 60% and 100% for BM Gem and Feige34, respectively, based on the slit width and the seeing size. Since Feige34 is a DA white dwarf having U=9.6, the brightness of BM Gem becomes U=15.0±0.5. Naive estimates of the distance to BM Gem are found by comparing its infrared flux densities with those of Y CVn, assuming that they have the same intrinsic observational properties except for the silicate features in the mid-infrared. A reliable parallax of Y CVn measured by Hipparcos is 4.59 mas, corresponding to a distance of 218pc. The flux ratios in I and K bands of Y CVn against BM Gem are 31.6 and 30.8, respectively, while those in the IRAS bands are 7.9, 3.6, 6.1 and 6.1 for 12, 25, 60, and 100 μm bands, respectively. We assume that the true distance ratio resides in between sqrt(3.6) and sqrt(31). Then the distance to BM Gem falls in between 400 pc and 1200 pc. We then find the absolute magnitude of BM Gem at U-band, M_U(BM Gem) in between +7.0 and +4.6 (no interstellar extinction correction applied). In a white dwarf which is hot enough to ionize the surrounding gas (DA white dwarf), M_U ranges from +9.1 to +11.1 for the effective temperature from 25000 K to 9700 K. It is therefore unlikely that a white dwarf is the direct source of the violet continuum light. The slope of the violet continuum does not match a white dwarf spectral energy distribution either. Though main sequence stars of G0, G5, and F5 have M_U of 5.03, 5.99, and 3.85, respectively, it is also unlikely that such a main sequence star is the source of the violet continuum. We do not see any spectral features typical for those spectral types in the observed spectrum. We therefore conclude that the source of the observed violet flux is neither a white dwarf nor a main sequence star of a spectral type earlier than K. The source of the continuum emission is probably the ionized region itself.

We now consider a consistent picture of BM Gem. The observational facts show that BM Gem is accompanied by an ionized, extended gas region which produces both the continuum emission and the Balmer lines with variable P-Cyg profile. It is natural now to assume the presence of an unseen, low-mass companion to BM Gem. The outflow from the primary carbon star, BM Gem is perhaps accreted by the companion,

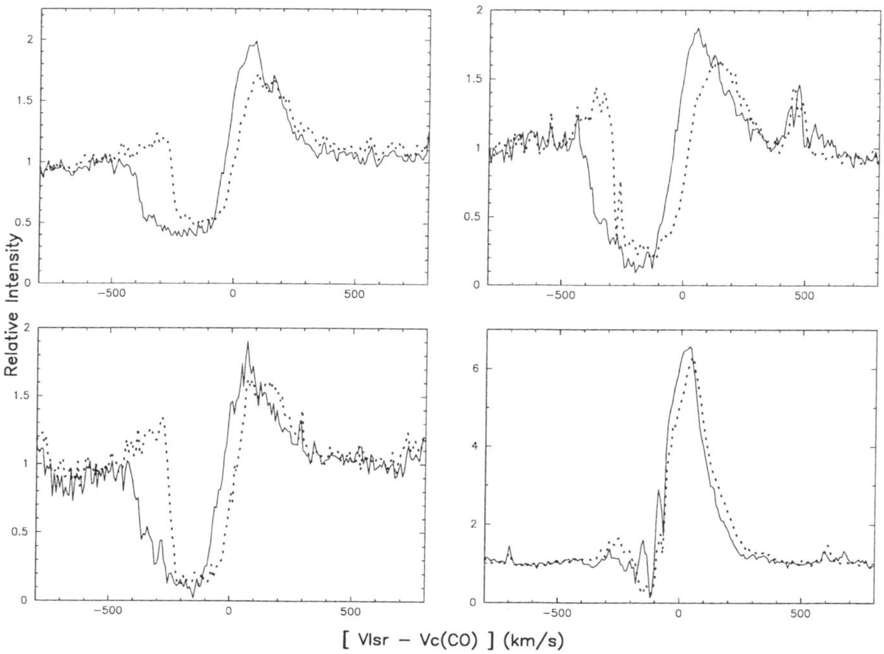

Figure 2. H_δ (top-left), H_ϵ (top-right), H_8 (bottom-left), and CaII K (bottom-right) lines in BM Gem observed on Jan 29, '01 (solid line) and April 14, '01 (dotted line). The spectra are normalized to the local continuum.

producing an accretion disk, of which inner part should be heated by the dissipation of the gravitational potential energy of the accreted matter. The hot region then emits ionizing photons, producing the observed ionized gas region. The accretion disk should also drive the high velocity, variable, rather spherical outflow discovered in this study. It perhaps happens very proximate to the center of the disk, like those jets from dwarf novae or infalls onto YY Ori type young stellar objects. It is, however, still very difficult to identify the precise location of the dust grains responsible for the silicate emission features, in spite of our discoveries here. Both a detailed analysis of our data and further observations at various wavelengths are necessary to pin down the location of the silicate emission region. We would like to notice that our picture does not rule out either a low luminosity main sequence star such as a M-dwarf or a white dwarf as the companion, at this moment.

3. Concluding Remarks

We observed the violet spectra of three silicate carbon stars using the high dispersion spectrograph on SUBARU telescope, as well as of a

Table 2. Comparison of the observed properties of the three silicate-carbon stars

Star Name	H_2O[1] maser	OH[2] maser	CO[3] (narrow)	Tc[4]	Violet[5] Cont.	Balmer[5] Lines	Vrad[6] Var.
BM Gem	N	N	Y	Y	Y	Y	6 km s^{-1}?
EU And	Y	N	Y	Y	N	N	5 km s^{-1}?
V778 Cyg	Y	Y	N	N	N	N	N

1: Benson&Little-Marenin 1987, Nakada et al.1987, 2: Little-Marenin et al.1988, 3: Kahane et al.1998, Jura and Kahane 1999, 4: Barnbaum et al.1991, 5: This work, 6: Barnbaum 1992

J-type carbon star, Y CVn. Balmer continuum as well as Paschen continuum emission has been detected in the region shortward of 4000Å in BM Gem. Balmer lines are also present up to H_{23} in the spectrum, all of which show P-Cyg type line profile. The gas outflow velocity amounts to 400 km s^{-1}. The absorption component of the P-Cyg profile shows a significant radial velocity variation in 75 days. Strong CaII K emission is also present while CaII H line absent. The overall spectra resembles to those of planetary nebulae and in particular, those of YY Ori stars, except that they show inverse P-Cyg profiles. These spectral features have never been observed in other cool AGB carbon stars. They suggest an ionized, high velocity outflow in a compact geometry. With the above findings combined with the previous studies in the literature we conclude that BM Gem is associated with an invisible low luminosity companion giving rise to an accretion disk which forms hot regions to ionize the surrounding Hydrogen gas, and at the same time causes a high velocity, perhaps rather spherical, gas outflows. The violet brightness does not rule out either possibility for the companion of a low luminosity dwarf star or a white dwarf. The nature of the outflow may be helpful to discriminate the two possibilities. In V778 Cyg and EU And both violet continuum and Balmer lines are absent. We summarize the observed properties of the three stars in Table 2. BM Gem, V778 Cyg, and EU And seem to have different circumstellar conditions, conforming a sequence of BM Gem -> EU And ->V778 Cyg. If it is real, it should be related to the difference in the binary parameters or the evolutionary status of those silicate carbon stars.

In order to further explore the ionized region of the circumstellar environment of BM Gem, ultraviolet spectroscopy with HST would be very interesting. Detecting radio continuum emission and recombination lines with VLA will also be intriguing, which will enable us to derive the density and temperature of the ionized gas. Direct imaging spectroscopy at 10 and 20 μm with COMICS on SUBARU may reveal the spatial structure of the emitting region of the silicate dust features. Further monitoring will also be fruitful of the radial velocities of silicate carbon stars in the optical.

Acknowledgments

This work is based on data collected at Subaru Telescope, which is operated by the National Astronomical Observatory of Japan. The author is grateful to the staff members of SUBARU telescope. We used the SIMBAD database. This work was supported in part by the Grant-in-Aid from the Ministry of Education, Culture, Sports, Science and Technology (No.13640247).

References

Barnbaum, C., Morris, M., Likkle, L., & Kastner, J. H. 1991, A&A, 251, 79
Barnbaum, C. 1992, AJ, 104, 1585
Benson, P.J., & Little-Marenin, I. R. 1987, ApJ, 316, L37
Jiang, B. W., Szczerba, R., & Deguchi, S. 2000, A&A, 362, 273
Jura, M., & Kahane, C. 1999, ApJ, 521, 302
Kahane, C., Barnbaum, C., Uchida, K., Balm, S.P., & Jura, M. 1998, ApJ, 500, 466
Lambert, D.L., Hinkle, K.H., & Smith, V.V. 1990, AJ, 99, 1612
Kaifu, N., Usuda, T., Hayashi, S. S. et al. 2000, PASJ, 52, 1
Kessler, M. F., Steinz, J. A., Anderegg, M. et al. 1996, A&A, 315, L27
Little-Marenin, I. R. 1986, ApJ, 307, L15
Little-Marenin, I. R., Benson, P. J., & Dickinson, D. F. 1988, ApJ, 330, 828
Lloyd-Evans, T. 1990, MNRAS, 243, 336
Molster, F.J., Yamamura, I., Waters, L. B. F .M., Nyman, L.-A., Kaufl, H.-U., de Jong, T., & Loup, C. 2001, A&A, 366, 923
Nakada, Y., Izumiura, H., Onaka, T., Hashimoto, O., Ukita, N., Deguchi, S. & Tanabe, T. 1987, ApJ, 323, L77
Noguchi, K., Aoki, W., Kawanomot, S., Ando, H., Honda, S., Izumiura, H., Kambe, E., Okita, K., Sadakane, K., Sato, B., Takada-Hidai, M., Tanaka, W., & Watanabe, E. 2002, PASJ, submitted
Shane, C.D. 1928, Lick Obs. Bull., 13, 123
Willems, F.J., & de Jong, T. 1986, ApJ, 309, L39
Willems, F.J., & de Jong, T. 1988, A&A, 196, 173
Yamamura, I., Dominik, C., de Jong, T., Waters, L.B.F.M., & Molster, F.J. 2000, A&A, 363, 629

MULTIPLE SHELLS AROUND IRC+10216

VLT/FORS1 observations

Patrick de Laverny
Observatoire de la Côte d'Azur, Dpt. Fresnel, UMR 6528
BP4229, F-06304 Nice cedex 4, France
laverny@obs-nice.fr

Abstract The shell-like dusty envelope of IRC+10216 has been observed in the visible band with the VLT/FORS1 instrument. The collected deep and high resolution images have allowed us to map simultaneously for the first time both the central (inner 10 arcsec) and the most external (up to 3 arcmin) regions of the envelope.

Keywords: AGB stars, circumstellar envelopes, IRC+10216

Introduction

IRC+10216 is the brightest and best-studied AGB carbon star. Due to its very large mass-loss rate, a huge dusty envelope extending up to several thousands stellar radii is formed. This envelope is externally illuminated by the interstellar radiation field (ISRF). Scattered ISRF photons can be detected in visible light revealing the almost spherical but non-uniform nature of the IRC+10216 envelope (see Mauron & Huggins, 1999, 2000). The ISRF as well as the dust properties in the IRC+10216 environment can then be studied precisely (see Mauron et al., 2002).

We present here new observations of this envelope in the V-band collected with the VLT/FORS1 instrument. Due to the deep sensitivity and good angular resolution (very good seeing conditions) of these images, the shell-like structure of IRC+10216 is very well detected. This work is presented with more details in de Laverny et al. (2002).

1. The rings and arclets in the IRC+10216 envelope

The new V-band image is shown in Fig 1. The observations have been collected in January 2000 under dark photometric sky and seeing

Y. Nakada et al. (eds.), Mass-Losing Pulsating Stars and their Circumstellar Matter, 197–200.

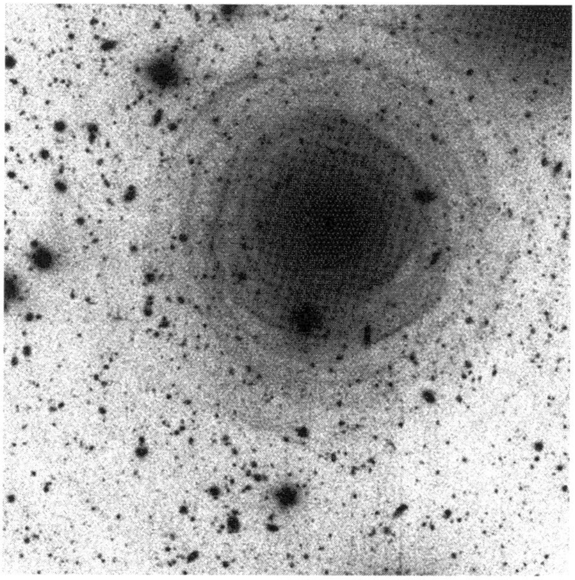

Figure 1. The dusty envelope of IRC+10216 in the V-band. The field of view is 3.2'×3.2', north is up and east to the left. The pixel size is 0.2 arcsec. The total observing time is 2 hours.

conditions around 0.6 arcsec. The dusty envelope is very well detected up to about 200 arcsec (by computing azimuthally averaged radial profiles). That corresponds to material ejected about 8 000 years ago.

We also show in Fig. 2 a map of the circumstellar shells. This image has been obtained by automatically removing most of the foreground and background stars and galaxies using a wavelet analysis. Then, a smooth radial brightness profiles has been subtracted to enhance the visibility of the structures seen in the envelope. Several remarks can then be done about these rings and arclets, as for example: (i) They are non-concentric and incomplete (ii) Some regions of the envelope have much less rings than others (iii) Several shells show some kind of interaction (iv) The intershell separation varies a lot in the envelope...

2. The inner central regions of the IRC+10216 envelope

The inner central regions are also very well detected in our data. Fig. 3 reveals a bipolar core with two very bright lobes. Their typical size is around 1-2 arcsec and the southern lobe is 40% brighter than the northern one. Such an asymmetrical structure has already been detected in

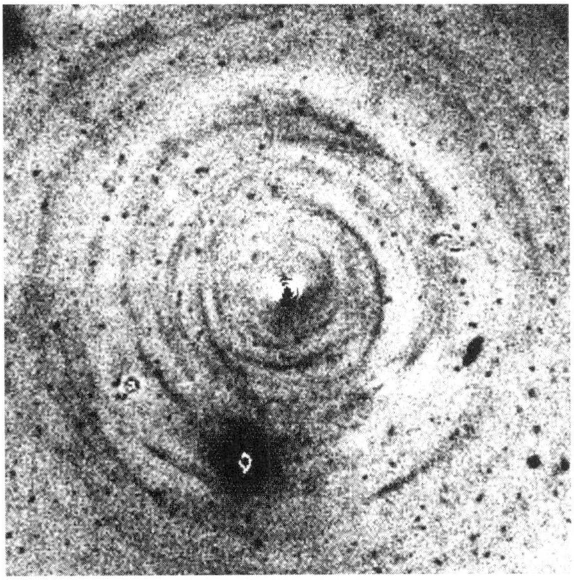

Figure 2. Arclets and incomplete rings seen in the envelope of IRC+10216. The field of view is 2.7'×2.7', north is up and east to the left.

much higher resolution observations but it is the first time that they are seen from the ground simultaneously with the very extended envelope. The orientation of this bipolar structure as well as its two bright lobes are consistent with previous work (see for instance Tuthill et al., 2000).

3. Conclusion

We have presented the first optical observations of the IRC+10216 envelope collected from the ground with high sensitivity and angular resolution. A map of the circumstellar structures reveals a rather heterogeneous pattern with incomplete rings, voids,... The bipolar shape of the most inner regions are also very well detected. These observations clearly demonstrated that the formation of winds in AGB stars is far more complex than assumed in current models.

Question by M. Feast:. When you see incomplete rings can you rule out the possibility that this is due to shadowing by other material in the circumstellar region rather than by a low density in the ring ?

Answer:. We detect incomplete rings almost everywhere in the envelope and particularly in the most external regions where the optical

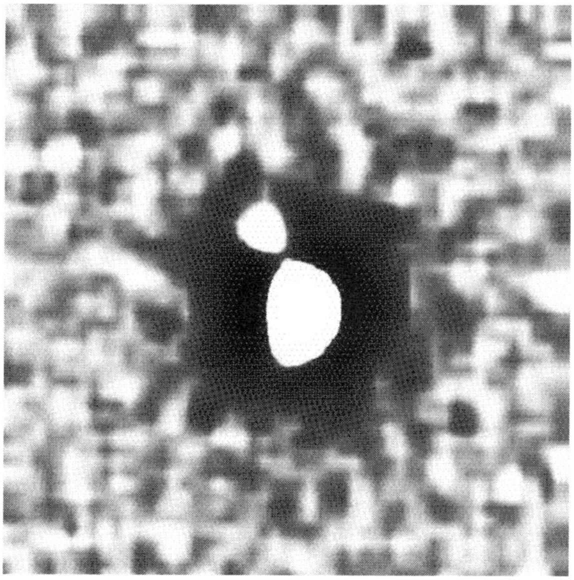

Figure 3. Magnification of the central regions of the IRC+10216 envelope in the V-band. The original image has been divided by a very smoothed one to enhance the most brightest features. The field of view is 10"×10", north is up and east to the left.

thickness between the outside and these rings is very low. Since illuminating photons come from the ambient galactic light, no material is thus present to cause such shadowing effects. We therefore think that detected incomplete rings do reflect lower density in some regions of these rings.

Acknowledgments

B. Vandame (ESO, Germany), B. Lopez (OCA, Nice, France) and N. Mauron (Montpellier University, France) are thanked for their help on this work which has been supported by CNRS/INSU and Ministère de l'Education Nationale et de la Recherche. The observations have been collected at ESO Paranal Observatory, Chile.

References

de Laverny, P., Vandame, B., Lopez, B., & Mauron, N. 2002, A&A, to be submitted
Mauron, N., & Huggins, P.J. 1999, A&A, 349, 203
Mauron, N., & Huggins, P.J. 2000, A&A, 357, 707
Mauron, N., de Laverny, P., & Lopez, B. 2002, A&A, submitted
Tuthill, P.G., Monnier, J.D., Danchi, W.C., & Lopez, B. 2000, ApJ, 543, 384

NIR MONITORING OF STAR CLUSTERS IN THE MAGELLANIC CLOUDS

Toshihiko Tanabé, Yoshifusa Ita, Noriyuki Matsunaga, Yoshikazu Nakada
Institute of Astronomy, The University of Tokyo, Mitaka, Tokyo, 181-0015, Japan
ttanabe@ioa.s.u-tokyo.ac.jp

Yasushi Nakajima, Chie Nagashima, Takahiro Nagayama, Tetsuya Nagata
Department of Astrophysics, Nagoya University, Furo-cho, Chikusa-ku, Nagoya, 464-8602, Japan

Ian S. Glass
South African Astronomical Observatory, PO Box 9, Observatory 7935, South Africa

Abstract From the near-infrared monitoring observations of intermediate-age clusters, we found that most AGB stars near the tip luminosity are variables in the infrared. It is also suggested that their amplitude may increase with their luminosities.

Keywords: AGB, variables, stellar evolution, Magellanic Clouds

Introduction

Globular clusters (GCs) are thought to be an ideal test field for the theory of stellar evolution. As is well known, the GCs in our Galaxy are all old and we are seeing in them the evolution of low mass stars (with a wide range in metallicity). The situation is however rather different in the Magellanic Clouds (MCs). GCs in the MCs span a wide range of ages, which enables us to examine the evolution of stars with various masses (although their metallicities are almost linearly dependent on the ages). Furthermore, GCs are numerous in the MCs and this richness

Y. Nakada et al. (eds.), Mass-Losing Pulsating Stars and their Circumstellar Matter, 201–204.

makes them particularly suitable for studying the short-lived later stages of stellar evolution.

We have started a near-infrared (NIR) monitoring observations of GCs in the MCs to make clear the evolution of AGB stars and especially to understand the characteristics of their variability. It is true that the amplitudes of variable stars become small in the NIR than in the optical wavelength region but NIR observations are vital in the following reasons. 1. AGB stars are inherently red and furthermore they are expected to become more red due to the heavy mass-loss. Indeed, we have so far discovered very red AGB stars which are very difficult to observe in the visible wavelengths (Tanabé, T. et al. 1997, 1998). 2. In the NIR, dwarf and sub-giant stars become faint and it is rather easy to do precise photometry of stars in the central part of rich clusters.

We plan to observe tens of GCs associated with the MCs repeatedly extending over 2 years to find AGB variables and to classify them according to their masses, metallicities, and periods. This catalog will consist of AGB stars with *known mass* and *metallicity* and thus can be used as a useful database for studying AGB evolutions.

Although the obtained data are still sparse, we will show preliminary results of our observations.

Observations

Observations are being carried out with the newly built 1.4-m telescope, IRSF (InfraRed Survey Facility) at the Sutherland station of the South African Astronomical Observatory. The telescope is equipped with a 3-channel NIR camera, SIRIUS (Simultaneous-3color InfraRed Imager for Unbiased Survey) (Nagayama et al. 2001) that allows J, H, K_s images to be obtained simultaneously. The field of view is 7.8 arcmin square with a scale of 0.45 arcsec/pixel.

Data Reduction

Images were processed by means of standard IRSF pipeline which is based on IRAF software package. A single image comprises 10 dithered 5 to 20-s exposures. A set of frames was combined to give an effective 50 to 200-s exposure in each of J, H, K_s. Photometry was carried out with the aid of DoPHOT (Schechter et al. 1993).

To find variable stars, we directly compare instrumental magnitudes of individual stars at different epochs. First, we calculate the median of the magnitude difference between 2 frames by comparing the magnitudes of identical stars on them. We regard it as a systematic magnitude difference and then re-calculate the "real" magnitude difference taking

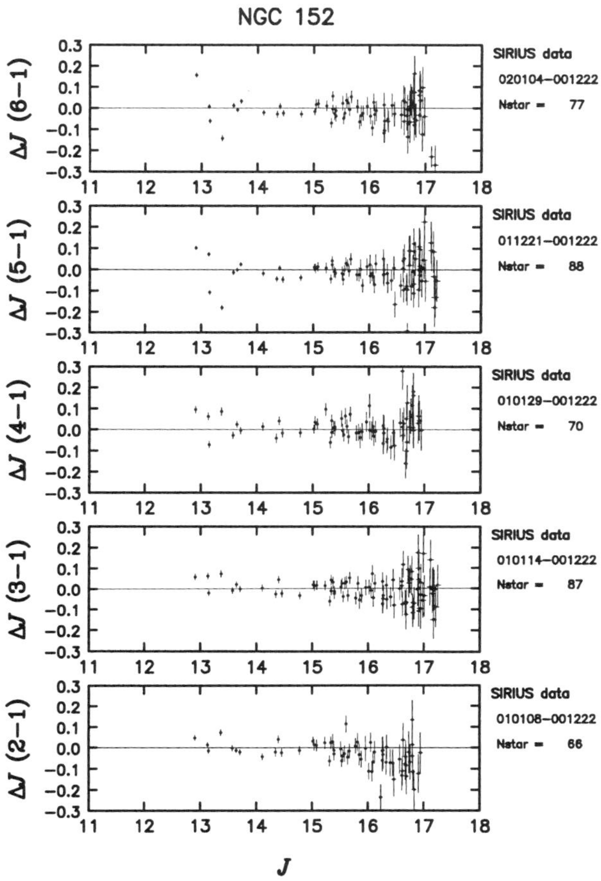

Figure 1. Magnitude differences of individual stars observed at different epochs. NGC 152 J-band.

account of it. Figure 1 shows an example of the results of this procedure in case of SMC cluster NGC 152 J-band. We can clearly see that at least several bright stars show variability.

Results and Discussion

In figure 2, we show the K_s vs. $J - K_s$ diagram of NGC 152 for reference. The star names are taken for MA stars after Mould & Aaronson (1982) and for the other three after Hodge (1981). From figure 1, we produced infrared light curves of several bright stars, shown in figure 3 for the case of the brightest AGB star in NGC 152. The similarity of the light curves of all the band suggests this star is a variable. NGC 152

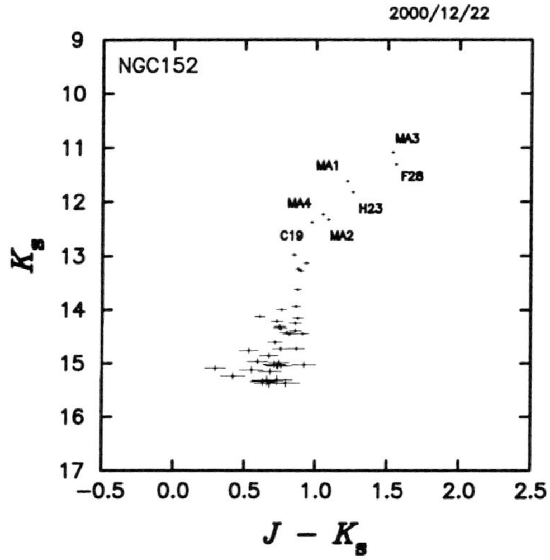

Figure 2. Color-Magnitude diagram of SMC cluster NGC 152.

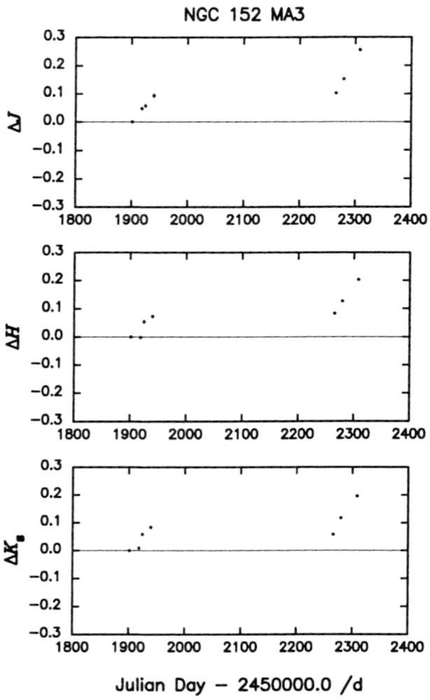

Figure 3. Infrared light curves of NGC 152 MA3.

F28, MA1, H23, MA2 also show similar light curves and are thought to be variables. Thus, we conclude that most, if not all, AGB stars near the tip of the AGB show variability. Although we have not been able to determine the amplitudes and periods of these stars, it seems that the amplitude increases with the luminosity. So far, we have analysed data of intermediate-age clusters, NGC 1783, NGC 1978 (LMC) and NGC 419 (SMC) besides NGC 152 and found that most bright AGBs have similar light curves and are thought to be variables.

References

Hodge, P. W. 1981, ApJ, 247, 894

Mould, J., & Aaronson, M. 1982, ApJ, 263, 629

Nagayama, T., et al. in The New Era of Wide Field Astronomy, ASP Conf. Series, Vol. 232, Eds. R. Clowes, A. Adamson, & G. Bromage, p.389

Schechter, P. L., Mateo, M., & Saha, A. 1993, PASP, 105, 1342

Tanabé, T., Nishida, S., Matsumoto, S., Onaka, T., Nakada, Y., Soyano, T., Ono, T., Sekiguchi, K., & Glass, I. S. 1997, Nature, 385, 509

Tanabé, T., Nishida, S., Nakada, Y., Onaka, T., Glass, I. S., & Sauvage. M. 1998, Ap&SS, 255, 407

THE BRIGHTEST AGB STARS OF THE LEO I DWARF SPHEROIDAL GALAXY

John Menzies
South African Astronomical Observatory, PO Box 9, Observatory 7935, South Africa
jwm@saao.ac.za

Abstract

The first results of a study of the dwarf spheroidal galaxy, Leo I, using the new Japanese-South African Infrared Survey Facility (IRSF) are presented. J, H, K_s observations show that most, if not all, of at least the top one magnitude of the AGB is populated by carbon stars. These objects form a tight sequence in the K_s,$(J-K_s)$ plane. In addition there are five very red objects which are believed to be dust enshrouded AGB stars. They have M_{bol} near -5.0 and are at the AGB tip. Comparison with carbon Miras in Magellanic Cloud clusters and taking into account other evidence on the ages and metallicities of Leo I populations suggests that these obscured stars belong to the youngest significant population of Leo I and have ages of ~2 Gyr. Some of them can be shown to be variable.

Keywords: Local group, Leo I, carbon stars, AGB, Mira variables

1. Introduction

A long-term programme has recently been started by a group of astronomers [1] from SAAO and the University of Tokyo to investigate Local Group galaxies in the near infrared (J,H,K_S). We are using the new Japanese-South African Infrared Survey Facility to study the stellar populations, evolution and structures of these galaxies, and in particular to detect and determine light curves for any Mira- and other red variables we find or which are already known. This report presents a summary of some early results obtained for Leo I.

[1]Michael Feast (U. Cape Town), Patricia Whitelock and John Menzies (SAAO), Yoshikazu Nakada and Toshihiko Tanabé (U. Tokyo).

Y. Nakada et al. (eds.), Mass-Losing Pulsating Stars and their Circumstellar Matter, 205–208.

2. Observations

The SIRIUS camera (Nagashima et al. 1999) was used to obtain simultaneous *(J,H,K_s)* images of a 7.8 arcmin square field at a scale of 0.45 arcsec/pixel. The central region of Leo I (field A) was observed at two epochs about 1 year apart, while a field 7 arcmin east of the centre (field B) was imaged only at the first epoch. A complete sample of stars to a limiting magnitude of about $K_s = 16.0$ was measured with the aid of the DoPHOT photometry program (Schechter, Mateo & Saha 1993). The *(K_s vs $(J - K_s)$)* colour-magnitude diagrams for the two fields are shown in Figure 1.

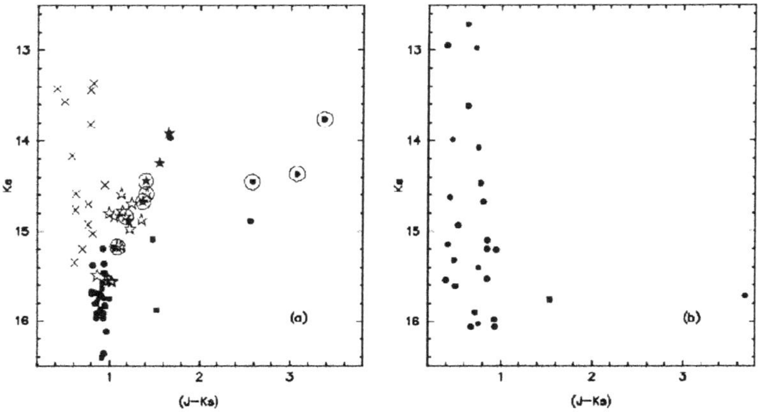

Figure 1. (a) K_s vs (J-K_s) for Leo I centre . Crosses are probable field stars, open star symbols denote ALW carbon stars, while filled star symbols refer to DB stars not measured by ALW. Filled squares represent probable background galaxies. Variables are circled. (b) K_s vs (J-K_s) for a field 7 arcmin east of the centre of the galaxy. The nearly vertical sequence is probably due to field stars, the square is probably a field galaxy, and star E (see text) is at the lower right.

3. Discussion

The bluest stars in field A and virtually all in field B are probably field stars. Most of the other stars in field A which form a tight sequence are AGB stars and most are known carbon stars (Azzopardi, Lequeux & Westerlund (1986 = ALW); Demers & Battinelli (2002 = DB)). Using bolometric corrections from Frogel, Persson and Cohen (1980), a distance modulus of 22.2 (Lee et al, 1993) and neglecting reddening, we find the carbon star sequence extends from $M_{bol} = -4.2$ at $(J - K_s) = 0.9$ to $M_{bol} = -5.1$ at $(J - K_s) = 1.7$. From work on galactic carbon stars (Whitelock 2000) the reddest of these stars are likely to be variable and are possibly Miras. Five of the carbon stars show evidence of variability from our data at the two epochs. The RGB tip is expected to be fainter than $M_{bol} = -4.0$, but without more work the tip of the AGB is not well defined by our observations.

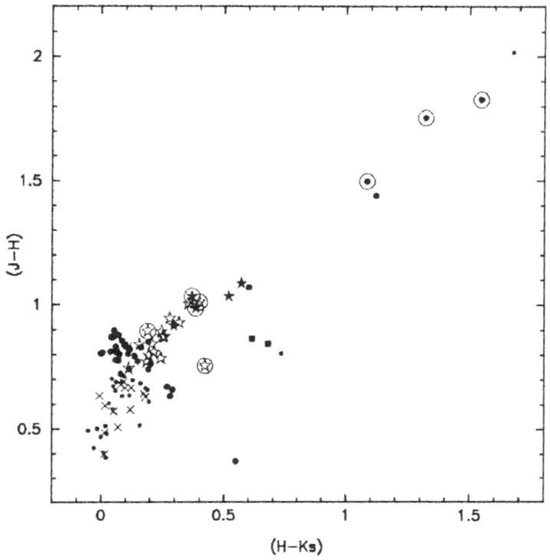

Figure 2 *(J-H),(H-K_s)* two-colour diagram for both fields. Symbols as in Fig 1 for field A; small filled circles for all stars in field B. The five obscured AGB stars are at upper right, star E being the reddest one.

There are 4 very red stars (A-D) in field A and one (E) in field B. Their positions in the colour-magnitude (Fig 1) and two-colour diagrams (Fig 2) are consistent with their being bright AGB stars obscured by circumstellar dust (see e.g. Whitelock 2000). In view of the fact that the top magnitude or more of the AGB sequence just discussed is heavily, and perhaps entirely, populated by carbon stars, it seems very likely that these obscured stars are also carbon stars. These five stars are the most extreme examples yet found of this type of obscured object in dwarf spheroidal galaxies. They are intermediate in colour between the two carbon Miras found in the Sagittarius dwarf spheroidal (Whitelock et al. 1999) and three obscured carbon Miras found by Nishida et al. (2000) in intermediate age clusters in the Magellanic Clouds. There is evidence of variability in stars A, B and C of the Leo I field A, and the reddest star (C) has varied by about 0.9 mag at K_s in 1 year. The Magellanic Cloud clusters are estimated to have ages in the range 1.6 to 2.0 Gyr, metallicities, [Fe/H] ~ -0.6, and turn-off masses of $\sim 1.5 M_\odot$. Gallart et al (1999) have suggested that major star formation in Leo I stopped about 2 Gyr ago, resulting in a population with a metallicity that could be as high as [Fe/H] ~ -0.6. The obscured stars and the most luminous unobscured AGB stars, at least, are possibly members of this youngest major stellar component of Leo I.

A small clump of stars in Fig 2 lies to the blue of the AGB carbon stars.

They are the faintest stars on the AGB in Fig 1. It would be interesting to know whether they are carbon or oxygen stars, and followup spectroscopic observations would be desirable.

The reddest obscured star, E, is interesting in that it is 8 arcmin from the centre of Leo I, whereas all of the known carbon stars are less than 4.5 arcmin from the centre. If it is an obscured AGB star, it is relatively young, and in a short-lived phase of its evolution; it probably traces a larger population of young stars. This supposedly young star though distant from the centre is still well within the tidal radius of Leo I (Irwin & Hatzidimitriou 1995). In dwarf spheroidals the younger populations are generally more concentrated to the centre than the older ones (e.g. Harbeck et al. 2001).

Further observations are being made to characterise the variable stars and to address some of the questions raised by our data so far.

References

Azzopardi, M., Lequeux, J., & Westerlund, B.E. 1986, A&A, 161, 232 (= ALW)
Demers, S., & Battinelli, P. 2002, AJ, 123, 238 (= DB)
Frogel, J.A., Persson, S.E., & Cohen, J.G. 1980, ApJ, 239, 495
Gallart, C. et al. 1999, ApJ, 514, 665
Harbeck, D. et al. 2001, ApJ, 562, L39
Irwin M., & Hatzidimitriou, D. 1995, MNRAS, 277, 1354
Lee, G.L. et al. 1993, AJ, 106, 1420
Nagashima, C. et al. 1999, in Star Formation 1999, ed. Nakamoto, T. (Nobeyama Radio Observatory).
Schecter, P.L., Mateo, M., & Saha, A. 1993, PASP, 105, 1342
Whitelock, P., et al. 1999, in IAU Symp. 192, The Stellar Content of Local Group Galaxies, eds. Whitelock, P., & Cannon, R. (San Francisco: ASP), 136
Whitelock, P. 2000, in IAU Symp. 177, The Carbon Star Phenomenon, ed. Wing, R.F. (Dordrecht: Kluwer), 179

MOLECULES AND DUST AROUND OXYGEN-RICH AGB STARS*

Jan Cami
Astronomical Institute "Anton Pannekoek", University of Amsterdam, The Netherlands
cami@astro.rug.nl

Abstract In this contribution, we present recent results from a study of the ISO/SWS spectra of a sample of O-rich AGB stars. Analysis of the molecular bands in the 2.3 - 8 μm region reveals an onion-like layering of the circumstellar matter in the extended atmosphere. The presence of an optically thick water layer for Mira variables is the most essential characteristic. We furthermore present an overview of the dust features observed in the 8 - 27 μm range of these spectra, tentatively identify one of the features with MgFeO and discuss the identification of other features with Al_2O_3 grains and spinel.

Keywords: Stars: AGB and post-AGB — Stars: atmospheres — circumstellar matter — Stars: late-type — Infrared:stars

1. Introduction

We studied the SWS spectra of a sample of 24 O-rich AGB stars with thin dust shells. We present here a summary of the recent results obtained by a detailed analysis of the molecular bands in the 2.3–8 μm region (Sect. 2) and an inventory of the dust features in these spectra in the 8–27 μm region (Sect. 3). A more in-depth discussion of the work presented here can be found in Cami (2002).

2. Molecular bands in the 2.3–8 μm region

In the wavelength range 2.3–8 μm, the SWS spectra are dominated by molecular bands due to H_2O, OH, SiO, CO, CO_2 and SO_2. Using

*Based on observations with ISO, an ESA project with instruments funded by ESA Member States (especially the PI countries: France, Germany, The Netherlands and the United Kingdom) with the participation of ISAS and NASA. The SWS is a joint project of SRON and MPE)

Y. Nakada et al. (eds.), Mass-Losing Pulsating Stars and their Circumstellar Matter, 209–212.

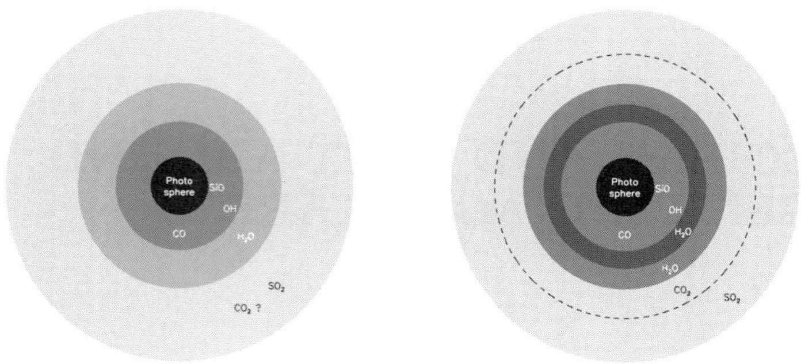

Figure 1. Cartoons depicting the location of the different molecules in Semi-Regulars *(left)* and Mira's *(right)*.

a new analysis method tailored to the molecular bands in the SWS observations (see Chapter 6 in Cami (2002) for a complete description), we derived excitation temperatures and column densities for each of the molecular bands. The excitation temperatures and column densities of SiO are systematically the highest, suggesting that the SiO bands are formed closest to the stellar atmosphere. The other molecular bands correspond to gradually lower excitation temperatures and column densities, indicating an onion-like layering of the various molecules in the extended atmosphere (see Fig. 1).

There are large differences however between the spectra of Semi-Regulars (SRs) and those of Mira variables. SRs show strong OH lines and deep (and hot) CO absorption bands, while the molecular bands due to CO_2 and SO_2 are regularly seen in emission. Their IR continuum in this wavelength range corresponds to color temperatures of 2300–3600 K. Mira's on the other hand do not show strong OH lines, only very weak (and cool) CO absorption and the bands due to CO_2 and SO_2 are most often observed as absorption bands. Their IR continuum is characterized by color temperatures of 1500–2000 K. Moreover, the Mira spectra show a pronounced absorption-like feature between 3.5 and 4 μm which cannot be identified by a molecular band.

This absorption-like feature in the Mira spectra actually corresponds to a lack of opacity from a hot and dense emitting water vapour layer (see also Yamamura et al., 1999). If the column density for this layer is high enough, the high opacity of water vapour would make this layer optically thick over a significant wavelength range, and hence mimic a blackbody function with a color temperature corresponding to the temperature of the layer. As a testcase, we modeled this layer for Mira, and found that

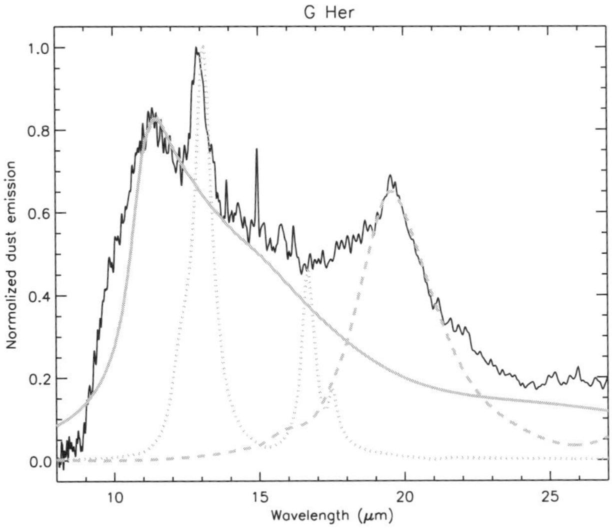

Figure 2. The dust emission from g Her compared to the extinction of natural spinel (dotted line), compact amorphous Al_2O_3 grains (full grey line) and $Mg_{(0.1)}Fe_{(0.9)}O$ (dashed line)

the entire SWS spectrum in this wavelength range is nicely reproduced when including a water layer extending to 2.2 $R_\star$ with a temperature of $\approx$1750 K and a column density of 10^{22} cm^{-2}. These parameters turn out to be very similar to the ones determined by Perrin (these proceedings).

The large spectral differences between the SRs and the Mira's can then be understood as follows (see also Fig. 1). The IR continuum for the SRs is determined by the stellar atmosphere. This is the background continuum, and depending on the relative size of the molecular layers with respect to this background continuum, the molecular bands will be seen in emission or absorption. The Mira's on the other hand have a water layer which is optically thick, and therefore we cannot observe the innermost molecules (mainly OH and CO). These molecular bands are therefore much weaker than in the SR spectra. This water layer then serves as the background continuum with a much lower color temperature against which the outermost molecular layers can absorb. As the background now has a much larger surface than for the SRs, the CO_2 and SO_2 bands are more likely to be observed as absorption bands than for the case of the SRs.

3. Dust Features in the 8–27 μm region

In the 8–27 μm region, the SWS spectra are dominated by circumstellar dust. The composition of the dust shells – determined by the spectral signatures of different dust components – shows remarkable differences from source to source and the richness of the dust spectra is often more bewildering than that of the gas.

Many spectra show a contribution from the well-known bands around 10 and 18 μm due to amorphous silicates, but only in 2 spectra these bands dominate the dust spectra. In some spectra, amorphous silicates are only marginally detected (see Fig. 2). The spectra also bear the signatures of other dust components; these are most obvious in the spectra with little amorphous silicates. Three dust features can be found in most spectra : a first feature peaking around 11 μm with a long red wing, another dust component is known as the 13 μm feature, and a third feature is a broad, almost triangular band with a peak near 19.5 μm. A few spectra also show a narrow ($\sim$ 1 μm) feature at 18 μm.

We tentatively identify the 19.5 μm feature with MgFeO, containing only a small fraction of Mg (see Fig. 2).

Miyata et al. (2000) suggested that the feature around 11 μm might be due to compact amorphous alumina grains, consistent with the results of an analysis by Speck et al. (2000). The SWS spectra corroborate this identification, as also the broad red wing is consistent with the observed spectra.

A lot of debate still exists about the carrier of the 13 μm feature. Posch et al. (1999) and Fabian et al. (2001) suggested spinel ($MgAl_2O_4$) to be the carrier of this feature. From the observational point of view, the peak position of the observed 13 μm feature is hard to reconcile with their laboratory measurements, even if grain shape effects are taken into account. Moreover, other strong bands in the laboratory data are absent even in those SWS spectra that have a very strong 13 μm feature and laboratory data from other groups show the strongest peak to be closer to 14 μm. Clearly, more work is needed to provide a solid identification for this carrier.

References

Cami, J. 2002, *Ph.D. Thesis*

Fabian, D., Posch, T., Mutschke, H., Kerschbaum, F., & Dorschner, J. 2001, A&A, 373, 1125

Miyata, T., Kataza, H., Okamoto, Y., Onaka, T., & Yamashita, T. 2000, ApJ, 531, 917

Posch, T., Kerschbaum, F., Mutschke, H., et al. 1999, A&A, 352, 609

Speck, A.K., Barlow, M.J., Sylvester, R.J., & Hofmeister, A.M. 2000, A&AS, 146, 437

Yamamura, I., de Jong, T., & Cami, J. 1999, A&A, 348, L55

DETECTION OF EXTENDED WATER VAPOR ATMOSPHERE OF MIRA BY NEAR-INFRARED SPECTROIMAGERY

Hideki Takami, Miwa Goto, Wolfgang Gaessler, Yutaka Hayano, Masanori Iye, David J. Saint-Jacques, Yukiko Kamata, Tomio Kanzawa, Naoto Kobayashi, Yosuke Minowa, Naruhisa Takato, Hiroshi Terada
National Astronomical Observatory of Japan,
650 North A'ohoku Place, Hilo, HI 96720
takami@naoj.org

Alan T. Tokunaga
Institute for Astronomy, University of Hawaii,
2680 Woodlawn Dr., Honolulu, HI 96822

Abstract

We present the first spectroimagery of extended molecular atmosphere of Mira spatially resolved by adaptive optics system. We found the molecular atmosphere of H_2O, CO and CO_2 is extended to about 40 mas. The presence of extended molecular layer has been revealed by recent infrared spectroscopic studies with space borne telescope ISO. The physical extent in our observation is consistent with their studies. The new technique that resolves objects in wavelength and in space simultaneously is proved to be very sensitive to the extended component respect to the continuum photosphere, because the time variable seeing effect is mostly canceled by relative measurement of the image size.

Keywords: AGB and post-AGB, circumstellar matter, adaptive optics

1. Introduction

Recent space-based spectroscopic observations by ISO/SWS have revealed the presence of extended (1 –100 R_*) and warm (a few 100 – 2000 K) molecular layers of the late type stars (Tsuji et al.1997, Yamamura et al.1999). This is related to the prediction of theoretical studies that H_2O, CO molecules contribute to the enhancement of the apparent stellar size in the near infrared at the absorption bands (Scholz & Takeda 1987, Watanabe & Kodaira 1979). To investigate the physical property

Y. Nakada et al. (eds.), Mass-Losing Pulsating Stars and their Circumstellar Matter, 213–216.

of the outer atmosphere, The state-of-art observation techniques such as optical interferometry and aperture masking have been employed to push the limit of spatial resolution (Mennesson et al.2000, Haniff et al.1995). We performed the spectroimaging with adaptive optics at the Subaru telescope. Although its spatial resolution is 70 mas at the K band, which is larger than the apparent size of late type stars, the technique turned out to be very sensitive to the difference in apparent stellar size between continuum and lines down to a few mas, because the effect of time variation of seeing is canceled out.

2. Observation and discussion

The observation was made on UT 2001 Dec 22 using the Infrared Camera and Spectrograph (IRCS; Kobayashi et al.2000, Tokunaga et al.1998) with the 8.2 m Subaru Telescope in conjunction with its adaptive optics system (Gaessler et al.2001, Takami et al.1998). The Subaru AO system equipped with 36-element curvature sensor is installed at the Cassegrain focus. The spectral resolution of the spectrograph was R=1000 in the K band. A SW Cet (M8III) is observed as a point source. The spectra and full width at the half maximum (FWHM) of apparent stellar size of Mira and SW Cet are shown in Figure 1. The spectra of Mira and SW Cet were divided by the spectroscopic standard.

The FWHM of Mira is enhanced at the molecular lines. We presume the enhancement is real and not the artifact of deep absorption bands because only Mira shows significant enhancement at the molecular lines, while both SW Cet and Mira have strong molecular absorption features such as CO bandhead at 2.3 μm. We derived the stellar diameter of Mira assuming that the stellar size is 18 mas (in FWHM) at the continuum with Gaussian function of center-limb-variation (Fig. 2). The continuum diameter is given by converting that of uniform disk model measured with interferometric techniques (Mennesson et al.2000). The opacity of H_2O, CO and CO_2 were taken from HITRAN/HITEMP database and convolved to the similar spectral resolution with the observation to identify the molecules responsible for the extended atmosphere (Rothman et al.1998). The profile of stellar diameter at each line is well reproduced by the synthetic optical depth curve (Fig. 2).

The enhancement in the stellar diameter at shorter than 2.07 μm and longer than 2.448 μm in the K band are dominated by H_2O opacity. At 2.28–2.40 μm region, the enhancement is well characterized by CO molecules. Excess enhancement at 2.005 μm and 2.020 μm is unaccounted by H_2O molecules only. We could assign it to CO_2 molecules based on the consistency of the wavelength and the spectral profile. The stellar size is enhanced as much as 40 mas at the molecular lines which corresponds to $2R*$. The extent is consistent with the results from spectroscopic observations (Yamamura et al.1999). However we should note that the spatial extent of the molecular atmosphere presented here only places the lower limit of the actual extent because the intrinsic physical dimension is smoothed out with the present modest wavelength resolution. The higher resolution spectroscopy is necessary to explore the outer edge of the molecular layers.

Acknowledgments

We thank all the staff and crew of the Subaru Telescope and NAOJ for their valuable assistance obtaining these data and their continuous support for the construction of the adaptive optics system and IRCS.

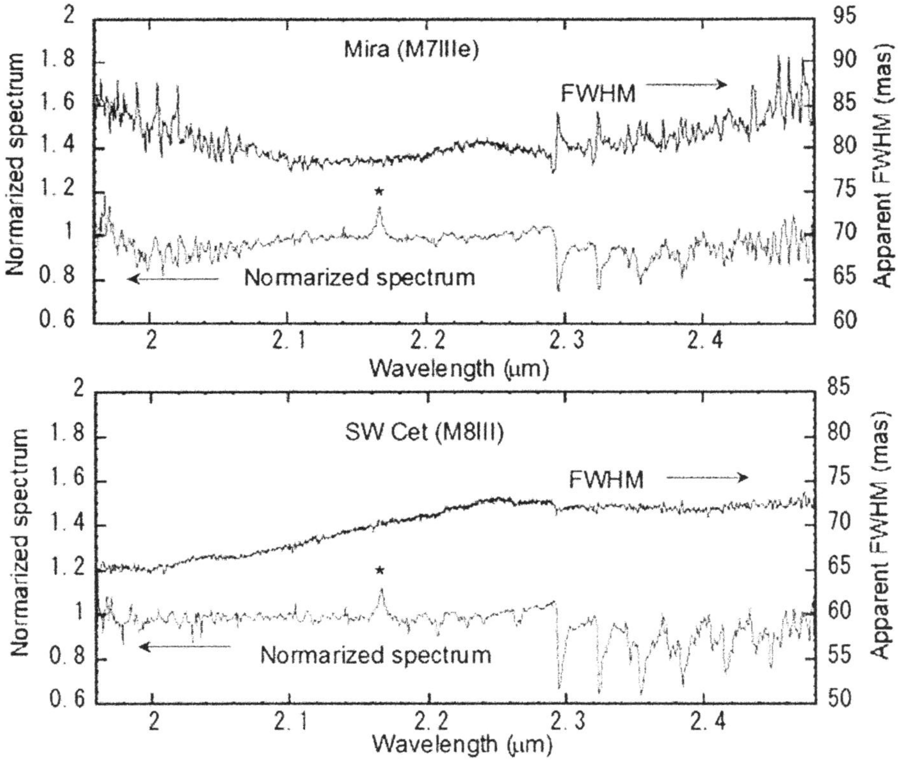

Figure 1. The FWHM of Mira (upper panel), SW Cet (lower panel) as functions of the wavelength overlaid with their spectra. Artificial features introduced by absorption lines of Brγ at 2.16 μm in the spectroscopic standard star is marked with an asterisk.

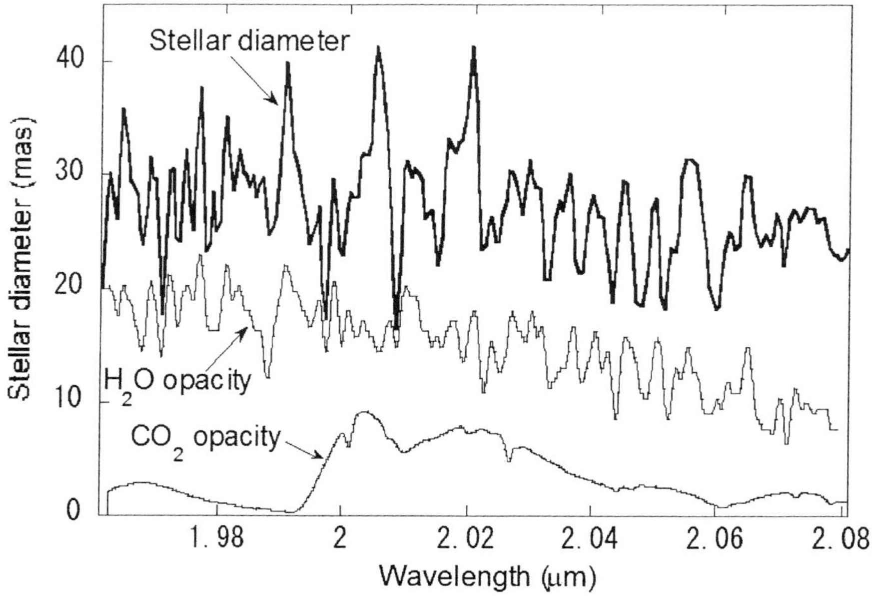

Figure 2. The stellar diameter of Mira as a function of the wavelength. The overlays are the opacity of H_2O and CO_2 calculated with HITRAN/HITEMP (Rothman et al.1998).

References

Gaessler, W., et al. 2001, Proc. SPIE, 4494, 30

Haniff, C.A., Scholz, M., & Tuthill, P.G. 1995, MNRAS, 276, 640

Kobayashi, N., et al. 2000, Proc. SPIE, 4008, 1056

Mennesson, B., Perrin, G., Coude du Foresto, V., Chagnon, G., Ruilier, C., Morel, S., Ridgway, S., Lopez, B., de Laverny, P., Traub, W., Carleton, N., & Lacasse, M. 2000, Proc. SPIE, 4006, 481

Rothman, L. S., et al. 1998, Journal of Quantitative Spectroscopy and Radiative Transfer, 60, 665

Scholz, M. & Takeda, Y., 1987, A&A, 186, 200

Takami, H., Takato, N., Otsubo, M., Kanzawa, T., Kamata, Y., Nakashima, K., & Iye, M. 1998, Proc. SPIE, 3353, 500

Tokunaga, A. T., et al. 1998, Proc. SPIE, 3354, 512

Tsuji, T., Ohnaka, K., Aoki, W., & Yamamura, I. 1997, A&A, 320, L1

Watanabe, T. & Kodaira, K. 1979, Pub. Astron. Soc. Japan, 31, 61

Yamamura, I., de Jong, T., & Cami, J. 1999, A&A, 348, L55

NIR SURVEY OF VARIABLE STARS IN THE LARGE MAGELLANIC CLOUD

Pulsation at the tip of the first giant branch?

Yoshifusa Ita[1]
[1]*Institute of Astronomy, School of Science, The University of Tokyo, 2-21-1 Osawa, Mitaka, Tokyo 181-0015, Japan*
yita@ioa.s.u-tokyo.ac.jp

T. Tanabé[1], Y. Nakada[1], N. Matsunaga[1], Y. Nakajima[2], C. Nagashima[2], T. Nagayama[2], D. Kato[2], T. Nagata[2], M. Tamura[3]
[2]*Department of Astrophysics, Nagoya University, Chikusa-ku, Nagoya 464-8602, Japan*
[3]*National Astronomical Observatory of Japan, Mitaka, Tokyo 181-8588, Japan*

Abstract The preliminary results of our ongoing NIR survey of the variable red giants in the Large Magellanic Cloud are presented. In the observed color-magnitude diagram $(J - K_s, K_s)$ and the stellar K_s magnitude distribution, the tip of the first giant branch (TRGB, $K_s \sim 12.3$ Nikolaev & Weinberg 2000) is clearly seen. Apart from the genuine AGB variables, we found many variable stars that populate just below the TRGB. From this result, we infer that most, if not all, of them are RGB variables.

Keywords: LMC, AGB, RGB, Variable star, Survey

Introduction

A monitoring survey toward the $1° \times 3°$ region along the LMC bar has been started since Dec. 2000, using the InfraRed Survey Facility (IRSF) at South African Astronomical Observatory Sutherland, to make complete catalog and sample of variable red giants.

Preliminary Results

So far, more than eight thousand variables have been detected and their JHK_s light curves have been accumulated. The photometric calibration

Y. Nakada et al. (eds.), Mass-Losing Pulsating Stars and their Circumstellar Matter, 217–218.

has been done, where we found that the survey is complete down to $K_s \sim 15.4$. In Figure 1 and 2, we show the stellar K_s magnitude distribution $N(m)$ of sources detected in all three wave bands (upper panel) and of variable stars (lower panel) and color magnitude diagram of variable stars, respectively. Because the plotted magnitudes are single-epoch

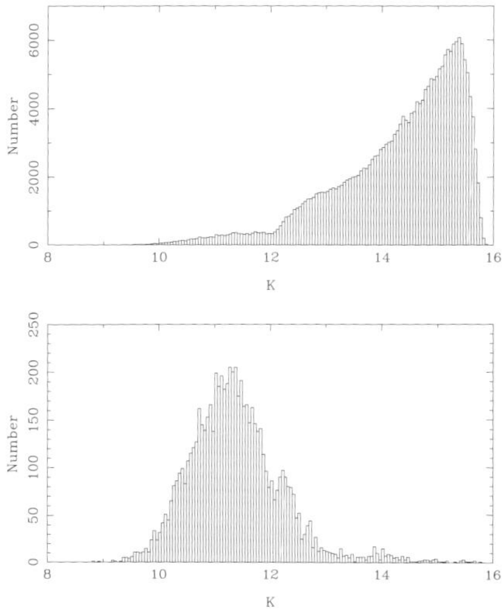

Figure 1. Magnitude distribution of sources detected in all three wave bands (upper panel, about 288000 sources), and of variable stars (lower panel, about 6800 variables).

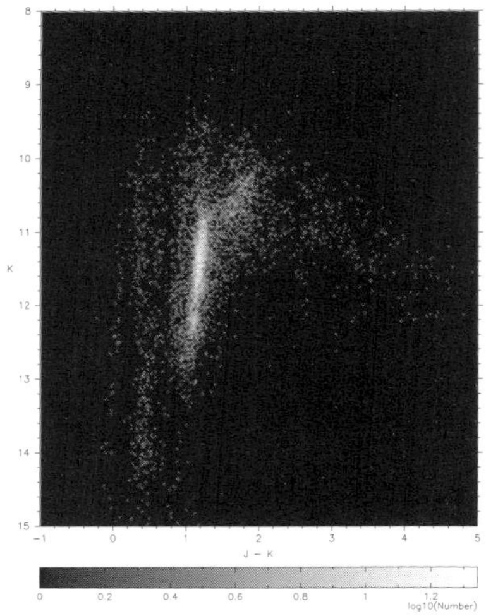

Figure 2. CM diagram of variable stars. (about 6800 variables)

ones, these are affected by the variability. Mean magnitudes will be used in the near future. In fig.1 upper, the discontinuity in $N(m)$ is clearly seen around $K_s \sim 12.2$, which corresponds to the TRGB. Also, as one can see in fig.1 lower and fig.2, there exist two peaks in number density along the red giant evolutionary track. One peaks well above the TRGB, which certainly consists of the genuine AGB variables, and the other peaks just below the TRGB. The members of the latter group show the short period and low amplitude light variation, quite different from those of genuine AGB variables (e.g., Mira). From these observational results, we infer the possibility that red giants pulsate at the TRGB, and the members of the latter peak are RGB variables.

The poster presented in Sendai can be downloaded from: http://www.ioa.s.u-tokyo.ac.jp/~yita/paper.html

References

Nikolaev, S., & Weinberg, M. D. 2000, ApJ, 542, 804

MASS LOSS HISTORY OF THE LOCAL MIRAS

Nakada, Y., Mito, H., Soyano, T., Tarusawa, K., Matsuyama, G., Matsunaga, N.
Kiso Observatory, Univ.Tokyo, Nagano 397-0101, Japan
nakada@kiso.ioa.s.u-tokyo.ac.jp

Abstract

Mira variables in the solar neighbourhood were selected from the IRAS Point Source Catalog with the help of the period-luminosity relation. The distribution of the IRAS color log(F25/F12) changes abruptly when the period exceeds 290 days. The implication of this change is discussed.

Keywords: Miras, IRAS, Period, Mass Loss

1. Mass loss and the AGB evolution

A general scenario of the stellar evolution for a mass-losing star is that it becomes a pulsating star near the AGB tip, strengthens the amplitude, starts a moderate mass-loss, accelerates the process to be hidden under a thick dust shell, and finally exhausts the envelope mass, and turns off the AGB to the planetary nebula. However, there are some evidence that the evolutionary path varies with the stellar mass. For example, few globular clusters contain planetary nebulae and no Mira variables have been found in the metal-poor globular clusters.

Knowledge of the age or mass for an individual star would be very important to study the mass dependence of the AGB evolution. Using the Magellanic clusters with known ages, Mould and Aaronson(1986) found a rather tight relation between the luminosity of the AGB tip and the age of the cluster older than 2 Gyr. Therefore, if we combine the well-known period-luminosity relation of Mira variables with the above luminosity-age relation we can obtain the ages of Mira variables from their periods. Although some correction might be needed to account

Y. Nakada et al. (eds.), Mass-Losing Pulsating Stars and their Circumstellar Matter, 219–220.

for the metallicity effect we will apply the same relation to the Galactic objects simply because we don't know the proper evaluation of this effect.

2. Local Miras

Local Miras have been selected using the IRAS Point Source Catalog (PSC) and the General Catalog of Variable Stars (GCVS). The distance D to the individual star was estimated by a simple formula, $L/(4\pi D^2)=F$, where L is the luminosity and F is the total flux. Van der Veen and Habing's formula was used to convert the IRAS flux F12 to F and Feast et al(1989)'s period-luminosity relation was applied to estimate L.

Since the lifetime of the Mira phase is much shorter than the nuclear burning time-scale, we assume that the luminosity remains the same during the Mira phase. Here, a star with 1 Mo runs along the line log(L/Lo)=3.6 while a star with 2 Mo is as bright as log(L/Lo)=3.9. Then, the color distribution of the Miras with a given period could be interpreted as the evolutionary path of a star with corresponding mass. The distribution of the IRAS color C12=log(F25/F12) shows a remarkable change at log(L/Lo)=3.7. Stars less luminous than this value are found only in the bluer side of the limit C12=0.1. It looks that they continue ejecting mass moderately until the end of the Mira phase, while more massive stars thrust into the heavy mass-loss range with the C12 exceeding 0.1.

However, there is still a possibility that the less massive stars jump into an extremely heavy mass-loss state where no optical data is available. Infrared monitoring should provide us a definite answer for this possibility. The H-band monitoring for selected sample of IRAS sources started at Kiso Observatory in September 2001.

References

Mould, J., & Aaronson, M. 1986, ApJ, 288, 551
Feast, M., Glass, I., Whitelock, P., & Catchpole, R. 1989, MNRAS, 241, 375
van der Veen, W., & Breukers, R. 1989, A&A, 213, 133.

NEAR-INFRARED MONITORING TO STUDY THE TRANSITION FROM AGB TO POST-AGB STARS

Takahiro Fujii
VERA project office, National Astronomical Observatory of Japan
Department of Physics, Kagoshima University
takahiro.fujii@nao.ac.jp

Yoshikazu Nakada
Kiso Observatory, Institute of Astronomy, University of Tokyo

Abstract We investigated light variability of late-AGB and post-AGB candidates in near-infrared H-band. A total of 113 objects were selected from IRAS PSC using their far-infrared colors. After four years monitoring observations, light variability was detected in 60 stars. These variables are classified into AGB or post-AGB stars according to their SED. We also carried out near-infrared spectroscopy to determine stellar temperature. We found the post-AGB variables in this study were classified into two group; one group contains F-G type post-AGB variables and the other group contains very cool(M-type) post-AGB variables.

Introduction

To study the transitional phase from AGB to post-AGB is important to understand a decrease in mass-loss which is believed to be associated with pulsation of its envelope and afterward planetary nebula formation. However, this phase is not well understood because of their rapid evolution and shortage of good observational candidates. In some case, this process undergo in the thick circumstellar envelope, therefore it is not visible in optical. Here we report our near-infrared monitoring of color-selected IRAS sources. Our sample contains both evolved AGB stars and post-AGB stars. There is very few previous study of variability in a sample of transitional objects (Fujii et al. 2001, Hrivnak 2001) , except for a few individual objects.

Y. Nakada et al. (eds.), Mass-Losing Pulsating Stars and their Circumstellar Matter, 221–222.

1. Observations and results

A total of 113 targets were selected from the IRAS PSC. The selection criteria were 1) $40° < l < 320°$ and $|b| > 2.5°$, 2) $Dec. > -30°$, 3) $F_{12\mu m} > 1.5$ Jy, 4) IRAS quality index is 333Any and 5) IRAS colors inside a box of dashed lines in Figure 1. The light variations were monitored with a near-infrared camera attached to the 105 cm Schmidt telescope at Kiso Observatory, University of Tokyo. We also carried out spectroscopy in the H and K-bands with the 188cm telescope at Okayama Astrophysical Observatory, NAOJ. A typical magnitude error is 0.03 mag. Out of 113 candidates, the variability was found in 60 objects. Figure 1 shows the distribution of the variables in IRAS two

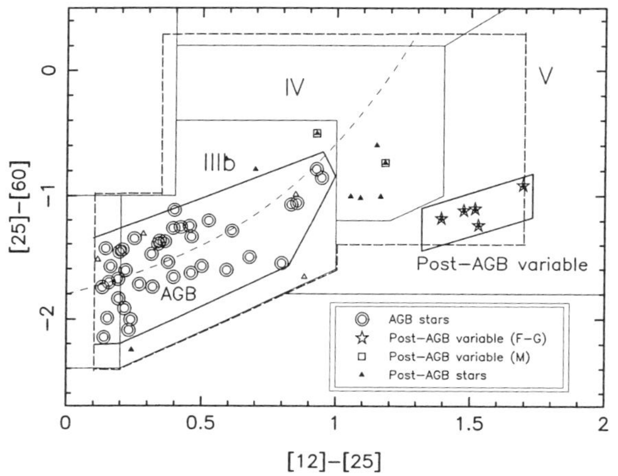

Figure 1 The positions of the variables in the IRAS color-color diagram. Circles, squares and stars denote AGB variables, cool post-AGB variables and F-G type post-AGB variables respectively.

color diagram. Objects are classified in AGB or post-AGB according to their SED. Post-AGB variables concentrate in a particular region, while AGB variables distribute leftward. These post-AGB variables are G-F type stars. In addition to them, we found two cool post-AGB variables (IRAS 05089+0459, 19511+3935) which are close to the AGB variable region. The peculiarity of dust shell of IRAS 05089 was pointed out by Fujii et al. (2002). Further investigation will be needed to understand the mechanism of variability of such cool post-AGB stars and its relationship with pulsation of AGB stars.

References

Fujii, T., Nakada, Y., & Parthasarathy, M. 2001, in Post-AGB Objects as a Phase of Stellar Evolution, ed. Szczerba, R. & Górny, S.K. (Kluwer), p.111

Fujii, T., Nakada, Y., & Parthasarathy, M. 2002, A&A, 385, 2002

Hrivnak, B.J. 2001, in Post-AGB Objects as a Phase of Stellar Evolution, ed. Szczerba, R. & Górny, S.K. (Kluwer), p.101

IDENTIFICATION AND NIR PHOTOMETRY OF THE ARECIBO SAMPLE OF OH/IR STARS

F. Jiménez-Esteban
Universidad Europea CEES, C/Tajo S/N, Villaviciosa de Odón, E-28670 Spain.
fje@laeff.esa.es

D. Engels
Hamburger Sternwarte, Gojenbersweg 112, D-21029 Hamburg, Germany.

P. García-Lario
ISO Data Centre, VILSPA, Apdo. de Correos 50727, E-28080 Madrid, Spain.

L. Agudo-Mérida
Universidad Alfonso X El Sabio, Apdo. de Correos 28691, Madrid, Spain.

Abstract In June 1999 we started a near-infrared (NIR) monitoring of the full Arecibo sample of OH/IR stars (385 sources) using the 1.23 m telescope at Calar Alto Observatory equipped with the MAGIC IR camera. Since then, we have completed 6 observing runs and obtained J, H and K images of almost all sources in the sample (97%) under good photometric conditions. NIR and optical counterparts have been identified and NIR photometric data has also been obtained.

Keywords: OH/IR stars, Near-Infrared

1. The sample

The Arecibo sample of OH/IR stars is the best sample available to study the statistical properties of OH/IR stars. This is because it is the most sensitive in terms of the maser flux and the only one complete (Lewis 1994). Note that the term "OH/IR star" is used here to designate IRAS selected AGB stars with OH maser emission, which comprises not

Y. Nakada et al. (eds.), Mass-Losing Pulsating Stars and their Circumstellar Matter, 223-224.

only the optically invisible objects, but also objects which would have been classified as Mira variables, if discovered optically. A minority of the Arecibo sample might actually be M-Supergiants or PPNe. Before our observations, images were available only for a small fraction of the stars in the sample in the NIR, and less than 5% are well studied variables with known periods.

2. Identification of the counterparts

The MAGIC camera uses a 256x256 pixel NICMOS3 array which provides an approximate field of view of 5'x5'. As of today, 75% of the Arecibo sources have been searched for their NIR and optical counterparts. In most cases a bright NIR counterpart was clearly identified. Interestingly, some of them are so red that they were found to be completely invisible in the optical. In the most extreme cases they are so heavily obscured that a bright counterpart was identified only in the K-band, being barely detectable in the H band and completely invisible at shortest wavelengths.

The search for optical counterparts was carried out by comparing the optical images extracted from the Digitised Sky Survey with the NIR images, improving this way the accuracy of their astronomical coordinates. 66% of objects studied so far show a clear optical counterpart, whereas 33% presented dubious or no optical counterpart.

3. Photometry

The mean flux of each source in each filter was derived from five individual frames where the star was located in different positions within the array. The associated error is the standard deviation of these five independent measurements. Typical errors are less than 0.2 magnitudes in all filters for well detected sources. Limiting magnitudes are 15.5 mag in J, 14.5 in H, and 13.5 in K. As of May 2002, 97% of the objects have already been observed under good photometric conditions at least once. Using a set of at least three non-variable reference stars in each field we will be able to perform accurate differential photometry under non-photometric conditions in future runs. The results obtained (including finding charts in the optical and in the NIR as well as the photometric data) will be made available soon in the World Wide Web.

References

Lewis, B.M. 1994, ApJS, 93, 549

INFRARED PROPERTIES OF O-/C-RICH MIRAS CLOSE TO THE GALACTIC PLANE

Lerothodi L. Leeuw*

South African Astronomical Observatory

P O Box 9, Observatory, 7935, South Africa

lleeuw@saao.ac.za or lerothodi@alum.mit.edu

Abstract Near- to mid-infrared photometric properties are preliminarily exhibited of *IRAS* Miras that lie within $-7.^\circ3 < b < +1^\circ$ and $282.^\circ5 < l < 285.^\circ5$. Tentative O-/C-rich classifications based on $K - L$ versus $[12\,\mu\mathrm{m}] - [25\,\mu\mathrm{m}]$ colours and pulsation period characteristics of the variable stars are presented. It is shown that O-/C-rich Miras come from two populations of statistically different period distributions.

Keywords: mira variables, circumstellar dust, colour diagrams, pulsation periods

1. Photometry and Classification of O/C Miras

Mira variables are mass-losing, pulsating late-type giants that are very red, hence infrared observations have a great importance in their study. Using the Mk III photometer attached to the South African Astronomical Observatory (SAAO) 1.9 m telescope, approximately 11 $JHKL$ photometry modules per star have been obtained for 101 Galactic *IRAS* sources that were expected to be Mira Variables (Leeuw 1997). For the 85 stars that were photometrically shown to be Miras, new periods were determined. Using similar two-colour diagrams as by Epchtein *et al.* 1987), tentative designations as O-rich or C-rich Miras were determined for 73 Miras with previously unpublished spectral classifications. Figure 1 (a) shows the two-colour positions of O-/C-rich stars, that were determined from demarcations based on plots of 11 O-rich and 10 C-rich Miras that had published spectral classifications. The physical reason for these separate regions lies in the different ratios of near to mid-infrared opacities for the O-/C-rich dusts (Le Bertre *et al.* 1994).

*Author acknowledges extensive collaboration with P.W. Whitelock and M.W. Feast.

Y. Nakada et al. (eds.), Mass-Losing Pulsating Stars and their Circumstellar Matter, 225–226.

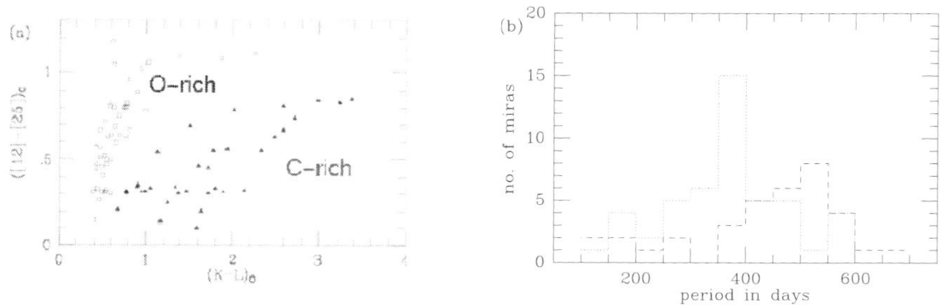

Figure 1. (a) The $K-L$ versus [12 μm]–[25 μm] diagram with O-rich (open squares) and C-rich (solid triangles) Miras that had published spectral classifications and those that were tentatively classified using this diagram. (b) The dotted-line and dashed-line histograms of number of O-rich and C-rich Miras respectively.

2. Pulsation Characteristics of O-/C-Rich Miras

Figure 1 (b) shows the histograms of number of Miras versus period in 50-day intervals, comparing the period distributions of the O-/C-rich Miras. The O-rich and C-rich Miras respectively have peaks at P=375 (15 stars) and P=525 days (9 stars), maximums of P=600 and P=716 days, and averages of P=386 and P=444 days. The minimum periods for the two types is about the same, just over 100 days. That the C-rich Miras have an average period longer than that of O-rich Miras is interesting and consistent with studies of Miras in the Large Magellanic Clouds (Feast *et al.* 1989), although the present sample is *IRAS* selected whereas the LMC sample was taken mainly from *I*-plate surveys. According to the Kolmogorov-Smirnov statistical test, there is only a very small probability of 0.00055 that the period distribution functions of the O-/C-rich Miras presented here come from the same populations.

3. Summary

Near- to mid-infrared O-/C-rich classifications and period distributions have been presented of 83 *IRAS*-selected Miras. The studies of the colours and periods respectively probe Mira circumstellar material and stellar pulsations and have shown that the O-/C-rich stars exhibited here come from two populations of significantly different period distributions.

References

Epchtein, N., Le Bertre, T., Lepine, J.R.D., Marques dos Santos, P., Matsuura, O. T., & Picazzio, E. 1987, A&AS, 71, 39

Feast, M.W., Glass, I.S., Whitelock, P.A., Catchpole, R.M. 1989, MNRAS, 241, 375

Leeuw, L.L. 1997, MSc. Thesis, University of Cape Town.

Le Bertre, T., Epchtein, N., Guglielmo, F., & Le Sidaner, P. 1994, A&SS, 217, 105

WATER VAPOR IN THE OUTER ATMOSPHERES OF RED GIANTS PROBED BY HIGH-RESOLUTION, MID-INFRARED OBSERVATIONS

Nils Ryde, John H. Lacy, Matthew J. Richter, David L. Lambert, Thomas K. Greathouse

McDonald Observatory and Department of Astronomy, University of Texas at Austin, TX 78712-1083, USA

ryde@astro.as.utexas.edu

Abstract Here, we report on our detection of water vapor in the mid-infrared (11-12 microns) spectrum of the K1.5 giant Arcturus (α Boo) obtained with the high-resolution, infrared spectrograph TEXES. The detection is presented in detail in Ryde et al. 2002 ApJ, submitted. We are successful in modeling several resolved, rotational lines of water vapor and vibration-rotational (1.5 microns) and rotational lines of OH with a modified MARCS photosphere.

Keywords: stars: individual (α Boo) – stars: AGB and post-AGB – stars: late-type – Infrared: stars

1. Observations and analysis

We have observed pure rotation, water vapor lines and OH (both rotation and vibration-rotation) in the disk averaged spectrum of Arcturus (α Boo) at 11-12 μm. The observations were obtained with the TEXES spectrograph mounted on the 3 meter NASA Infrared Telescope Facility telescope on Mauna Kea. The data presented here were observed at a resolving power of approximately $R = 80,000$. TEXES, the Texas Echelon Cross Echelle Spectrograph, a mid-infrared (5-25 microns) spectrograph, is presented in Lacy et al. (2002 PASP 114, 153).

In analysing the data we used model photospheres calculated with the MARCS code. This version of the MARCS code is the final major update of the code and its input data in the suite of MARCS model-photosphere programs first developed by Gustafsson et al. (1975, A&A 42, 407) and further improved in several steps, e.g. Plez et al. (1992, A&A 256, 551),

Y. Nakada et al. (eds.), Mass-Losing Pulsating Stars and their Circumstellar Matter, 227–228.

Jørgensen et al. (1992, A&A 261, 263), and Edvardsson et al. (1993, A&A 275, 101).

We have calculated a model photosphere for Arcturus with $T_{\rm eff} = 4300\,{\rm K}$, $\log g = 1.5\,({\rm cgs})$, $[{\rm Fe/H}] = -0.5$, and $\xi_t = 1.7\,{\rm km\,s^{-1}}$. The OH (v=2-0) 1.5 micron lines and the OH (v=1-1), which are formed deep in the atmosphere, are predicted well in the standard MARCS model. However, a standard model fails to predict the observed water-vapor lines and the OH (v=0-0) lines.

We produced a semi-empirical model for the outer photosphere, by modifying the temperature structure by lowering it in the very outer parts of the photosphere compared to a flux-constant, hydrostatic, standard MARCS model photosphere. From this temperature structure we consistently recalculate the model. Based on this new modified model, we are able to predict the observed strengths of all OH and water vapor lines satisfactorily.

2. Discussion

We have made an unexpected discovery of absorption lines of water vapor in the disk-averaged spectrum of the K1.5III red-giant star Arcturus (see Ryde et al. 2002 ApJ, submitted). Arcturus is the hottest star yet to have shown water-vapor features. Based on kinematic information, line widths and excitation temperature, we argue that the water vapor is photospheric and not circumstellar. Until now, H_2O has neither been observed nor expected to exist in photospheres of stars of Arcturus' effective temperature. By modifying the outer layers of standard model photospheres (which may not describe these regions properly) we have succeeded in finding a model which yields a synthetic spectrum matching the observations. Thus, it is possible to achieve a photospheric spectrum containing water vapor also for early K giants if the outer parts are cooled for some reason. Possible reasons for the unexpected water vapor include photospheric inhomogeneities (such as convective flows, molecular catastrophes, or even star spots) and departures from LTE in the photospheric structure or line formation in the boundary layers.

Acknowledgments

This work was supported by the Swedish STINT Foundation and the Robert A. Welch Foundation of Houston, Texas. The construction of the TEXES spectrograph was supported by grants from the NSF and observing with TEXES was supported by the Texas Advanced Research Program. We acknowledge the support of the IRTF staff.

THE MASS-LOSS HISTORY OF CARBON STARS PROBED BY CO LINE AND DUST CONTINUUM EMISSION

Fredrik Larsen Schöier
Leiden Observatory, the Netherlands
fredrik@strw.leidenuniv.nl

Nils Ryde
McDonald Observatory and Department of Astronomy, University of Texas at Austin, USA
ryde@astro.as.utexas.edu

Hans Olofsson
Stockholm Observatory, Sweden
hans.olofsson@astro.su.se

Abstract An extensive modelling of CO line emission from the circumstellar envelopes around a number of carbon stars is performed. By combining radio observations and infrared observations obtained by ISO the circumstellar envelope characteristics are probed over a large radial range. The combined data set gives better determined envelope parameters, and puts constraints on the mass loss history of these carbon stars. Moreover, analysis of the dust emission puts further constraints on the mass-loss rate history. The stars presented here are not likely to have experienced any drastic long-term, mass-loss rate modulations, at least less than a factor of about 5, over the past thousands of years.

Keywords: stars: AGB and post-AGB – circumstellar matter – late-type – mass-loss – Infrared: stars

1. Introduction

Mass loss is associated with many stages of stellar evolution. Of particular interest for understanding the underlying mechanism(s) are observational results on its temporal changes. Mass loss in late stellar evo-

Y. Nakada et al. (eds.), Mass-Losing Pulsating Stars and their Circumstellar Matter, 229–230.

lution can be both highly episodic with modulations of the mass loss rate by several orders of magnitude over a few thousand years (e.g., Olofsson et al. 1996, A&A 311, 587). But also less dramatic and more regular modulations on time scales of a few hundred years have recently been observed (Mauron & Huggins 1999, A&A 349, 203; Mauron & Huggins 2000, A&A 359, 707).

The ISO satellite detected FIR CO line emission towards only three carbon stars IRC+10216, IRAS 15194–5115, and CIT 6. Other sources were not observed long enough by ISO to allow detection. By combining the ISO observations with complimentary radio data the envelope characteristics are probed over a large radial range typically corresponding to a time scale of several thousand years.

2. Results and discussion

The CO millimetre and FIR rotational transitions are modelled using a detailed, spherically symmetric, NLTE radiative transfer code based on the Monte Carlo method (Schöier & Olofsson 2001, A&A 368, 969). The code simultaneously solves the energy balance equation and the statistical equilibrium equations allowing for a self consistent treatment of CO line cooling. The modelling of CO rotational line emission at millimetre and FIR wavelengths put constraints on the physical properties of a CSE over its full radial extent. The full data set can be consistently modelled assuming a constant mass loss rate. The publicly available dust radiative transfer code DUSTY (Ivezic & Elitzur 1997, MNRAS 287, 799) was used to model the observed spectral energy distributions. The dust modelling provides a complimentary estimate of the mass loss rate. The effect of the dust emission on the excitation of CO was further tested and found to be only of minor importance.

The CO data in combination with the analysis of the dust emission constrain the mass loss rate history and we find that any longer-term mass loss rate modulations are likely to have been smaller than about a factor of 5 over the past several thousand years for the three carbon stars where FIR CO line emission was detected by ISO. For further information see Schöier, Ryde & Olofsson (2002, accepted by A&A; astro-ph/0206078).

Acknowledgments

F.L.S. is supported by the Netherlands Organization for Scientific Research (NWO) grant 614.041.004. N.R. acknowledges support from the Swedish Foundation for International Cooperation in Research and Higher Education (STINT).

AN ANALYSIS OF THE 2.6 – 3.7 MICRON SPECTRUM OF R DOR

Nils Ryde
McDonald Observatory and Department of Astronomy, University of Texas at Austin, TX 78712-1083, USA
ryde@astro.as.utexas.edu

Abstract Here, I present our modelling (Ryde & Eriksson, A&A 2002, 386, 874) of the 2.6 – 3.7 microns spectrum of the red semiregular variable R Doradus observed with the Short-Wavelength Spectrometer on board the Infrared Space Observatory.

Keywords: stars: individual (R Dor) – stars: AGB and post-AGB – stars: late-type – Infrared: stars

1. Observations and analysis

Our goal is to discern whether a moderately varying M giant, such as R Doradus, could be modelled satisfactory with a hydrostatic model photosphere. We observed the $2.6-3.7$ micron region with the ISO SWS 06 at a resolution of $R = 2000 - 2500$ and the full $2.4 - 45$ micron region with the SWS 01 at $R = 30$. The low-resolution SWS01 observation is used to ensure that the different SWS06 observations in this region were properly aligned.

We have calculated hydrostatic model photospheres with the MARCS code in spherical geometry. Synthetic spectra were calculated using these model photosphere. The synthetic spectra, also calculated in spherical geometry, include lines from H_2O (Partridge & Schwenke, 1997, J. Chem. Phys. 106, 4618), CO (Goorwitch, 1994, ApJS 95, 535), SiO (Langhoff & Bauschlicher, 1993, Chem. Phys. L. 211, 305), CH (Jørgensen et al., 1996, A&A 315, 204), CN (Jørgensen & Larsson, 1990, A&A 238, 424 and Plez, 1998, priv. com.), OH (Goldman et al., 1998, JQSRT 59, 453), C2 (Querci et al. 1971 and Jorgensen, 2001, priv. com.), and CO_2 (Hitemp; Rothman et al., 1992, JQSRT 48, 469).

Y. Nakada et al. (eds.), Mass-Losing Pulsating Stars and their Circumstellar Matter, 231–232.

We are able to fit the observed range well with a hydrostatic model photosphere, see Ryde & Eriksson (A&A 2002, 386, 874). Our best fit model have the fundamental parameters of $T_{\rm eff}$ = 3000 K, log(g)=0, and solar metallicity. We find that the wavelength region investigated is sensitive to the effective temperature used in the modeling of the photosphere. The agreement between the synthetic spectrum and the ISO observations is encouraging, especially in the wavelength region of 2.8 – 3.7 microns, suggesting that a hydrostatic model photosphere is adequate for the calculation of synthetic spectra in the near infrared for this moderately varying red giant star. However, an additional absorption component is needed at 2.6-2.8 microns. Thus, we also find that the model does not account for this extra absorption component at the beginning of the interval. For a discussion of this see Ryde & Eriksson. The spectral signatures are dominated by water vapour in the stellar photosphere, but several photospheric OH, CO, and SiO features are also present.

2. Conclusion

R Dor is a moderately varying M Giant. We have modeled its 3 micron ISO spectrum (resolving power is ca 2000) successfully with a hydrostatic model photosphere. Dynamical models are, however, certainly needed for the Mira stars, which show larger variations with time. Line opacities contributing at 2.6 – 3.7 microns are H_2O, CO, OH, and SiO. The region is found to be very temperature sensitive, and it is found that an extra absorption is needed at 2.6 - 2.8 microns.

Acknowledgments

We should like to thank B. Gustafsson, K. Eriksson and D. L. Lambert for inspiration and enlightening discussions. This work was supported by the P.E. Lindahl Foundation Fund of the Royal Swedish Academy of Sciences and the Swedish Foundation for International Cooperation in Research and Higher Education.

ANNUAL PARALLAX MEASUREMENTS OF MIRA-TYPE VARIABLES WITH PHASE-REFERENCE VLBI OBSERVATION

Tomoharu Kurayama
Department of Astronomy, School of Science, the University of Tokyo, Japan
VERA Project Office, National Astronomical Observatory, Japan
kuraymtm@cc.nao.ac.jp

Tetsuo Sasao
VERA Project Office, National Astronomical Observatory, Japan
sasao@miz.nao.ac.jp

Abstract This study is to measure an annual parallax of water masers around a Mira-type variable, AW Tau, with phase-referencing VLBI observation. We have obtained preliminary values of the annual parallax of 9.7 ± 0.4 mas.

Keywords: Mira-type variables, parallaxes, distances, phase-referencing VLBI

1. Introduction

The final goal of this observation is to establish the period-luminosity relation of solar-neighborhood Miras.

The period-luminosity relation of Miras are clearly seen in the Large Magellanic Cloud (Feast et al. 1989), but is not clear in the solar-neighborhood (van Leeuwen et al. 1997). This is because distances to Miras in the solar-neighborhood are highly uncertain.

So if we can measure annual parallaxes of Miras with enough accuracy, we can establish a good period-luminosity relation in solar-neighborhood Miras. This will lead to determination of the distance modulus of Large Magellanic Cloud and also a restriction of physical properties of late-type stars.

Y. Nakada et al. (eds.), Mass-Losing Pulsating Stars and their Circumstellar Matter, 233–234.

2. Observation and Results

Observation was carried out with following parameters (Table 1). We have finished the analysis of AW Tau, one of the five observed Miras. Movements of maser spots and fitting results are shown in Figure 1. Note that the maser spot in the 4th epoch observation is separated from the others, so we consider that it is another maser spot and omit from the fitting. From the fit, we have obtained a parallax of 9.7 ± 0.4 mas.

For error estimate in a phase-referencing VLBI observation, here we consider three sources of systematic errors: (i) Error from reference source structure, (ii) Error from airmass correction uncertainties, (iii) Error from baseline error. (i) is almost negligible. This is seen from closure phases. (ii) is less than 0.1 mas at airmass estimation errors 1 cm and at 2000 km baselines. (iii) is 0.04 mas at baseline-length errors 1 cm and at the 200 km baseline. Error estimations of (ii) and (iii) are in the worst case.

3. Conclusion and Future

We have obtained preliminary values of the annual parallax (9.7 ± 0.4 mas) and proper motions for AW Tau. We have evaluated systematic errors, and demonstrated the possibility of the measurement of annual parallaxes at 0.1-mas-order error level. In order to establish the period-absolute-magnitude relation, we are going to analyze the other 4 sources, and willing to observe more sources by VLBA and VERA.

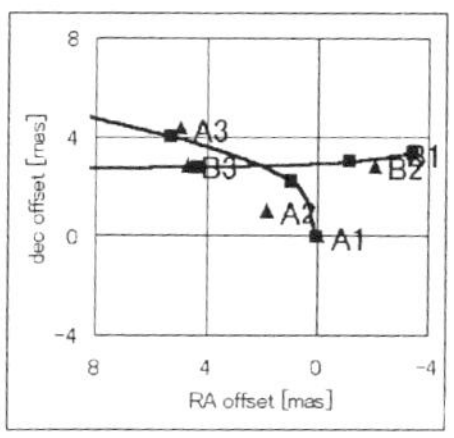

Figure 1. Movements of maser spots. Triangles show observed maser positions. Squares and lines show fitted results.

Array	VLBA 10 stations
Freqency	22 GHz
Nodding cycle	40 sec
Freq. set	32 Msps × 2 bit × 2 ch
Bandwidth	16 MHz
5 hours × 4 epochs × 5 sources	

Table 1. Observation setups.

References

Feast, M., et al. 1989, MNRAS, 241, 375
van Leeuwen, F., et al. 1997, MNRAS, 287, 955

MAPPING THE CIRCUMSTELLAR ENVELOPE OF OH26.5+0.6

D. Fong

Department of Astronomy, University of Illinois, 1002 W. Green Street, Urbana, IL 60801, USA

K. Justtanont

SCFAB, Stockholm Observatory, Department of Astronomy, SE-106 91 Stockholm, Sweden

kay@astro.su.se

M. Meixner, M.T. Campbell

Department of Astronomy, University of Illinois, 1002 W. Green Street, Urbana, IL 60801, USA

Abstract We report the detection of CO J=1-0 of the circumstellar envelope of OH 26.5+0.6. From modelling both the dust and CO, we deduce that the envelope structure is as followed : a superwind with a mass loss rate of 5×10^{-4} $M_\odot$ yr^{-1} surrounded by a tenuous wind of 10^{-6} $M_\odot$ yr^{-1}.

The circumstellar envelope around the OH/IR star, OH 26.5+0.6, was detected by Berkeley-Illinois-Maryland-Association (BIMA) Millimeter Array in the CO J=1-0 transition using the B-, C- and D-configurations. We notice a "missing flux" between the velocity of 12-16 km s^{-1}, most likely due to interstellar absorption along the line of sight, compared to profiles from higher CO transitions (Justtanont et al. 1996) and OH maser observations (Bower & Johnson 1990).

We modelled the infrared spectral energy distribution of the star and derive a dust mass loss rate of $\sim 2 \times 10^{-6}$ $M_\odot$ yr^{-1} and the outer dust

Y. Nakada et al. (eds.), Mass-Losing Pulsating Stars and their Circumstellar Matter, 235–236.

envelope of 10^{16} cm, using archived data from the spectrometers (SWS and LWS) aboard the Infrared Space Observatory (ISO).

We calculated a self-consistent model of the gas mass loss rate, using the dust to drive the gas assuming that the gas and dust components are fully coupled, and the gas is heated by the dust and cooled by molecular rotational transitions of H_2O and CO, as well as by adiabatic cooling (Justtanont et al 1996). We derive a gas mass loss rate of 5×10^{-4} $M_\odot$ yr^{-1} for the superwind. In order to explain the observed BIMA data, we need to include a second component, a tenuous AGB wind of 10^{-6} $M_\odot$ yr^{-1} (Fig. 1).

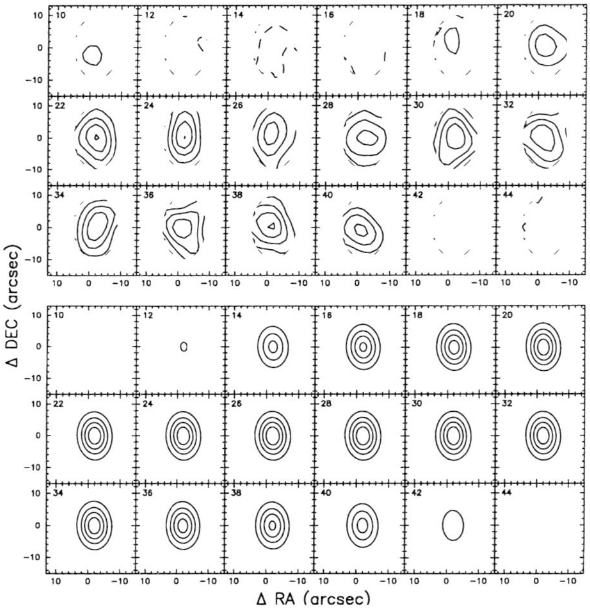

Figure 1. The observed channel maps with the emission outside 2× the beam area masked to separate out the intervening interstellar clouds (top panels) and the model CO maps (bottom panels). Our model shows that CO emission is expected to range from 12 to 42 km s^{-1} while the observations reveal missing flux between 12-20 km s^{-1} due to interstellar absorption.

References

Bower, P.F., & Johnson, K.J. 1990, ApJ, 354, 676

Justtanont, K., Skinner, C.J., Tielens, A.G.G.M., Meixner, M., & Baas, F. 1996, ApJ, 456, 337

A STUDY OF OH AND H_2O MASERS IN THE CIRCUMSTELLAR ENVELOPES AROUND MIRAS

Koji Murakawa
Subaru Telescope, HI USA
murakawa@naoj.org

Yates, J. A.
University College London, UK

Richards, A. M. S., Cohen, R. J.
Jodrell Bank observatory, Manchester, UK

van Langevelde, H. J.
Joint institute for VLBI in Europe, the Netherlands

Keywords: OH and H_2O masers, EVN/global VLBI and MERLIN, Miras

Introduction

We study the mass loss phenomenon which is though to be generally seen in the evolution of late type stars. We are using radio interferometry European VLBI Network (EVN) and Multi-Element Radio Lined Network (MERLIN) to study maser emission from gas molecules of H_2O and OH. This enable us to obtain 10-30 mas resolution images of mass-loss in AGB stars.

1. U Ori

U Orionis is one of the brightest long-period variables belonging to the Mira Ceti type with luminosity period of 368 days. The mass loss rate have been estimated $2.9\times10^{-7}M_{\odot}yr^{-1}$. A recent distance estimate for U Ori is $\sim$300 pc (Knapp et al. 1998).

Y. Nakada et al. (eds.), Mass-Losing Pulsating Stars and their Circumstellar Matter, 237–238.

The left frame of Fig.1 shows a composite image of OH and H_2O maser emissions of U Ori. A ring-like structure in the H_2O maser emission line is seen in this epoch. This structure could be in result of latitude dependent mass losing. However this cannot seen in previous work (Bains et al. 1994). Two spots appear to the North and the South in the OH maser emission line. This could be due to that its driven source of the star light can reach through in direction perpendicular to the equatorial plane.

2. VX Sgr

VX Sagittari associates with the Sgr. OB1 cluster and is thought to be a red supergiant with mass of $\sim 10 M_\odot$ at 1.7kpc or an AGB star at 400pc. Its mass loss rate is $\sim 1.5 \times 10^{-5} M_\odot yr^{-1}$. The luminosity period is $\sim$732 days.

The velocity structure of H_2O maser emission line ranges between -18 and 29 kms^{-1}. Typical radial expansion velocity of $\sim$20 kms^{-1}. The mass loss wind makes ellipsoidal envelope structure with 150 (NS)$\times$300 (WE) mas^2. Distribution of H_2O maser clouds suggests that they are part of a radially latitude dependent mass loss outflow.

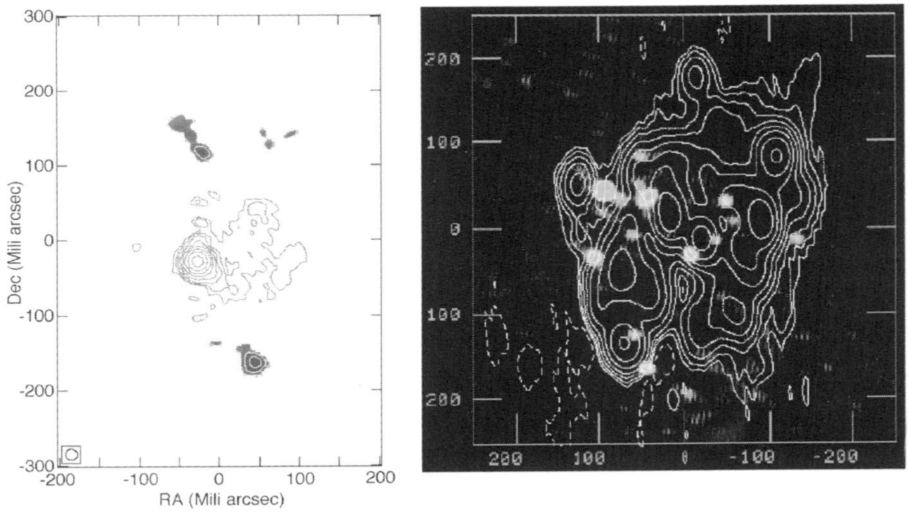

Figure 1. OH and H_2O masers maps of U Ori (left) and VX Sgr (right). In the left frame, the contour at the center is for H_2O maser and gray scale with contour is for OH maser. In the right frame, the contour is for H_2O maser and gray scale is for OH maser.

VARIATIONS OF PHYSICAL PARAMETERS IN CIRCUMSTELLAR ENVELOPES AROUND CARBON-RICH MIRA VARIABLES

Hyun-il Sung[1,2], Sang-gak Lee[2]
[1] Korea Astronomy Observatory, Taejeon, KOREA
[2] Astronomy Program, SEES, Seoul National University, Seoul, KOREA
hisung@kao.re.kr

Abstract We have studied variations of physical parameters in the circumstellar envelope of carbon-rich Mira variables at different pulsation phases. After comparing the ISO/SWS spectra of carbon-rich Miras with spectral energy distributions(SEDs) calculated in models, we could find the best fitted parameters for the physical state of the dust envelope at each phase.

Keywords: Circumstellar matter, Mira variables, pulsation

We studied physical parameters of the dust forming region around Carbon-rich Mira variables at various pulsating phases. These Miras have long period with light variations of 4 magnitude on the visual band and large mass loss rates which were known from the radio observations.

Observations were carried out with the Infrared Space Observatory (ISO) Short Wavelength Spectrometer (SWS; de Graauw et al. 1996) for three Carbon-rich Mira Variables, T Dra, V CrB, and V Cyg, at 6$\sim$8 phases. SWS covers the wavelength range between 2.4 to 45μm with a resolution of R=300$\sim$1000. To reduce the SWS spectra, we used the OSIA and ISAP software.

In order to derive the properties of the dust grains and central star, the SWS spectra were fitted with radiative-transfer models. The radiative-transfer calculations are performed with the code DUSTY developed by Ivezic et al.(1997), which is publicly available. This code is based on exact solutions of a self-consistent equation for the radiative energy

Y. Nakada et al. (eds.), Mass-Losing Pulsating Stars and their Circumstellar Matter, 239–240.

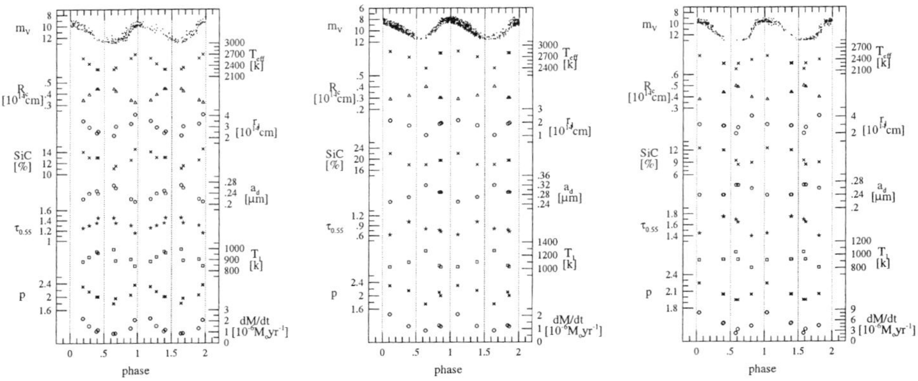

Figure 1. Physical parameters as a function of pulsational phase for Carbon-rich Mira Variables, T Dra, V CrB, and V Cyg.

density, including dust scattering, absorption, and emission in spherical shell geometry.

Figure 1 Shows optical light curve obtained from the AAVSO and phase dependence of physical parameters determined from model fitting. Each point is plotted twice to show clearly the variability.

The results found from the model fitting of the IR SEDs are summarized below.

1. The variations of T_{eff}, r_1, SiC/AMC ratio and $\dot{M}$ are found to have nearly the same trend that of light curve. The other parameters, R_c, a_d, $\tau_{0.55}$, and T_1 are inversely varing to the light curve.

2. The maximum to minimum ratios of T_{eff}, SiC/AMC, a_d, $\tau_{0.55}$, and T_1 are more or less around 1.3. In the case of R_c and r_1, these ratios are about 1.5 and 2, respectively.

3. The variation of grain size would followed in the change of two rates, evaporation rate and growth rate as the variation of T_{eff}.

4. The slope of SED at the long wavelength region ($\lambda > 15\mu$) was dependent the index p of power-law density distribution.

References

de Graauw, Th., Haser, L. N., Beintema, D. A. et al., 1996, A&A, 315, L49

Ivezic, Z., Nenkova, M., Elitzur, M., 1997, User Manual for DUSTY, available on the Internet : http://www.pa.uky.edu/~moshe/dusty

RADIATIVE TRANSFER MODELING OF PERIODIC MASS-LOSS ENHANCEMENT DURING THE AGB PHASE

Angela Speck, Margaret Meixner
Univ. of Illinois at Urbana-Champaign
1002 W. Green St., Urbana, IL 61801, USA.
akspeck@astro.uiuc.edu

Maia Nenkova, Moshe Elitzur
University of Kentucky

Keywords: Thermal pulses; mass loss; Egg nebula; dust

1. Introduction

Thermally pulsing asymptotic giant branch (AGB) stars suffer mass loss which leads to the formation of a circumstellar shell of gas and dust. At the end of the AGB phase, mass-loss stops and the circumstellar shell begins to drift away from the star. Assuming the velocity of the AGB wind has been relatively constant, the history of mass loss during the AGB phase is imprinted on the dust shell of the post-AGB envelope. By studying the distribution of matter in these circumstellar shells we can gain a better understanding of the mass-loss processes involved in the evolution of these stars. Using far-infrared (IR) ISOPHOT images of the circumstellar shells of the Egg nebula and AFGL 618, we showed that dust shells can be imaged out to a radius of 2-3 pc, and that these dust shells show evidence for episodic enhancements in the mass loss with a period of $\sim 10^4$ years. Here, we present radiative transfer (RT) models of the dust shell around the Egg nebula using the 1-d RT code DUSTY which has been modified to include heating by the ISRF.

2. The models

The observed surface brightness distributions at 120 and 180μm show two well-defined features (bumps) on a slope that gradually flattens out. We have used various density profiles together with heating from the ISRF to try to match the intensity distributions. Our only variables are

Y. Nakada et al. (eds.), Mass-Losing Pulsating Stars and their Circumstellar Matter, 241–242.

the dust density distribution and the normalization factor of the ISRF. The external heating from the ISRF is *always* necessary.

In order to get bumps in the dust shell we have included spikes of increased dust density in the density profiles. The bumps must be separated by sufficient space so that they are resolved. The best fits so far are shown in Fig. 1, which includes the input density profiles.

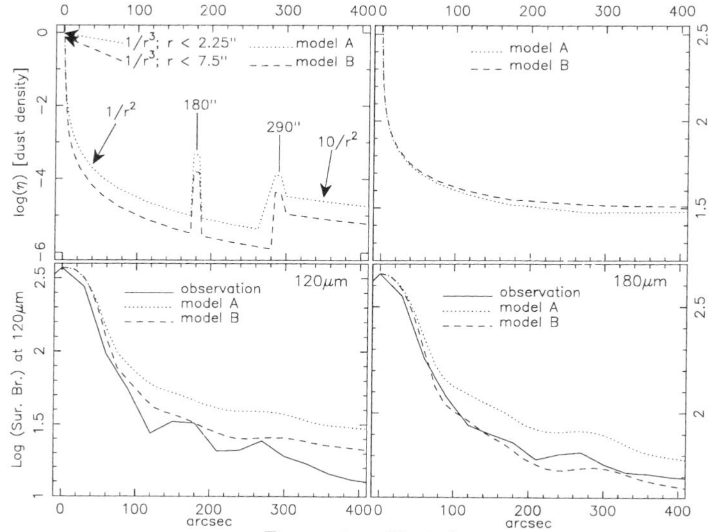

Figure 1. Best fit.

3. Interpretation

The modeled density distribution implies that the mass-loss rate was higher earlier in the progenitor AGB star's evolution and that its mass-loss rate decreased $\sim 7.5 \times 10^5$ years ago, possibly after a thermal pulse. However, the increased density in the outer shell may be the result of the outflowing matter piling up as it reaches the interface with the ISM. The need for a $1/r^3$ fall-off in the innermost part of the shell is indicative of the superwind region where the density reflects the dramatically increasing mass-loss rate at the end of the AGB phase. The spikes in the density distribution correlate with the timescales for thermal pulses during the AGB phase.

References

Gillett, F.C., et al. 1986, ApJ, 310, 842

Mauron, N., & Huggins, P.J. 1999, A&A, 349, 203

Nenkova, M., Ivezic, Z., & Elitzur, M. 2000, in ASP Conf. Ser. Vol. 196, p77

Olofsson, H., et al. 2000, A&A, 353, 583

Speck, A.K., Meixner, M., & Knapp, G.R. 2000, ApJL, 545, L145

Young, K., Phillips, T.G., & Knapp, G.R. 1993, ApJ, 409, 725

SPATIALLY RESOLVED 3 μm SPECTROSCOPY OF IRAS 22272+5435

Miwa Goto, W. Gaessler, Tomio Kanzawa, Naoto Kobayashi, Hideki Takami, Naruhisa Takato, Hiroshi Terada,
Subaru Telescope, 650 North A'ohoku Place, Hilo, HI 96720, USA
mgoto@duke.ifa.hawaii.edu

Yutaka Hayano, M. Iye, Yukiko Kamata, D. J. Saint-Jacques, Y. Minowa
National Astronomical Observatory of Japan
Mitaka, Tokyo, 181-8588, Japan

Abstract

We present spatially resolved 3 μm spectroscopy of hydrocarbon dust in a proto-planetary nebula IRAS 22272+5435. The spectroscopy with adaptive optics system at the 8.2 m Subaru Telescope revealed intriguing spatial distribution of hydrocarbon molecules and dust. The intensity of aromatic hydrocarbon feature at 3.3 μm relative to the aliphatic feature at 3.4 μm increases with the distance from the central star at $0''.35$–$0''.82$ (560-1300 AU). The spectral variation is well reproduced by a sequence of spectra of a laboratory analog of hydrocarbon dust subjected to thermal annealing. It indicates that a similar thermal process takes place at the formation of hydrocarbon dust in IRAS 22272+5435.

Keywords: circumstellar matter, hydrocarbon dust, IRAS 22272+5435

1. Observation

IRAS 22272+5435 is a extremely carbon rich proto-planetary nebula with peculiar infrared spectral features (Kwok, Volk, & Hrivnak 1989; Geballe et al.1992). The simultaneous presence of 3.3 μm aromatic and 3.4 μm aliphatic features in IRAS 22272+5435 produces a good laboratory for the study of transition between the two types of hydrocarbon bonds.

The spectroscopic observation was made on UT 2001 July 13 using the Infrared Camera and Spectrograph (IRCS) (Tokunaga et al.1998;

Y. Nakada et al. (eds.), Mass-Losing Pulsating Stars and their Circumstellar Matter, 243–244.

Kobayashi et al.2000) with the 8.2 m Subaru Telescope in conjunction with the AO system (Takami et al.1998; Gaessler et al.2002). A medium resolution grism was used with a $0''.30$ slit in the 58 mas camera section to provide spectra from 2.84 to 4.18 μm with resolving power of 600–800. The visible central star was used as the reference source for the AO system. The position angle of the slit was $56°$ along the elongation in the mid-infrared image that corresponds to a dusty torus (Ueta et al.2001).

2. Result

The absorption bands at 3.0 μm in the central star spectra are identified with the combination of P and R branches of C_2H_2 and HCN. The molecular absorption is conspicuous close to the central star, and abruptly disappear at $0''.35$ or 560 AU. The broad emission bands at 3.3 and 3.4 μm become strong in turn at the distance. This is the first observation of the hydrocarbon gas and dust clearly separated in location. It indicates the hydrocarbon molecules are indeed the source of the solid material, and that the leftover gas of the grain formation is being observed near the central star.

The relative intensity of the aliphatic feature at 3.4 μm to aromatic feature at 3.3 μm decreases with the distance from the central star. We suggest the stochastic heating of very small grains ($a < 6$Å) or the grain surface composition at the end of grain growth would be responsible for the changing aliphatic and aromatic spectral features. The time dependent mass loss should be relevant to account for the spatial variation.

Acknowledgments

We thank all the staff and crew of the Subaru Telescope and NAOJ for their valuable assistance obtaining these data and continuous support for IRCS and Subaru AO construction. Special thank goes to A. T. Tokunaga for many inspiring discussions with him. M. Goto is supported by a Japan Society for the Promotion of Science fellowship.

References

Kwok, S., Volk, K. M., & Hrivnak, B. J. 1989, ApJ, 345, 51

Geballe, T. R., Tielens, A. G. G. M., Kwok, S., & Hrivnak, B. J. 1992, ApJ, 398, L89

Tokunaga, A. T., et al. 1998, Proc. SPIE, 3354, 512

Kobayashi, N., et al. 2000, Proc. SPIE, 4008, 1056

Takami, H., et al. 1998, Proc. SPIE, 3353, 500

Gaessler, W., et al. 2002, Proc. SPIE, 4494, in press.

Ueta T., Meixner, M., Hinz, P. M., Hoffmann, W. F., Brandner, W., Dayal, A., Deutsch, L. K., Fazio, G. G., & Hora, J. L. 2001, ApJ, 557, 831

FORMATION AND EVOLUTION OF DUST CLUMPS AROUND COOL STARS

Gilles Niccolini
Observatoire de la Côte d'Azur, Dept. Fresnel, UMR 6528
BP 4229, F-06034 Nice Cedex 4, France
nicolin@obs-nice.fr

Peter Woitke
Woitke@physik.TU-Berlin.DE

Bruno Lopez
lopez@obs-nice.fr

Abstract In order to model the possibly cloudy circumstellar envelopes of AGB and post-AGB stars we developed a code describing the coupling between radiative transfer and time-dependent heterogeneous growth and evaporation of carbon grains.

Introduction

High angular observations of the carbon star IRC+10216 (Weigelt et al. 1998, Hannif & Buscher 1998, Tuthill et al. 2000) revealed the presence of inhomogeneous structures in the inner regions of its circumstellar dust envelope. This might be interpreted by the presence of dust clumps. In order to study the temporal evolution and the formation of such clumps we have developed a code describing the coupling between radiative transfer and dust grain processing. We present some first test calculations illustrating the feasibility of such a study.

Transfer/dust processing code

A first code handles the radiative transfer through the dust envelope using a Monte Carlo algorithm for 2D axisymmetric geometries

Y. Nakada et al. (eds.), Mass-Losing Pulsating Stars and their Circumstellar Matter, 245–246.

(Niccolini et al. 2002). The second code describes the time-dependent heterogeneous growth and evaporation of carbon grains (Gail & Sedlmayer 1988). The temperatures of the dust grains are computed by the transfer code and provided to the dust code which computes new dust opacities at the next time step. The temporal evolution of the clump is followed by repeating this sequence.

Test Calculation

In Fig. 1 a dust clump is initially (upper plot) located at $2\,R_*$ from a central carbon star ($T_{\rm eff} = 2650\,{\rm K}$, $L_* = 10^4\,L_\odot$) in which 10 % of the available carbon atoms are condensed. The clump starts evaporating after $\sim$ 80 days (lower plot), while a new dust shell is forming on both sides of it.

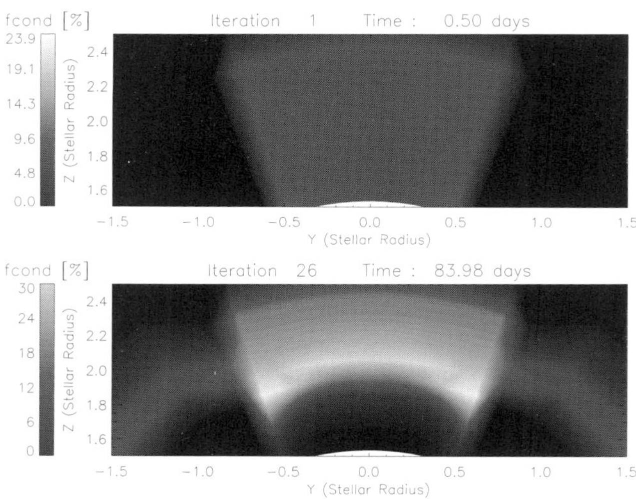

Figure 1. Evolution of a dust clump. The degree of condensation $f_{\rm cond}$ (percentage of C atom not blocked in CO and condensed into dust grain) is shown for two selected time steps.

References

Gail, H.-P., & Sedlmayr, E. 1988, A&A, 206, 153
Haniff, C.A., & Buscher, D.F. 1998, A&A, 334, L5
Niccolini, G., Woitke, P., & Lopez, B. 2002, submitted to A&A
Tuthill, P.G., Monnier, J.D., Danchi, W.C., & Lopez, B. 2000, ApJ, 543, 284
Weigelt, G., Balega, Y., Blöcker, T., Fleischer, A.J., Osterbart, R., & Winters, J.M. 1998, A&A, 333, L51

THE POLARIMETRIC EVIDENCE OF THE CIRCUMSTELLAR DUST IN SYMBIOTIC STAR Z ANDROMEDAE

M. Isogai[1], Y. Ikeda[2], K.S. Kawabata[3], M. Seki[1]
[1] Astronomical Institute, Tohoku University, Japan
[2] Genesia Corporation, Japan
[3] SUBARU Telescope Project Office, National Astronomical Observatory, Japan
iso@astr.tohoku.ac.jp

Keywords: symbiotic star, binary, circumstellar dust, polarimetry

1. Introduction

Symbiotic stars are interacting binary systems consisting of a red giant and a hot star which are embedded in circumstellar matter. Though it is well known that many symbiotic stars show intrinsic polarization (e.g. Schulte-Ladbeck et al. 1990), the origin is not specified partly because their observations have been based on multi-color polarimetry whose wavelength resolution is too low for separating the intrinsic polarization from the observed one. Therefore we have started to carry out spectropolarimetry for some symbiotic stars.

Z And is a prototypical symbiotic star and is one of our target stars. Fortunately, we could carry out spectropolarimetry of Z And both in the quiescent phase and in the outburst phase which started in September 2000 (Mattei et al. 2000).

2. Observations and Results

We observed Z And on nine nights between December 1998 and January 2001, using a low-resolution spectropolarimeter, HBS (Kawabata et al. 1999). Polarizations observed in both phases clearly show the time variation, which evidently indicates the presence of intrinsic polarization. Their variability in polarization, however, is quite different in both phases.

Y. Nakada et al. (eds.), Mass-Losing Pulsating Stars and their Circumstellar Matter, 247–248.

2.1 Quiescent phase

Since the time variation in quiescent phase seems to be correlated with the orbital motion, we fitted observed polarization to a simple scattering model and extracted the intrinsic polarization in quiescent phase. The simple model is that a point source (a hot star) moves around a spherical scatterer (a cool star) along the circular orbit at an inclination i. We assumed that the scattering phase function is the same as in Rayleigh scattering. Fig.1 shows the derived intrinsic polarization in quiescent phase. The wavelength dependence of $P(\lambda) \propto \lambda^{-1.88\pm0.34}$ indicates that the polarization is due to Mie scattering by circumstellar dust around the cool star.

2.2 Outburst phase

The intrinsic polarizations in outburst phase are derived using the same interstellar component in quiescent phase (Fig.2). The wavelength dependences of $P(\lambda) \propto \lambda^{-2.70\pm0.20}$ on 10/15 and $\lambda^{-3.51\pm0.16}$ on 10/18 indicate the presence of the circumstellar dust which may be formed in the outburst phenomenon. On the other hand, the polarizations from November 23 are different in their wavelength dependence and position angle from those on October. These may suggest that the circumstellar dust has already been evaporated or diffused out in a month or so.

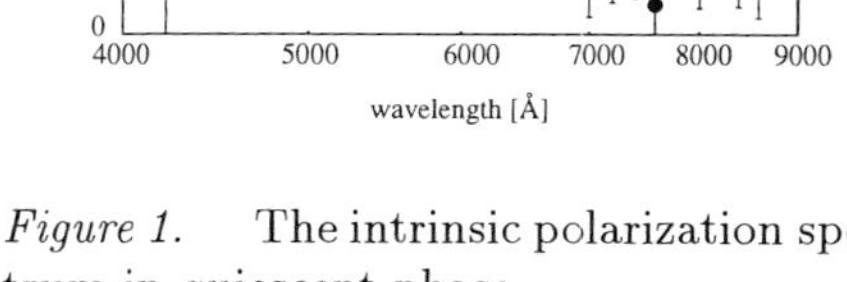

Figure 1. The intrinsic polarization spectrum in quiescent phase.

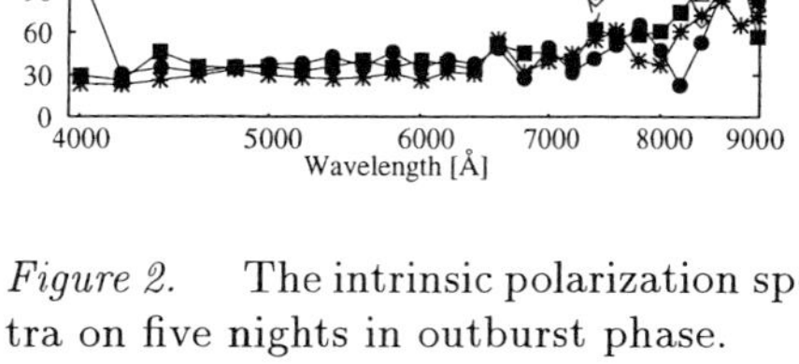

Figure 2. The intrinsic polarization spectra on five nights in outburst phase.

References

Kawabata, K.S., Okazaki, A., Akitaya, H., et al. 1999, PASP, 111, 898

Mattei, J.A. 2000, IAU Circ., No. 7489

Schulte-Ladbeck, R.E., Aspin, C., Magalhaes, A.M., et al. 1990, A&AS, 86, 227

BISPECTRUM SPECKLE INTERFEROMETRY OF THE CARBON STAR IRC+10216

G. Weigelt, T. Blöcker, K.-H. Hofmann, A. Men'shchikov, J.M. Winters
Max-Planck-Institut für Radioastronomie, D-53121 Bonn, Germany

Y. Balega
Special Astrophysical Observatory, Nizhnij Arkhyz, 35147, Russia

The prototypical carbon Mira IRC +10 216 is an Asymptotic Giant Branch (AGB) star suffering from strong mass loss (several 10^{-5} $M_{\odot}$/yr; Loup et al. 1993) which has led to an almost complete obscuration of the star by dust. Due to the high mass-loss rate, long period of $P = 649$ d and carbon-rich chemistry of the dust-shell, IRC +10 216 is obviously in a very advanced stage of its AGB evolution. High-resolution near-infrared imaging of IRC +10 216 has revealed that on sub-arcsecond scales (100 mas) its dust shell is clumpy, bipolar, and changing on a time scale of only ~1 yr (Weigelt et al. 1997, 1998, Haniff & Buscher 1998, Osterbart et al. 2000, Tuthill et al. 2000). Since most dust shells around AGB stars are known to be spherically symmetric, whereas most proto-planetary nebulae (PPN) show an axisymmetric geometry (Olofsson 1996) it appears likely that IRC +10 216 has already entered the transition phase to the PPN stage.

New bispectrum speckle-interferometry observations of IRC +10 216 were carried out with the Russian 6 m SAO telescope in the J, H, and K band in Sep. 1999, Oct. 2000, and March 2001 continuing the monitoring of Osterbart et al. (2000) which covered five epochs between 1995 and 1998. Figure 1 shows the reconstructed K-band images of the innermost region of IRC +10 216 in 1996, 1998 and 2001. The resolution varies between 87 and 73 mas. The dust shell consists of several compact components, at the beginning within a radius of 200 mas, which steadily change in shape and brightness. For instance, the apparent separation of the two initially brightest components A and B increased from 201 mas in 1996 to 351 mas in 2001. At the same time, component B is fading and has almost disappeared in 2000 whereas the initially faint components

Y. Nakada et al. (eds.), Mass-Losing Pulsating Stars and their Circumstellar Matter, 249–250.

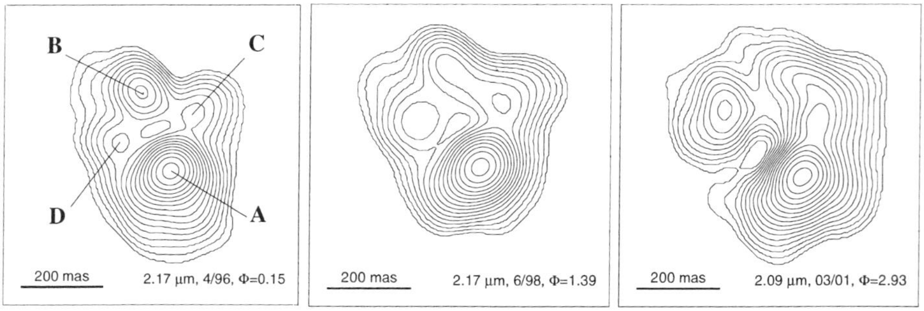

Figure 1. K-band speckle reconstructions of IRC +10 216 in 1996 (left), 1998 (middle), and 2001 (right). The resolution is 82 mas, 87 mas, and 73 mas, resp. Contour levels are plotted from 0.1 mag to 3.1 mag relative to the peak intensity in steps of 0.2 mag. North is up and east is to the left.

C and D have become brighter. In 2001, the intensity level of component C has increased to almost 40% of the peak intensity of component A. Both components appear to have started merging in 2000. These changes of the dust-shell appearance can be related to changes of the optical depths caused, e.g., by mass-loss variations. The present monitoring, covering more than 3 pulsation periods, shows that the structural variations are not related to the stellar pulsation cycle in a simple way. This is consistent with the predictions of hydrodynamical models that enhanced dust formation takes place on a timescale of several pulsation periods (Fleischer et al. 1995). Our recent 2-D radiative transfer modeling (Men'shchikov et al. 2001) suggests that the bright and compact component A is not the direct light from the central star but the southern lobe of a bipolar structure dominated by scattered light. Instead, the carbon star is at the position of the fainter northern component B.

References

Fleischer, A.J., Gauger, A., & Sedlmayr, E. 1995, A&A, 297, 543
Haniff, C.A., & Buscher, D.F. 1998, A&A, 334, L5
Loup, C., Forveille, T., Omont, A., & Paul, J.F. 1993, A&AS, 99, 291
Men'shchikov, A.B., Balega, Y., Blöcker, T., et al. 2001, A&A, 368, 497
Olofsson, H. 1996, Ap&SS, 245, 169
Osterbart, R., Balega, Y., Blöcker, T. et al. 2000, A&A, 357, 169
Tuthill, P., Monnier, J., Danchi, W., & Lopez, B., 2000, ApJ, 543, 284
Weigelt, G., Balega, Y., Hofmann, K.-H., et al. 1997, in Science with the VLT Interferometer, (Berlin: Springer), 206
Weigelt, G., Balega, Y., Blöcker, T., et al. 1998, A&A, 333, L51

GROUND BASED MIR OBSERVATIONS OF C-RICH PROTO PLANETARY NEBULAE

Miyata, Takashi
Kiso Observatory, Institute of Astronomy, University of Tokyo, Japan
miyata@kiso.ioa.s.u-tokyo.ac.jp

Kataza, H.
Institute of Space and Astronautical Science, Japan

Okamoto, Y. K.
Kitasato Univ., Japan

Sako, S., Honda, M., Onaka, T., Okada, Y.
School of Science, University of Tokyo, Japan

Yamashita, T., Murakawa, K.
Subaru Telescope, National Astronomical Observatory of Japan

Introduction

IRAS22272+5435 and IRAS07134+1005 are known as the 21 micron feature proto planetary nebulae (PPNe). Their spectra show a strong feature at 21 micron and some emission features in the near- and the mid-infrared regions. These dust features are unique to C-rich PPNe, and the carrier materials of them have not been well understood so far. For investigating the carrier materials and their formation process mid infrared observations with high spatial resolution are quite useful since the spatial distribution of the dust emissions record history of the dust formation.

Y. Nakada et al. (eds.), Mass-Losing Pulsating Stars and their Circumstellar Matter, 251–252.

Observations

We carried out N-band imaging (8.7, 10.3, and 12.4μm), Q-band imaging (18.5, 20.8, and 24.5μm), and N-band low-resolution spectroscopy of IRAS22272+5435 and IRAS07134+1005 with SUBARU/COMICS on June 2001 and January 2002. Diffraction limited images and spectra were obtained both in the N-band (~ 0.3") and the Q-band (~ 0.6").

Results

1. N-band images of IRAS22272+5435 and IRAS07134+1005 consist of an unresolved core, a circular halo, and arc like structures. The core is quite compact (r < 0.16"), and it may be caused by a strong mass ejection in the proto-planetary phase.

2. Spatially resolved N-band spectra of IRAS22272+5435 (Figure-1) show two prominent emission features peaked at 8 μm and 11 μm. The spatial distribution of these features differs from each other. The 8 μm feature is mainly concentrated on the core and the inner edge of the south-east arc, while the 11 μm feature is widely distributed in the core, the arcs, and the halo.

3. The "21 micron feature only" map has been reconstructed from the Q-band images of IRAS07134+1005 (Figure-2). The 21 micron feature is relatively prominent on the arcs. This suggests that its carrier material may be effectively formed in the super wind phase at the end of the AGB evolution.

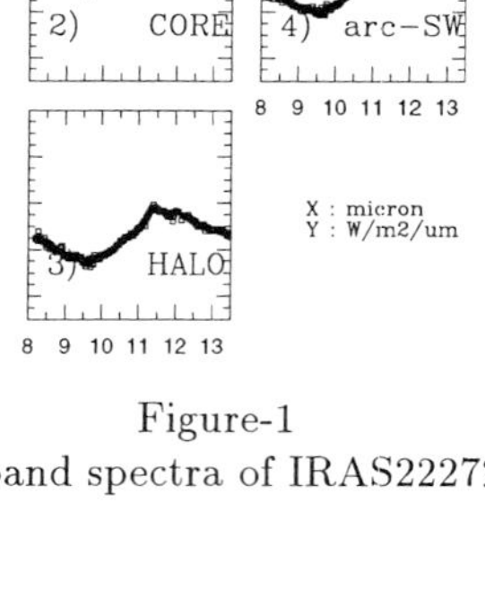

Figure-1
N-band spectra of IRAS22272

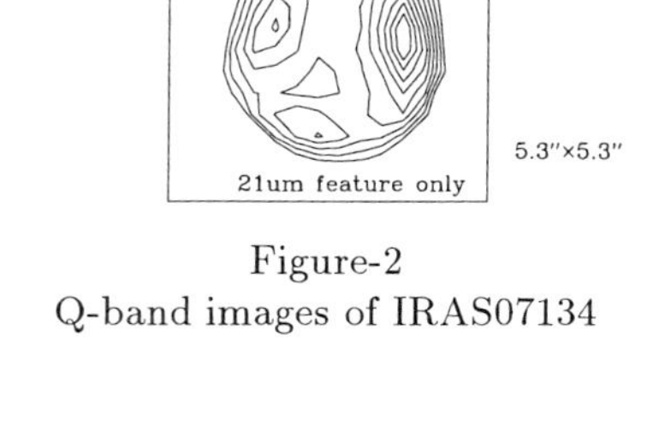

Figure-2
Q-band images of IRAS07134

MID-INFRARED POLARIMETRY OF AXI-SYMMETRIC MASS LOSS AND THE SHAPING OF NGC 7027

Colby A. Jurgenson, Robert E. Stencel
University of Denver Observatories
cjurgens@du.edu

Abstract We obtained the first ever mid-IR polarimetry maps of the axi-symmetric planetary nebula NGC 7027. The observations were taken with the University of Denver's TNTCAM2 at the Wyoming Infrared Observatory's (WIRO) 2.3 m aperture telescope, providing 0.78 arcsecond resolution. Polarization maps in the mid-IR can help constrain the properties and alignment of dust grains in the circumstellar environment.

Keywords: Polarimetry, infrared, NGC7027

1. Introduction

The role of mass loss in stellar evolution is a topic of considerable interest due to its effect upon the structure of the circumstellar envelope as well as the star's lifetime on the AGB. Much work has been done to better our understanding through spectroscopy and imaging, but very little has been done in the mid-IR using polarimetry as a technique in studying mass loss. Johnson & Jones (1991 AJ) studied polarization characteristics in the near-IR from a sample of evolved objects from red giants to planetary nebulae. They found polarization increases with age from the AGB to the post AGB phase and decreases with increasing planetary nebula age. They also note that aspherical mass loss is a continual feature in the late stages of stellar evolution and that the axi-symmetric morphology we observe in planetary nebulae most likely begins as early as the AGB. Thermal emission in the IR from evolved stars is dominated by a warm dust continuum with grains that scatter radiation into preferred directions. Imaging polarimetry can provide a glimpse into grain orientations and alignment. It is in this effort that we produced mid-IR polarimetry maps in oneμm-wide bandpasses of NGC 7027 between 9-13 μm.

Y. Nakada et al. (eds.), Mass-Losing Pulsating Stars and their Circumstellar Matter, 253–254.

2. Observations & Discussion

The University of Denver's TNTCAM2 is a mid-IR imaging polarimeter in which all the optical components (filters, lenses, retarder and polarizer) are at liquid helium temperatures (Klebe et al.1998 SPIE). The observations were obtained at WIR0 2001June6. The composite image, summed over the four bandpasses, closely resembles the NICMOS image by Latter et al.2000 ApJ. The polarization is measured from rotation of the waveplate through four positions separated by 45 degees. A stationary wire grid functions as the polarizer. From the four intensity images Stoke's parameters are calculated to determine magnitude and direction of polarization at each pixel. Each polarization vector plotted has a signal to noise ratio of 40 or higher. The mid-IR is dominated by dust emission, which in the case of NGC 7027, obscures the source of radiation. From spectral information obtained by ISO-SWS (Bernard-Salas et al.2001A&A, Hony et al.2001A&A): 10.6μm is continuum dominated plus weak [SIV] 10.5μm emission (5%); 11.6μm shows continuum plus PAH emission at 11.2μm ($\sim$25%); 12.6μm shows continuum plus PAH emission at 12.8μm (5%). The 10.6 and 12.6μm maps exhibit maximum polarization along a NW-SE axis, while the 11.6μm map with the stronger PAH contribution shows polarization oriented along the NE-SW axis with reductions 30 degrees from the NW-SE axis, corresponding to bipolar outflows noted by several authors. Details are being prepared for publication elsewhere and at website: www.du.edu/~cjurgens.

3. Conclusion

NGC 7027 is a $\sim$600 year old planetary nebula that exhibits axisymmetric structure in the form of an ellipsoidal photo-dissociation region (PDR)/ionized shell surrounded by a bi-conical H_2 envelope. In the near/mid-IR there are disruptions running along the NW-SE direction and the NE-SW direction. These disruptions are believed to have been once active jets. A Chandra x-ray image shows similar bi-lobed structure with a prominent jet running NW-SE. Since the central source is not the dominant source in the mid-IR, the polarization cannot be interpreted as simple, single scattering of the central stellar emission. There are then two polarization components, polarization associated with dust-scattered ambient mid-IR emission (from star plus surrounding grains) and polarization associated with thermal emission of dust grains. The structured nature of the polarization maps leads us to conclude that the grains are aligned. Whether the alignment mechanism is due to a magnetic field or stellar wind is being tested with new models (Carciofi, private communication). Reference details available on request.

HIGH RESOLUTION IMAGING AND POLARIMETRY OF THE RED RECTANGLE

Koji Murakawa
Subaru Telescope, HI, USA
murakawa@naoj.org

T. Miyata
Kiso Observatory, Institute of Astronomy, University of Tokyo, Japan
miyata@kiso.ioa.s.u-tokyo.ac.jp

M. Tamura
National Astronomical Observatory of Japan
hide@subaru.naoj.org

H. Suto
Subaru Telescope, HI, USA
suto@naoj.org

Keywords: C-rich PPN, near-infrared high resolution imaging, polarimetry

Introduction

The red rectangle (AFGL 915) is a C-rich proto-planetary nebula (PPN) with a x-shape structure at the center. The red rectangle has a A0 type post-AGB star HD44179 which is thought to be a carrier source of UIR emission (Cohen 1975 and Miyata et al. 2000). It is thought to be that there is a companion of M type star as a carrier of flux (Rowan-Robinson & Harris 1983).

1. Observation

The HKL' band direct imaging has been carried out using a near-infrared high resolution camera (Coronagraphic Imager with Adaptive

Y. Nakada et al. (eds.), Mass-Losing Pulsating Stars and their Circumstellar Matter, 255–256.

Optics; CIAO) mounted on the 8m Subaru telescope. The Subaru 36 element adaptive optics was used to get nearly diffraction image (FWHM of 0.07-0.1"). The *H* band imaging-polarimetry has been carried out using a wire grid polarizer on CIAO.

2. Discussion

The position of the local minimum in Fig.2-b is consistent with that of L' band image. This suggests that this peak could be due to HD 44179. Polarization map in Fig.2-a shows the center of illumination is consistent with the center of 2 lobes. (a) HD 44179 is not the dominant source to illuminate the nebulae. (b) there is unseen source at the gap between the Northern and Southern lobes which illuminates the nebulae. This illumination source could be M type star.

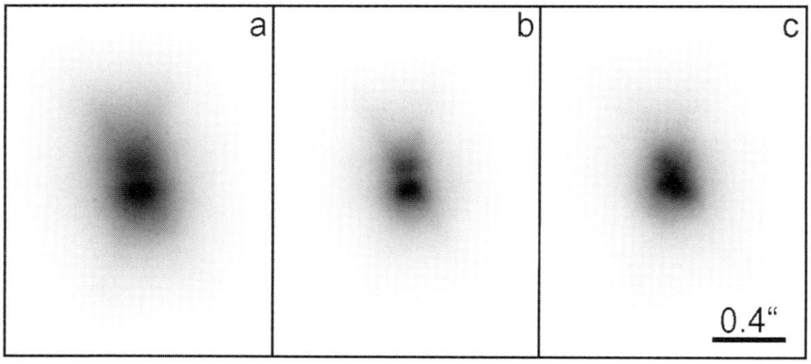

Figure 1. HKL' band images.

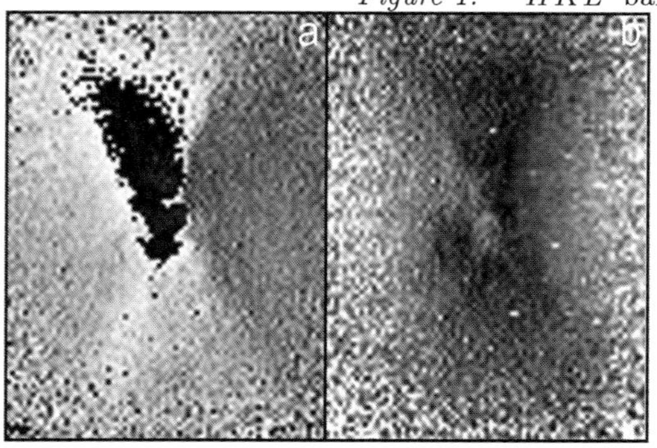

Figure 2. The *H* band imaging-polarimetry. (a) and (b) shows the position angle and degree of polarization, respectively. In (a), position angle is expressed in gray scale. Black and white are negative and positive position angle, respectivel.

TIME VARIATION OF MID-INFRARED SPECTRA OF M-TYPE MIRA VARIABLES*

Takashi Onaka
Department of Astronomy, University of Tokyo, Bunkyo-ku, Tokyo 113-0033, Japan
onaka@astron.s.u-tokyo.ac.jp

Teije de Jong
SRON-Utrecht, 3584 CA Utrecht, The Netherlands
Astronomical Institute, University of Amsterdam, 1098 SJ Amsterdam, The Netherlands

Issei Yamamura
Institute of Space and Astronautical Science, Sagamihara, Kanagawa 229-8510, Japan

Abstract Time variation of mid-infrared (MIR) spectra of M-type Mira variables has been investigated by ISO/SWS observations of Z Cyg and T Cep. The variation in the Z Cyg spectrum can be accounted for by the variation in the stellar luminosity, if appropriate dust emissivity parameters are adopted. No apparent variation in the spectral shape despite of the intensity variation suggests that emission from high-temperature grains dominates in the MIR spectrum of T Cep.

Keywords: Circumstellar dust, Mira variables, Infrared spectroscopy

M-type Mira variables, Z Cyg and T Cep, were observed with the Short Wavelength Spectrometer (SWS) on board the Infrared Space Observatory (ISO) 7 times over more than one visual variability cycle in approximately equal intervals. The variation in the near-infrared region has been investigated by Matsuura et al. (2002). In this paper we report the variation in the mid-infrared (MIR) dust features in both stars.

*Based on observations with *ISO*, an ESA project with instruments funded by ESA Member States (especially the PI countries: France, Germany, the Netherlands and the United Kingdom) and with the participation of ISAS and NASA.

Y. Nakada et al. (eds.), Mass-Losing Pulsating Stars and their Circumstellar Matter, 257–258.

Z Cyg shows the prominent 10 and 20 μm silicate emission and a large variation in its SWS spectrum, which is well synchronized with the visual light variation. Details of the observational results and analysis of Z Cyg are given in Onaka et al. (2002). Based on the model analysis we have found: (i) All the observed spectra at 7 different epochs can be fitted with a single silicate dust emissivity fairly well apart from minor emission features, (ii) the observed variations (both in the spectral shape and intensity) can be interpreted in terms of the stellar luminosity variation, *if the proper dust emissivity is adopted.* Conversely, the circumstellar silicate dust emissivity can be derived from the variability observations of the MIR spectra, (iii) the derived dust emissivity has a relatively large 20 μm to 10 μm band ratio (0.54 ± 0.08) and suggests the inner dust shell temperature of 700 ± 100 K, (iv) the derived emissivity accounts for the circumstellar dust emission of some AGB stars, but not all of them, and (v) there are varieties in the 20 μm emissivity of circumstellar dust, possibly related to the presence of another dust component.

T Cep shows unusual silicate features with fine structures in 10 μm, indicating that the usual silicate is not dominant dust in T Cep. The time variation of the T Cep spectrum is not completely synchronized with the visual variability phase though the amplitude is small. There are variations also in the molecular emission features and they are reported in Cami et al. (1999) for CO_2 of 13–17 μm and in Yamamura et al. (1999) for SO_2 of 7.3 μm. There seems no apparent correlated variation with the phase in the MIR color $[f(12\,\mu)/f(25\,\mu\text{m})]$ despite of the variation in its intensity. Little variation in the color suggests that the MIR emission of T Cep is dominated by relatively high-temperature dust grains. In addition the 20 μm feature peak is located at a wavelength longer than the Z Cyg dust and a dust feature complex is seen in 28–32 μm. The feature at 32 μm has been observed in other evolved stars (cf. Molster et al. 2002). The variation in the 32 μm of T Cep seems to not correlate with other part of the spectrum variation. We have investigated possible carriers of this feature, such as diopside or aluminum oxide. All of them should have strong features at other wavelengths, which are not seen in the T Cep spectrum, and thus the carrier material remains unidentified.

References

Cami, J., Yamamura, I., de Jong, T., et al. 1999, ESA SP-427, 281
Matsuura, M., Yamamura, I., Cami, J., et al. 2002, A&A, 389, 972
Molster, F., Waters, L. B. F. M., & Tielens, A. G. G. M. 2002, A&A, 382, 222
Onaka, T., de Jong, T., & Yamamura, I. 2002, A&A, 388, 573
Yamamura, I., de Jong, T., Onaka, T., et al. 1999, A&A, 341, L9

THREE MICRON SPECTRA OF CARBON STARS IN THE LARGE MAGELLANIC CLOUD

M. Matsuura, A.A. Zijlstra, A.J. Markwick
Department of Physics, UMIST, P.O. Box 88, Manchester M60 1QD, UK
(M.M.) m.matsuura@umist.ac.uk

J.Th. van Loon
School of Chemistry and Physics, Keele University, Staffordshire ST5 5BG, UK

I. Yamamura
The Institute of Space and Astronautical Science, Kanagawa 229-8510, Japan

Paul M. Woods
European Southern Observatory, Alonso de Cordova 3107, Casilla 19001, Santiago 19, Chile & UMIST, UK

Abstract Infrared spectra of six carbon-rich stars were obtained with the Very Large Telescope (VLT). We found a metallicity dependence in the equivalent width of the molecular bands, and we discuss the causes of this.

Keywords: stars: AGB – stars: atmospheres – Galaxy: Magellanic Clouds

Introduction

Stellar evolution models predict that third dredge-up enhances the number of ^{12}C atoms in the atmosphere of the AGB stars. The relative ^{12}C enhancement increases at low metallicity (e.g. Vassiliadis and Wood 1993). This suggests that at lower metallicity, the molecular abundances may also be different. To investigate the metallicity dependence of the atmospheric content of AGB stars, we obtained infrared spectra of carbon-rich stars in the Large Magellanic Cloud (LMC), the metallicity of which is about half of that of the solar neighbourhood.

Y. Nakada et al. (eds.), Mass-Losing Pulsating Stars and their Circumstellar Matter, 259–260.

Results and discussion

Three micron spectra of six carbon stars in the LMC were obtained with ISAAC at the VLT on 12th–14th of December, 2001. Example spectra are shown in Fig. 1. Several molecular bands are detected; the 3.1 μm HCN and C_2H_2 band, the 3.5 μm HCN band, and a 3.8 μm band, which is probably due to C_2H_2.

The equivalent width of the 3.1 μm HCN and C_2H_2 band is compared for LMC and Galactic carbon stars (Fig. 2). The additional data are taken from van Loon et al. (1999) and the ISO data archive. The LMC carbon stars show a larger equivalent width. This tendency is found in the 3.5 μm HCN band and in the 3.8 μm band as well. At low metallicity, these molecules are expected to be low abundance because less carbon and nitrogen atoms in the atmosphere. Our results are opposite way.

This metallicity dependence can be caused by a systematic difference in the C/O ratio in the atmosphere. Chemical models assuming local thermodynamic equilibrium predict that C_2H_2 and HCN are more abundant in gas with a higher C/O ratio. Since the amount of ^{12}C enhancement at the third dredge-up increases with lower metallicity, the LMC AGB stars may have higher C/O ratio.

The authors are grateful to Drs. P.A. Whitelock, M.W. Feast, L.B.F.M. Waters, and the ESO support astronomers for their help in this project.

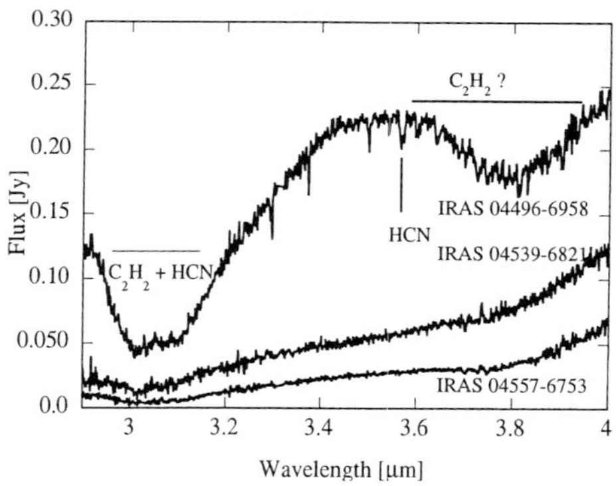

Figure 1. An example of the spectra of LMC carbon-rich stars

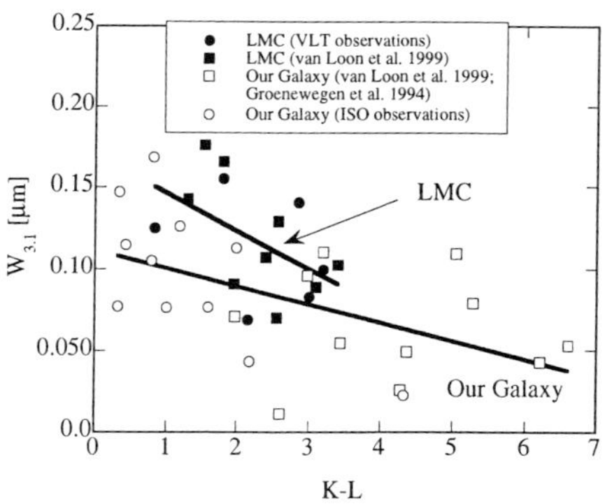

Figure 2. The equivalent width of the 3.1μm HCN and C_2H_2 band for LMC and Galactic carbon stars

References

van Loon, J.Th., Zijlstra, A.A., & Groenewegen, M.A.T. 1999, A&A, 346, 805

Vassiliadis, E., & Wood, P.R., 1993, A&A, 413, 641

IV

RADIO AND MM OBSERVATIONS OF CIRCUMSTELLAR MATTER

THE STRUCTURE OF WINDS IN AGB STARS

Moshe Elitzur
Department of Physics & Astronomy
University of Kentucky, Lexington, KY 40506
moshe@uky.edu

Željko Ivezić
Department of Astrophysical Sciences
Princeton University, Princeton, NJ 08544
ivezic@astro.princeton.edu

Dejan Vinković
Department of Physics & Astronomy
University of Kentucky, Lexington, KY 40506
dejan@pa.uky.edu

Abstract Most dusty winds are described by a set of similarity functions of a single independent variable that can be chosen as τ_V, the overall optical depth at visual. The self-similarity implies general scaling relations among the system parameters, in agreement with observations. Dust drift through the gas has a major impact on the structure of most winds.

Keywords: stars: AGB and post-AGB, mass-loss, winds, outflows, dust, hydrodynamics, radiative transfer

1. Introduction

The complete description of a dusty wind should start with a full dynamic atmosphere model and incorporate the processes that initiate the outflow and set the value of $\dot{M}$. These processes are yet to be identified with certainty, the most promising are stellar pulsation (e.g. Bowen 1989) and radiation pressure on the water molecules (e.g. Elitzur, Brown & Johnson 1989). Proper description of these processes should be followed by grain formation and growth, and subsequent wind dynamics. Two ambitious program attempting to incorporate as many aspects

Y. Nakada et al. (eds.), Mass-Losing Pulsating Stars and their Circumstellar Matter, 265–273.

of this formidable task as possible have been conducted over the past few years by groups at Berlin (see J.M. Martin, these proceedings) and Vienna (see Dorfi et al 2001). While much has been accomplished, the complexity of this undertaking necessitates simplifications such as a pulsating boundary. In spite of continuous progress, detailed understanding of atmospheric dynamics and grain formation is still far from complete.

Fortunately, the full problem splits naturally to two parts, as recognized long ago by Goldreich & Scoville (1976). Once radiation pressure on the dust exceeds all other forces, the rapid acceleration to supersonic velocities decouples the outflow from the earlier phases—*the supersonic phase would be exactly the same in two different outflows if they have the same mass-loss rate and grain properties even if the grains were produced by entirely different processes.* Furthermore, these stages are controlled by processes that are much less dependent on detailed microphysics, and are reasonably well understood. And since most observations probe only the supersonic phase, models devoted exclusively to this stage should reproduce the observable results while avoiding the pitfalls and uncertainties of dust formation and the wind initiation.

2. Dusty Winds: Problem Setup and Solution

A steady-state dusty wind is controlled by three forces. First, radiation pressure drives the outflow. When $\tau_V > 1$, the radiation is continually reddened and its coupling to the dust is degraded; this weakening of the force can be described by a radial profile $\phi(r)$. Radiation pressure acts mostly on the dust grains while the wind mass is mostly in the gas particles, and collisional coupling between these two components introduces the second force, drag. Gilman (1972) has shown that the dust-gas relative velocity reaches steady state within a distance much shorter than all other scales. The drag force can then be eliminated, because the dust fully mediates to the gas the radiation pressure force. But the drift still has an important effect, since the radiative force per unit mass is proportional to n_d/n, and separate mass conservation for the dust and gas implies $n_d/n \propto v/v_d$.[1] Finally, the gravitational pull of the central star opposes the expansion and must be overcome in any outflow.

[1] Even with prompt dust formation and no further grain growth or destruction, the dust abundance varies in the shell because of the difference between the dust and gas velocities.

Accounting for the three forces, the equation of motion in terms of dimensionless velocity w and radial distance y is (Elitzur & Ivezić 2001)

$$\frac{dw^2}{dy} = \frac{P^2}{y^2}\left(\phi(y)\zeta(y) - \frac{1}{\Gamma}\right). \tag{1}$$

The constants P and Γ are the ratios of radiation pressure to the drag and gravity, respectively, $\zeta = v/v_d$ is the drift profile and ϕ is the reddening profile, obtained from the separate, but coupled, equation of radiative transfer. The outflow problem is fully described by this simple mathematical equation with only three free parameters—the constants P and Γ and the initial value $w(y = 1)$. There are no other free parameters. The wind structure will be exactly the same in systems that have different mass (M), luminosity (L) or mass loss rate ($\dot{M}$) when these properties combine to produce the same P and Γ.

The complete solution of equation 1 is presented in Elitzur & Ivezić (2001), including the analytic solution for optically thin winds ($\tau_V < 1$). Here we present observational implications.

2.1 Physical Domain

The relation $\dot{M}v_\infty \leq L/c$ has often been used as a physical bound on radiatively driven winds, even though the mistake in this application when $\tau_V > 1$ has been pointed out repeatedly (e.g. Ivezić & Elitzur 1995). Instead, the proper form of momentum conservation is $\dot{M}v_\infty = \tau_F L/c$, where τ_F is the flux-averaged optical depth, and it yields

$$\dot{M}v_\infty = \frac{L}{c}\left(Q_*/Q_V\right)\tau_V\left(1 + \tau_V\right)^{-0.36}. \tag{2}$$

Here Q_V is the dust efficiency coefficient at visual and Q_* is its Planck-average with the stellar temperature. Typical values of Q_*/Q_V are 0.1 for silicate and 0.25 for carbon dust. Since the ratio $\dot{M}v_\infty/(L/c)$ can exceed unity and increase without bound, momentum conservation does not constrain the wind. Instead, the constraints come from force considerations—the outward force must exceed gravity (positive derivative in eq. 1), leading to the two phase space boundaries shown in figure 1. The liftoff bound, shown in the left panel, sets a lower limit on $\dot{M}$, proportional to M^2/L; below this minimal $\dot{M}$ the grains are ejected without dragging the gas because the density is too low for efficient dust–gas coupling. As a condition on the wind origin, this bound is the most uncertain part of the solution. The result $\dot{M}_{min} \propto M^2/L$ is reasonably secure (a similar relation was noted by Habing et al 1994), but the proportionality constant involves a fudge factor. This factor

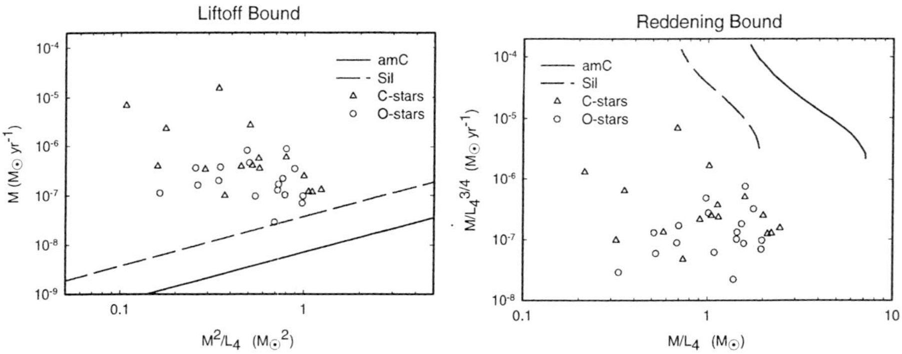

Figure 1. Winds powered by radiation pressure on dust should fall above the liftoff bounds shown in the left panel and below the reddening bounds in the right one. L_4 is the stellar luminosity in 10^4 $L_\odot$ (Ivezić & Elitzur 2002).

can only be determined from a more complete formulation that handles properly grain growth. The bound shown in the right panel reflects the weakening of radiative coupling in optically thick winds because of the radiation reddening, setting an upper limit on $\dot{M}$.

2.2 Scaling

Most stars are located well inside the allowed region of phase space, as is evident from figure 1. The dependence on Γ and $w(1)$ disappears rapidly with distance from the boundaries, therefore most dusty winds are described by a set of similarity functions of just a single independent variable P, which is equivalent to τ_V. The self-similarity implies general scaling relations. Systems with the same combination $\dot{M}/L^{3/4}$ necessarily have also the same $\dot{M}v_\infty/L$ because both are uniquely related to τ_V.

The velocity profile for all winds can be summarized with the simple expression

$$v = v_\infty \left(1 - \frac{r_a}{r}\right)^k, \qquad \text{where} \begin{cases} k = \frac{2}{3} & \text{when } \tau_V \lesssim 1 \\ k \simeq 0.4 & \text{when } \tau_V \gtrsim 1 \end{cases} \tag{3}$$

and where $r_a = r_1[1 - (v_1/v_\infty)^{1/k}]$, with v_1 the wind velocity at its starting radius r_1. The profile shape of v/v_∞ is nearly the same for all winds, a weak dependence on τ_V enters only in the power k. The small-τ_V value $k = \frac{2}{3}$ reflects the effect of the drift, which dominates the dynamics in that regime, the switch to a more moderate profile at large τ_V is caused by reddening.

The dimensionless solution involves only properties of individual grains. The dust abundance is irrelevant and affects only the velocity scale v_∞.

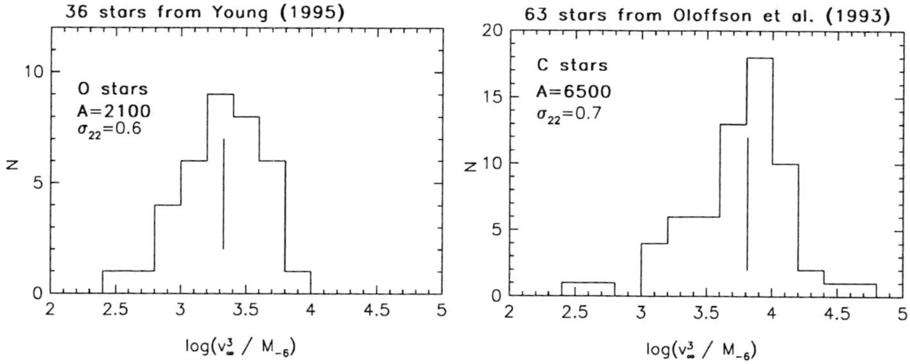

Figure 2. Histograms of $v_\infty^3/\dot{M}$ for oxygen- and carbon-rich stars with $\tau_V < 1$ outflows. In each data set, both v_∞ (in km s^{-1}) and $\dot{M}$ (in 10^{-6} $M_\odot$ yr^{-1}) were determined from CO observations. A vertical line marks the mean of each distribution. The corresponding coefficient A (see eq. 4) is listed together with the value of σ_{22} it implies (Ivezić & Elitzur 2002).

When only the radiation pressure force is taken into account, v_∞ increases with L and is independent of $\dot{M}$ (Goldrecih & Scoville 1976). Drift effects change this behavior fundamentally, producing instead

$$v_\infty^3 = A\,\dot{M}\,(1+\tau_V)^{-1.5}, \tag{4}$$

where the proportionality coefficient A contains properties of the grain material and $n_d\sigma_d/n$. Therefore, outflows with $\tau_V < 1$ should obey $v_\infty^3 \propto \dot{M}$ if they have the same dust properties. Remarkably, *even though the wind is driven by radiation pressure, its velocity is independent of luminosity.* The dependence on L enters only in optically thick winds because $\tau_V = A^{1/3}(Q_V/Q_*)\dot{M}^{4/3}/L$.

The behavior implied by eq. 4 had been noticed by Young (1995) from 36 nearby Mira variables with low mass-loss rates. Young's data is presented in figure 2 as a histogram of $v_\infty^3/\dot{M}$ together with a similar histogram for C-rich stars with small optical depths (data from Olofsson et al 1993). Each histogram shows a pronounced peak—as implied by eq. 4 when the dust properties do not vary among members of the set. The value of $n_d\sigma_d/n = 10^{-22}\sigma_{22}$ cm^{-2} determined for each set from its average A and properties of silicate and carbon grains is listed in the figure. Remarkably, the two samples of different types of stars produce the same σ_{22}. In spite of the large differences in atmospheric and grain properties between O- and C-rich stars, the fraction of material channelled into dust is such that σ_{22} turns out to be the same in both.

Drift and reddening play a major role in the dynamics. As $\dot{M}$ decreases, the dust and gas decouple and the velocity decreases too. As $\dot{M}$

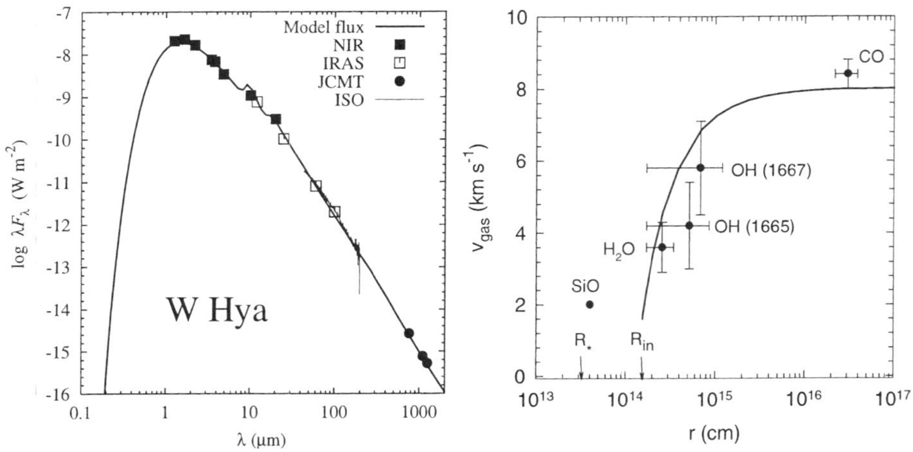

Figure 3. Modeling the wind in W Hya (Zubko & Elitzur 2000) with outflow calculations by the code DUSTY (Ivezić et al 1999). Only one free parameter was varied, τ_V, determined to be 0.83 from the SED fit in the left panel. Right: Model prediction for the gas velocity profile. Data points show masers and CO thermal emission from the wind. The SiO maser emission originates in the extended atmosphere and is not part of the outflow.

increases and the wind becomes optically thick, reddening degrades the efficiency of the radiation pressure force and the velocity again decreases. In between, the velocity reaches maximum $v_{\max} \simeq 20 L_4^{1/4}$ km s^{-1}, obtained at $\dot{M}(v_{\max}) \simeq 2\times10^{-6} L_4^{3/4}\ M_\odot\ \mathrm{yr}^{-1}$. The proportionality coefficients are slightly different for carbon and silicate dust.

2.3 Individual Sources

Since τ_V controls both the dynamics and the radiative transfer, in any given source the spectral energy distribution (SED) can be fitted with only one free parameter and the corresponding solution should provide a fit also for the radial velocity profile. Figure 3 shows the SED fit for W Hya. A byproduct of the solution that produced this SED is the shape of the dimensionless velocity profile. There is no freedom to modify this profile, and all the data points (from observations of CO thermal emission and various masers) indeed follow it when scaled to dimensionless values. Figure 3 shows the results with actual physical variables, setting the dimensional scales for r and v from the Hipparcos distance of 115 pc and the terminal velocity of 8 km s^{-1} from CO emission.

Significantly, neither L, $\dot{M}$ or dust-to-gas ratio were input parameters of the model. Instead, these quantities are derived afterwards when the SED fitting results are supplemented by the distance and velocity scales.

The bolometric flux is a direct product of the fitting process and together with the distance to the source determines the overall luminosity. And once L, v and τ_V are known, $\dot{M}$ and the dust abundance are calculated from general self-similarity relations.

3. Water Thermal Emission

Water is an abundant component of the outflow in oxygen-rich stars and the dominant coolant at radial distances up to $\sim 10^{15}$ cm. But because of atmospheric absorption, the discovery of its thermal emission had to wait for ISO. With the wind properties in W Hya determined from SED analysis, Zubko & Elitzur (2000) successfully fitted all 21 water lines observed by ISO using H_2O abundance as the only free parameter. Harwitt & Bergin (2002) subsequently found the same line ratios in VY CMa. This highly luminous supergiant has $L = 5\times10^5\ L_\odot$ and $\dot{M} \simeq 2\times10^{-4}\ M_\odot\ \mathrm{yr}^{-1}$ while in W Hya the corresponding values are only $1.1\times10^4\ L_\odot$ and $2.3\times10^{-6}\ M_\odot\ \mathrm{yr}^{-1}$. In spite of these large differences, scaling the model results for W Hya by an overall factor of 10 produces adequate fit for the H_2O lines in VY Cma too.

Thanks to the multitude of observed transitions, in H_2O line analysis the abundance is derived rather than entered as input. This is a significant advantage over the common determination of $\dot{M}$ from CO observations, which involves only one transition and must assume an abundance beforehand.

4. Deviations from Spherical Symmetry

Many AGB shells display spherical symmetry, but there are numerous exceptions. A peculiar case is the source CIT3, one of the most extreme infrared AGB objects. Observations by Hofmann et al (2001) show a 1.24μm image elongated along a symmetry axis (fig. 4, top left panel). The elongation disappears at slightly longer wavelengths, and the image becomes almost perfectly circular at 2.12μm. This peculiar behavior has been successfully explained by Vinković et al (2002) with the aid of a newly developed 2D radiative transfer code, LELUYA (see www.leluya.org). In the model geometry, the wind spherical symmetry is broken by two bi-polar cones and the observing viewpoint is inclined to the axis. The density distributions in the conical regions and the bulk of the outflow are deduced directly from the observed brightness of the J-band image. The model results, shown in the lower panels of fig. 4, are hardly distinguishable from the observations.

The wavelength-variation of the images reflects interplay between scattering and emission. Scattered light dominates at $\lambda \lesssim 1.5\ \mu$m and

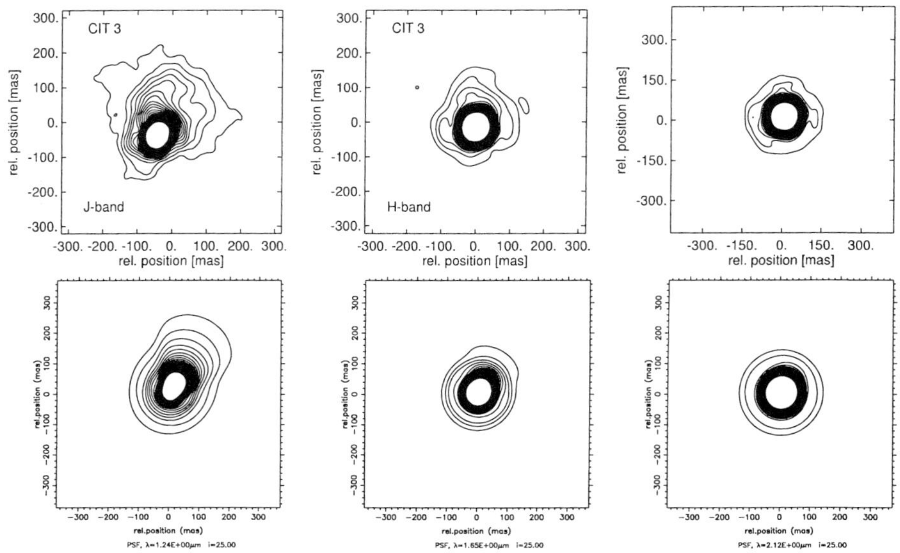

Figure 4. Top Row: Contour plots of the observed CIT3 images at 1.24 μm (J-band), 1.65 μm (H-band), and 2.12 μm (K-band) from left to right (Hofmann et al 2001). Bottom: Model results of Vinković et al, 2002.

the images trace the large-scale contours of the density distribution. This explains the elongated shape of the J-image. At longer wavelengths emission takes over and the images are affected by the shape of the dust isotherms, becoming circular due to the central heating of the dust.

5. Conclusions

The "standard model" seems to be working satisfactorily. Detailed fits are successfully obtained with a single free parameter, τ_V, as predicted. Dust drift is a major ingredient of the dynamics and the reason for the peculiar correlation discovered by Young (1995).

Two fundamental issues remain open and can be addressed only by more complete models that incorporate the wind origin (such as those of the Berlin and Vienna groups). The first is a more reliable determinations of the phase space lower boundary shown in fig. 1. What is the minimal $\dot{M}$? The second involves the dust abundance, which sets the scale of terminal velocities in AGB outflows to ~ 10 km s^{-1}; were that abundance 100 times higher, typical velocities would be ~ 100 km s^{-1} instead. What is so special about $n_d\sigma_d/n \sim 10^{-22}$ cm^{-2} and why is it the same in both C- and O-rich stars? It is entirely possible that the answers to these two open issues are related.

References

Bowen, G.H. 1989, in *Evolution of peculiar red giant stars*, ed. H.R. Johnson & B. Zuckerman (Cambridge University Press: Cambridge), p. 269

Dorfi, E.A., Höfner, S., & Feuchtinger, M.U. 2001, in *Stellar pulsation - nonlinear studies*, ed. M. Takeuti & D.D. Sasselov (Dordrecht: Kluwer) p. 137

Elitzur, M., Brown, J.A. & Johnson, H.R. 1989, ApJ, 341, L95

Elitzur, M., & Ivezić, Z. 2001, MNRAS, 327, 403

Gilman, R.C. 1972, ApJ, 178, 423

Goldreich, P., & Scoville, N. 1976, ApJ, 205, 144 (GS)

Habing, H.J., Tignon, J. & Tielens, A.G.G.M. 1994, A&A 286, 523 (HTT)

Harwit, M. & Bergin, E. A. 2002, ApJ, 565, L105

Hofmann, K.-H., et al. 2001, A&A, 379, 529

Ivezić Ž., & Elitzur M. 1995, ApJ, 445, 415

Ivezić Ž., & Elitzur M. 2002, in preparation

Ivezić, Ž., Nenkova, M., & Elitzur, M. 1999, User Manual for DUSTY, University of Kentucky Internal Report, accessible at http://www.pa.uky.edu/~moshe/dusty

Olofsson, H., Eriksson, K., Gustafsson, B., & Carlstrom, U. 1993, ApJS, 87, 267

Vinković, D., Bloecker, T., Elitzur, M., Hofmann K.-H., & Weigelt, G. 2002, in preparations

Young, K. 1995, ApJ, 445, 872

Zubko, V., & Elitzur, M. 2000, ApJ, 544, L137

MOLECULAR DATA ON CIRCUM-STELLAR ENVELOPES AROUND AGB STARS

Valentín Bujarrabal
Observatorio Astronómico Nacional
Ap. 1143, E-28800 Alcalá de Henares, Spain
bujarrabal@oan.es

Abstract Observations of molecular line emission in circumstellar envelopes around AGB and post-AGB stars are the best tool to measure a number of circumstellar gas parameters: total mass, present or past mass-loss rate, expansion velocity, shape, chemical composition, etc. In this review, I discuss basic results and relevant recent work on this topic. Both observational data and theoretical analysis are reviewed. The reliable determination of some parameters and recent work on SiO masers are in particular discussed.

Keywords: AGB and post-AGB stars, circumstellar gas, molecules, mass-loss rates

1. Introduction

Observations of molecular lines, particularly low-J rotational transitions, are probably the best tool to study circumstellar envelopes around AGB stars (CSEs).

The first reason is that the low excitation requirements of such transitions are well adapted to the low kinetic temperatures present in CSEs, typically $\sim 1000 - 10$ K. At such temperatures one expects the gas to be molecular (except in the outer layers, where photodissociation by the interstellar UV field is important) and that the lowest energy levels of molecules, usually at a few K from the ground, are easily populated. The second reason is that such lines can be efficiently observed using ground based radiotelescopes, able to produce data with high spectral and spatial resolution, as we will see in this contribution.

We have mentioned the role of photodissociation in the chemistry of CSEs, which in particular defines the radius of the molecular envelope. A molecule exposed to the interstellar UV is photodissociated in typically 10^{10} seconds; in actual cases, this time is somewhat larger due to

Y. Nakada et al. (eds.), Mass-Losing Pulsating Stars and their Circumstellar Matter, 275–282.

shielding by dust grains or selfshielding. Since the expansion velocity in CSEs is of the order of 10 km s^{-1}, the molecular radius must be $\gtrsim 10^{16}$ cm. In some important cases, like CO in relatively thick envelopes, one gets molecular radii of about 10^{17} cm. Photodissociation is not only important in order to estimate the molecular radius. The dissociation of stable molecules produce a high number of radicals that, before being dissociated themselves, give raise to an active (photoinduced) chemistry.

2. CO

Rotational lines from ^{12}CO are particularly intense, mainly because of the high abundance of this molecule. The J=1–0 and the J=2–1 transitions (λ = 2.6 mm and 1.3 mm) are the most commonly observed ones. The analysis of such lines is particularly easy because of the favorable conditions from both the points of view of the chemistry and excitation.

CO is a very stable molecule, and its abundance is expected to be practically the maximum allowed by the C and O nuclei abundance, see e.g. Lafont et al. (1982). (For that reason, in C-rich envelopes, in which carbon is more abundant than oxygen, very few O atoms remain out of CO to form other O-bearing molecules, that are then rare while many C-bearing molecules appear; the opposite holds in O-rich CSEs.) Only photodissociation, in the outer layers, acts on the CO abundance.

The treatment of excitation is particularly easy for CO. First, its rotational levels are placed in a single ladder, defined by a rotational constant equal to 2.8 K. Second, the CO rotational transitions are easily thermalized, due to its particularly low permanent dipole moment (and Einstein coefficients).

Under these conditions one can easily deduce approximate expressions of the net photon production rate for the CO low-J transitions. In the optically thin case, we get:

$$n_{\rm ph}({\rm s}^{-1}{\rm cm}^{-3}) \sim A n_{\rm u} \overset{\propto}{\sim} n_{\rm co} e^{-E_{\rm u}/kT_{\rm rot}}/F_{\rm par} \overset{\propto}{\sim} n_{\rm CO} e^{-E_{\rm u}/kT_{\rm rot}}/T_{\rm rot} \quad .$$

Note that, for the single CO ladder, the partition function ($F_{\rm par}$) is approximately proportional to the rotational temperature ($T_{\rm rot}$, the typical excitation temperature for the rotational transitions). $n_{\rm u}$ is the upper level of the transition, with energy equal to $E_{\rm u}$, and $n_{\rm co}$ is the CO total abundance per cubic centimeter.

Assuming that the rotational transitions are thermalized to the kinetic temperature ($T_{\rm k}$), overexcitation ($E_{\rm u} \ll kT_{\rm rot}$), and that the CO density decreases with the square radius ($n_{\rm co} \propto 1/r^2$, which holds for constant mass loss rate and constant CO relative abundance), we deduce:

$$n_{\rm ph} \overset{\propto}{\sim} 1/r^2/T_{\rm k} \quad .$$

Let us write the emission rate per unit shell (for a simple law $T_{\rm k} \propto 1/r$):

$$n_{\rm ph,shell}({\rm s^{-1}cm^{-1}}) \overset{\propto}{\sim} 1/T_{\rm k} \overset{\propto}{\sim} r \quad .$$

CO lines are sometimes opaque, but opacity can be treated using ^{13}CO lines or an escape probability formalism:

$$n_{\rm ph} \sim A\beta n_{\rm u} \sim n_{\rm u}/\tau \overset{\propto}{\sim} \frac{1}{e^{-E_{\rm u}/kT_{\rm rot}} - 1}\frac{1}{r} \quad ,$$

$$n_{\rm ph} \overset{\propto}{\sim} T_{\rm k}/r, \quad n_{\rm ph,shell}({\rm s^{-1}cm^{-1}}) \sim \text{constant}$$

We have taken into account that, for high values of the optical depth τ, the escape probability, β, is $\propto 1/\tau$.

These formulae are just crude approximations, but they show that the calculation of the CO emissivity is simple and its dependence on the physical conditions is smooth, at least under the overexcitation regime (for underexcitation, the term $e^{-E_{\rm u}/kT_{\rm rot}}$ cannot be neglected and the dependence on $T_{\rm rot}$ becomes sharp). Therefore, actual calculations of the CO emissivity are expected to be robust and the determination of the envelope parameters from fitting of the observations must be reliable. (On the other hand, calculations of the CO emissivity including all the ingredients of the excitation reproduce the behavior predicted by the above formulae, for both optically thin and thick cases; in particular they show that the underexcitation regime would only appear in shells farther than the photodissociation radius, at least for the mm-wave lines.)

2.1 CO: results

Because of these properties of the CO line emission, its observations have provided reliable information on basic properties (mostly, global properties) of CSEs. The main result has been the measurement of the mass loss rate in more than 500 stars. The method is relatively simple, and results from different authors and models in general yield quite compatible values, within a factor 2–3. We can see $\dot{M}$ calculations in Knapp & Morris (1985), Bujarrabal et al. (1989), Margulis et al. (1990), Kastner (1992), Loup et al. (1993), Olofsson et al. (1993), Young (1995), Knapp et al. (1998), Groenewegen et al. (1998), Skinner et al. (1999), Schöier & Olofsson (2001), etc. Note that in estimations older than the photodissociation radii calculations by Mamon et al. (1988), the mass loss rates tend to be smaller since the new accurate radii are shorter than previously believed. Particularly interesting are the systematic $\dot{M}$ estimations by Loup et al., who used just simple formulae and got values that are not very different from those obtained from complex modeling (except perhaps for the lowest $\dot{M}$, for which, as mentioned by Loup et al., their formulae may be not very accurate).

Typically, $\dot{M}$ values between $\sim 10^{-8}$ and a few 10^{-5} $M_\odot$ yr^{-1} are found. The lower limit probably holds, at least, for nearby Miras, for which relatively complete surveys exist and yield just a few nondetections (Margulis et al. 1990, Olofsson et al. 1993, Schöier & Olofsson 2001). The upper limit probably has also a meaning for nearby objects, since it is difficult to believe that we have missed a strong emitter similar to IRC+10216 (a well known carbon star placed at 150–200 pc and with a mass loss rate $\sim 2\ 10^{-5}$ $M_\odot$ yr^{-1}).

However, these $\dot{M}$ values are systematically lower than those found from data on post-AGB objects (see Bujarrabal et al. 2001 and references therein). At least one half of protoplanetary nebulae (PPNe) were formed by a process of mass ejection with rates higher 10^{-4} $M_\odot$ yr^{-1}, apparently during the last AGB phases. However, these high mass-loss rates are not observed in AGB stars (indeed, the total mass in PPNe is often larger than 1 $M_\odot$, more than 10 times larger than the masses of AGB CSEs). The reason is not clear, perhaps it could due to the short duration of this superwind (a few thousand years) or to the high obscuration expected at that moment. The missing link between the AGB and post-AGB properties may be the very red OH/IR star OH26.5+0.6, for which Justtanont et al. (1996) deduced a strong increase of $\dot{M}$ (up to $\sim 5.5\ 10^{-4}$ $M_\odot$ yr^{-1}) in the last 200 yr; there is a poster on this topic in this symposium. If such a strong increase in $\dot{M}$ in the last AGB phases is confirmed to be systematic, we would have to conclude that the formation of a planetary nebula is in fact not the result of a slow, smooth process of mass ejection during the whole AGB, but mostly of a single, quasi-explosive event: the star would eject most of its mass during a few thousand years at the end of this phase.

We note that these works have also estimated the ^{12}CO abundance (relative to the total number of particles), typical values between 2 10^{-4} and 10^{-3} are given (lower values being found for PPNe and O-rich sources). The abundance of ^{13}CO is found to be $\sim$ 1–2 10^{-5} (see also Kahane et al. 1992, 2000).

The expansion velocity of CSEs has been measured from CO data, from profile modeling or just by inspection of the spectra, see the above references. In general, values between 5 and 30 km s^{-1} are found. (Again these values are significantly smaller than those found for the fast bipolar flows in PPNe, but they are thought to be a post-AGB phenomenon.)

Mapping of the CO emission has been used to study the shape of CSEs, at least at large scale, see Bujarrabal & Alcolea (1991), Stanek et al. (1995), Neri et al. (1998). From their comprehensive observations,

Neri et al. were able to show that at least 80% of the AGB envelopes show a significant spherical symmetry (difference between major and minor diameters smaller than $\sim$ 15%) and are isotropically expanding. This is not in contradiction with the existence of multiple shells, in particular of 'detached envelopes', which will not be further discussed in this review (see Knapp et al. 1998, Lindqvist et al. 1999, Olofsson et al. 2000). The spherical symmetry of AGB CSEs is again very different to the prominent and widespread bipolar geometry found in PPNe, which appears even in young PPNe (objects that apparently have recently left the AGB). Signs of significant axial symmetry have been found in the inner shells of a few AGB envelopes from IR or optical images, but existing systematic studies in molecular emission attained a resolution $\sim 5''$, not enough to study these features.

3. Other molecules

Many molecules (about 120) have been detected in CSEs. Together with the well know ones, like CO, SiO (thermal and maser emission), HCN, HNC, CN, CS, HC_3N, H_2O (thermal and maser emission), OH (thermal and maser emission), etc, we can mention new interesting detections, like C_6H_2, C_6H_6 (benzene), AlNC, SiCN, C_3, SiC_3, etc (see Cernicharo et al. 2000, 2001, Guélin et al. 2000, Ziurys et al. 2002, etc). Note that many of the recent detections have been done in the submm and FIR wavelength ranges and in PPNe or young PNe.

The first results one expects from the observation of these molecules are their relative abundances, the basic input to circumstellar chemistry models. Abundances are found to strongly vary from O-rich to C-rich objects. As expected, C-bearing molecules are only abundant in C-rich envelopes, while O-bearing molecules are abundant in O-rich sources. See observational results in Bujarrabal et al. (1994), Olofsson et al. (1998), Bieging et al. (2000), etc. Theoretical predictions can be found in Lafont et al. (1982), Millar & Herbst (1994), Willacy & Millar (1997) and Millar et al. (2001). We can see examples of average relative abundances and abundance ratios in Bujarrabal et al. (1994); molecules like HCN and SiO attain relative abundances close 10^{-5} in, respectively, C- and O-rich envelopes; but its abundance drops by more than a factor ten in O- and C-rich objects respectively. S-type stars present intermediate abundances. Other species are of course less abundant; for instance, some of the recent detections mentioned above have relative abundances smaller than 10^{-9}. These values are qualitatively explained by theoretical models; although, for instance, theory in general predicted still stronger differences between O-rich and C-rich gas.

Another important result obtained from these observation is the measurement of the molecular envelope extent. Usually, stable molecules are found to be abundant in the inner shells, though many radicals appear as a result of photodissociation or photoinduced chemistry. See observations by Dayal & Bieging (1995), Lucas & Guélin (1999), Lindqvist et al. (2000). Theory is usually compatible with observations, see calculations by Mamon et al. (1998), Millar & Herbst, Willacy & Millar (1997). As an example, HCN in IRC+10216 occupies an inner region surrounded by the CN distribution (Dayal & Bieging, Lucas & Guélin). The radius of the CN shell is found to be $\sim 4\ 10^{16}$ cm; in 1982, Lafont et al. predicted a radius for this shell equal to $3\ 10^{16}$ cm, quite in agreement with observations.

4. Maser emission: SiO

Maser emission by molecules has also been an important tool to study CSEs. In this review I will only summarize results on SiO maser emission, because of the very interesting recent works on it and also as introduction to several contributions to this symposium.

Maser emission appears in SiO rotational transitions, up to J=8–7, in vibrationally excited states, from v=1 to v=4 (see e.g. Pardo et al. 1998). The most intense and often observed ones are the v=1 and 2 J=1–0 (7mm) and the v=1 J=2–1 (3mm) lines. SiO maser lines have been detected in about 1000 stars, mainly O-rich objects, including bright AGB stars (Miras, semiregulars and irregulars), OH/IR stars, supergiants and a few PPNe. The large number of stars emitting in SiO masers has even allowed interesting galactic studies (e.g. Deguchi et al. 2000). Let us note, among the classical results, the good correlation found for large star samples between the SiO intensity and the stellar IR flux (particularly at 8 μm, the wavelength of the Δv=1 vibrational transition), see Bujarrabal et al. (1987), Jiang (2002). It is also remarkable that the maser flux found in large-amplitude pulsators, i.e. Miras and large-amplitude semiregulars and supergiants, is significantly larger than that observed in other red stars (see e.g. Alcolea et al. 1990).

VLBI observations have been particularly revealing, showing the structure of the inner CSE with spatial resolution $\sim$ 1 mas (which is equivalent to about 3–5 10^{12} cm for typical distances, i.e. about 1/10 of the stellar radius; see Diamond et al. 1994, Kemball & Diamond 1997, etc). In the observed stars (about 10 objects), the masers systematically appear in a ring around the star, at a few stellar radii. It has been also found that the maser lines predominantly show linear polarization in the tangential direction. The possible coincidence of the J=1–0 v=1 and v=2 spots has been carefully studied. Miyoshi et al. (1994) claimed that spots of both lines were almost always coincident, concluding that this demonstrated that the maser pumping is collisional (since radiative models tend to require slightly different conditions to pump both lines); but those pioneering observations just attained a resolution of 7 mas. Recent observations (Desmurs et al. 2000), with a resolution about ten times better, found however systematic shifts between both lines of about 1 mas. This result has been confirmed by several recent observations presented in this symposium.

SiO masers are known to vary. For regular variables, they show the same period as the star, but their maxima systematically appear at phase ~ 0.1 with respect to optical maxima (e.g Nyman et al. 1985), being therefore in phase with IR maxima. After 11 years of tight monitoring of the J=1–0 v=1 and 2 lines in 21 stars, Pardo et al. (2003, in preparation) conclude that the average phase lag for Miras is ~ 0.12; for two red objects in which IR light curves exist, the SiO vs. IR average shift was found to be ~ -0.01, just a few days.

Finally, I briefly mention that the pumping mechanism of SiO masers is not still well understood. The population inversion in a v–state is thought to be due to the presence of high optical depths in the radiative deexcitation v -> v-1 (Kwan & Scoville 1974); but the excitation mechanism, i.e. the energy source, is still debated. Radiative models (e.g. Bujarrabal et al. 1994) naturally reproduce most observations, but require a high velocity gradient or a shell geometry to be efficient. Collisional models (Lockett & Elitzur 1992, Humphreys et al. 2002) seem simpler from the theoretical point of view, but have problems to explain the tight observational relation between the IR and SiO emission in evolved stars.

References

Alcolea, J., Bujarrabal, V., & Gómez-González, J. 1990, A&A, 231, 431
Bieging, J.H., Shaked, S., & Gensheimer, P.D. 2000, ApJ, 543, 897
Bujarrabal, V., Planesas, P., & del Romero, A. 1987, A&A, 175, 164
Bujarrabal, V., Gómez-González, J., & Planesas, P. 1989, A&A, 219, 256
Bujarrabal, V., & Alcolea, J. 1991, A&A, 251, 536
Bujarrabal, V. 1994, A&A, 285, 953
Bujarrabal, V., Fuente, A., & Omont, A. 1994, A&A, 285, 247
Bujarrabal, V., Castro-Carrizo, A., Alcolea, J., & Sánchez Contreras, C. 2001, A&A, 377, 868
Cernicharo, J., Goicoechea, J.R., & Caux, E. 2000, ApJ, 534, L199
Cernicharo, J., Heras, A.M., Tielens, A.G.G.M., et al. 2001, ApJ, 546, L123
Dayal, A., & Bieging, J.H. 1995, ApJ, 439, 996
Deguchi, S., Fujii, T., Izumiura, H., et al. 2000, ApJS, 130, 351
Desmurs, J. F., Bujarrabal, V., Colomer, F., & Alcolea, J. 2000, A&A, 360, 189

Diamond, P.J., Kemball, A.J., Junor, W., et al. 1994, ApJ, 430, L61
Groenewegen, M., van der Veen, W., & Matthews, H. 1998, A&A, 338, 491
Guélin, M., Muller, S., & Cernicharo, J. 2000, A&A, 363, L9
Humphreys, E.M.L., Gray, M.D., Yates, J. A., et al. 2002, A&A, 386, 256
Jiang, B.W. 2002, ApJ, 566, L37
Justtanont, K., Skinner, C.J., Tielens, A.G.G.M., et al. 1996, ApJ 456, 337
Kahane, C., Cernicharo, J., Gómez-González, J., & Guélin, M. 1992, A&A, 256, 235
Kahane, C., Dufour, E., Busso, M., et al. 2000, A&A, 357, 669
Kastner, J.H. 1992, ApJ, 401, 337
Kemball, A.J., & Diamond, P.J. 1997, ApJ, 481, L111
Knapp, G.R., & Morris, M. 1985, ApJ, 292, 640
Knapp, G.R., Young, K., Lee, E., & Jorissen, A. 1998, ApJS, 117, 209
Kwan, J., & Scoville, N. 1974, ApJ, 194, L97
Lafont, S., Lucas, R., & Omont, A. 1982, A&A, 106, 201
Lindqvist, M., Olofsson, H., Lucas, R., et al. 1999, A&A, 351, L1
Lindqvist, M., Schöier, F.L., Lucas, R., & Olofsson H. 2000, A&A, 361, 1036
Lockett, P., & Elitzur, M. 1992, ApJ, 399, 704
Loup, C., Forveille, T., Omont, A., & Paul, J.F. 1993, A&AS, 99, 291
Lucas, R., & Guélin, M. 1999, in IAU Symp. 191, Asymptotic Giant Branch Stars, eds. T. Le Bertre, A. Lèbre, & C. Waelkens (San Francisco: ASP), 305
Mamon, G.A., Glassgold, A.E., & Huggings, P.J. 1988, ApJ, 328, 797
Margulis, M., van Blerkom, D.J., Snell, R.L., & Kleinmann, S.G. 1990, ApJ, 352, 615
Millar, T.J., & Herbst, E. 1994, A&A, 288, 561
Millar, T.J., Flores, J.R., & Markwick, A.J. 2001, MNRAS, 327, 1173
Miyoshi, M., Matsumoto, K., & Kameno, S., et al. 1994, Nature, 371, 395
Neri, R., Kahane, C., Lucas, R., et al. 1998, A&AS, 130, 1
Nyman, L.-Å. 1985, PhD Thesis, Chalmers University of Technology
Olofsson, H., Eriksson, K., Gustafsson, B., & Carlstrom, U. 1993, ApJS, 87, 267
Olofsson, H., Lindqvist, M., Nyman, L.-Å., & Winnberg, A. 1998, A&A, 329, 1059
Olofsson, H., Bergman, P., Lucas, R., et al. 2000, A&A, 353, 583
Pardo, J.R., Cernicharo, J., et al. 1998, A&A, 329, 219
Schöier, F.L., & Olofsson, H. 2001, A&A, 368, 969
Skinner, C.J., Justtanont, K., Tielens, A.G.G.M., et al. 1999, MNRAS, 302, 293
Stanek, K.Z., Knapp, G.R., Young, K., & Phillips, T.G. 1995, ApJS, 100, 169
Willacy, K., & Millar, T.J. 1997, A&A, 324, 237
Young, K. 1995, ApJ, 445, 872
Ziurys, L.M., Savage, C., Highberger, J.L., et al. 2002, ApJ, 564, L45

Question, by E. Humphreys: Our current models include both collisional and radiative pumping, but not a variation of the radiative pumping during the stellar cycle. I have not included any special geometrical schemes. I have not attempted to discern whether the radiative/collisional pump dominates in the SiO maser as yet.

Answer: I am sure that you actually include collisional and radiative excitation, but I think that, as far as the geometry and velocity do not satisfy the requirements I mentioned, no efficient radiative maser pumping will appear.

RADIO PHOTOSPHERES: IMAGING AND LIGHT CURVES

Mark J. Reid
Harvard–Smithsonian Center for Astrophysics
60 Garden Street, Cambridge, MA 02138, USA
reid@cfa.harvard.edu

Karl M. Menten
Max-Planck-Institut für Radioastronomie
Auf dem Hügel 69, Bonn, D–53121, Germany
kmenten@mpifr-bonn.mpg.de

Abstract Radio continuum observations of a sample of oxygen-rich Miras indicate temperatures and densities at radii near 2 R_* average $\approx$ 1500 K and $\sim 10^{-12}$ gm cm^{-3}. The nearly constant radio light curves rule out strong shocks in this region, and SiO maser proper motions indicate convective-like motions with speeds of $\approx$ 4 km s^{-1}. The large amplitudes of visual light curves can be explained by the variable formation of molecular oxides (i.e., TiO). Simple models can reproduce the visual light curves and these models confirm the densities and low temperatures derived from the radio observations.

Keywords: Mira variables; light curves; shocks; masers

1. Introduction

A typical oxygen-rich Mira variable has a radius, R_*, (defined by line-free regions of the spectrum) of about 2 AU. Surrounding the star is an extensive "molecular photosphere" which can be opaque across portions of the visible and infrared spectrum out to nearly 2 R_*. Also at about 2 R_* one detects a "radio photosphere," surrounded by an SiO maser emitting region. Beyond the SiO region dust grains form, radiation pressure accelerates the dust, transferring outward momentum to the gas and leading to significant mass loss.

Y. Nakada et al. (eds.), Mass-Losing Pulsating Stars and their Circumstellar Matter, 283–290.

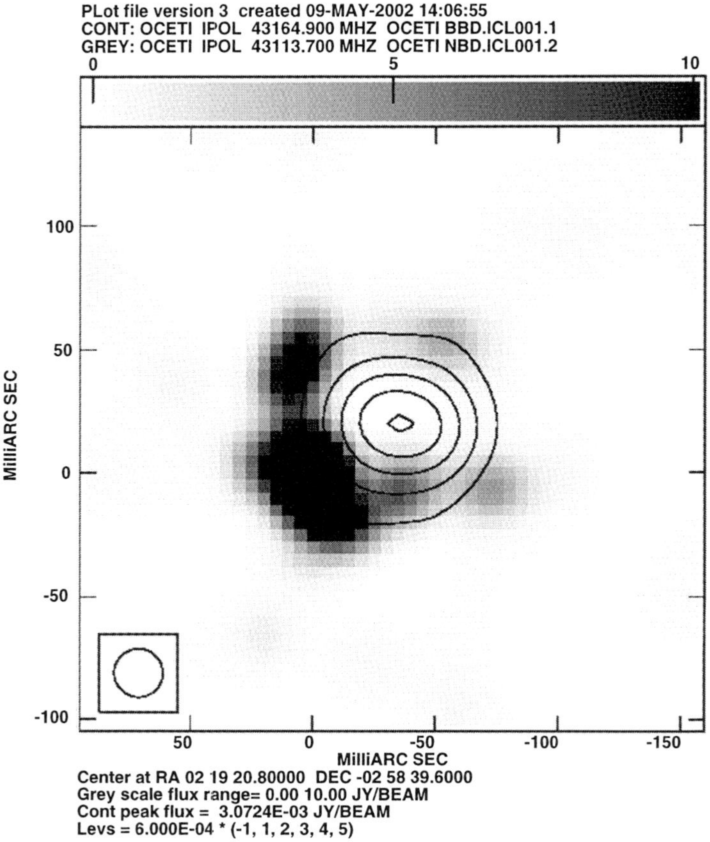

Figure 1. Radio continuum at 43 GHz (contours) and SiO (v=1, J=1-0) masers (grey-scale) for *o* Ceti observed with the VLA in November 2000.

In this paper, we summarize recent observations at radio wavelengths that characterize the physical conditions in the radio photosphere near 2 R_*. These include simultaneous imaging observations of the radio continuum and SiO maser emission from some Mira variables. We also summarize what is known about radio "light curves" and how a simple physical model can explain not only the radio light curves, but also the infrared and extremely large amplitude visible light curves. These observations and supporting models constrain the physical and dynamical conditions between 2 and 5 R_*

2. Radio Photospheres

Reid & Menten (1997a) published measurements made with the Very Large Array [1] (VLA) for a small sample of low mass-loss rate Miras (*o* Ceti, R Leo, W Hya, R Aql, χ Cygni, and R Cas). Their results reveal the 8 to 22 GHz spectral index for these stars and the 8 GHz light curves for three (*o* Ceti, W Hya, R Leo) of them. Also, they imaged the radio continuum of W Hya at 22 GHz (Reid & Menten 1990). Based on these observations, it appears that such Miras display the following characteristics:

- flux density proportional to the square of the observing frequency between 8 and 22 GHz;
- brightness temperature of $\approx$ 1500 K for W Hya;
- observed size of $\approx$ 4 AU ($\approx$2 R_*) for W Hya; and
- nearly constant light curves (variation $\lesssim \pm 15\%$ about the mean).

Recent results (Reid & Menten, in preparation) of simultaneous observations of the radio continuum and SiO masers at 43 GHz with the VLA resolve and image the radio photosphere better than before and locate precisely the SiO masers with respect to the star. For example, Fig. 1 shows the radio continuum and the SiO masers superposed for *o* Ceti. The data were taken at the beginning of November, 2000, when the longest VLA baselines afforded a resolution of about 45 mas (although a "CLEAN" restoring beam of 30 mas was used in the figure). The star is resolved, and modeling the visibility data indicates a (uniform disk) diameter of $\approx$ 52 mas and a brightness temperature of $\approx$ 1670 K. The stellar image shows signs of elongation in the east-west direction. These results are comparable to those for W Hya obtained by Reid & Menten (1990, 1997a) and in new data taken in November 2000 for W Hya.

The detection of a "radio photosphere" near $2R_*$ provides a new observational tool to increase our understanding of the region between the star and the dust formation zone. For the physical conditions expected in the radio photosphere, free electrons, obtained mostly from the ionization of potassium and sodium, provide the dominant opacity through free-free interactions with neutral H and H_2. Reid & Menten (1997a) proposed a simple physical model which can explain the radio frequency observations. At cm-wavelengths, unity optical depth is achieved at a

[1]The VLA is operated by the National Radio Astronomy Observatory under a cooperative agreement with the National Science Foundation.

radius of about 5×10^{13} cm, where the density and temperature are $\sim 10^{-12}$ gm cm^{-3} and ≈ 1600 K, respectively. This produces a flux density, which is roughly twice that expected for black-body emission from the underlying star, as observed.

Fig. 2 shows radio light curves measured at 8.4 GHz for W Hya and o Ceti. The radio emission from these stars (and R Leo, not shown) are consistent with a constant flux density. Only for o Ceti is there a hint of a periodic variation in the 8.4 GHz light curve, possibly peaking at optical phase of about 0.6. The lack of strong variability of the cm-wavelength flux density for low mass-loss Miras like o Ceti, R Leo, and W Hya is an important observational constraint.

Varying the temperature or radius of the unity optical depth point in in "static" model of the radio photosphere by more than about ± 150 K and $\pm 4 \times 10^{12}$ cm results in greater observed variability than observed (Reid & Menten 1997a). Models with strong disturbances (e.g., shocks) propagating through the radio photosphere at $\gtrsim 5$ km s^{-1} also produce variability greater than observed (Reid & Menten 1997b). Below about 5 km s^{-1}, however, such disturbances could "pile up" within a thin region in the radio photosphere and produce nearly constant radio emission as observed. Thus, the lack of strong variability at cm-wavelengths indicates that periodic shocks are probably strongly damped in amplitude and/or speed by the time they reach $2R_*$.

3. SiO Masers

The SiO masers in o Ceti (Fig. 1) form an incomplete ring with a diameter of about 90 mas centered on the star. Since the continuum and maser data were taken simultaneously and "cross-calibrated," they are properly registered to better than 1 mas accuracy. The simplest interpretation for the SiO masers is that they form in a spherical shell between radii of $\approx 3-5$ R$_*$. The inner boundary of the SiO masers is determined either by densities $\gtrsim 10^{-14}$ gm cm^{-3}, leading to thermalization of the level populations (e.g., Lockett & Elitzur 1992), or by high radio-continuum opacity from H$^-$ free-free interactions (Reid & Menten 1997a). The outer boundary occurs when SiO densities become too low for strong maser amplification. Note that in addition to the rapid decrease in total density with increasing radius in the extended atmosphere, the SiO density can fall even faster owing to silicate grain formation.

Proper motions of SiO masers in TX Cam indicate complex dynamics (Diamond & Kemball 1999). Expansion at a speed of ≈ 4 km s^{-1} is the dominant motion. However, some SiO maser spots are also observed

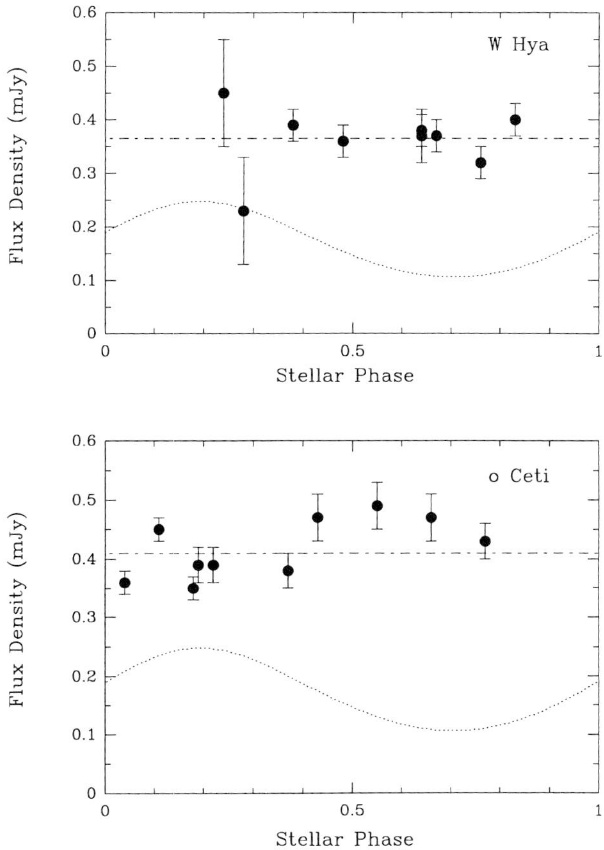

Figure 2. Radio light curves measured by Reid & Menten (1997a) at 8.4 GHz (points with error bars) for W Hya and *o* Ceti. The dotted line is the flux density predicted from a pulsating stellar photosphere. The dash-dotted line indicates the best fitting constant flux density.

to move inwards at a few km s^{-1}, suggesting a possible large-scale convective structure. In any event, there are no indications of fast moving ($\gtrsim 10$ km s^{-1}) shocks in the SiO masing region. This complements a similar conclusion, that strong, fast shocks are not found in the underlying radio photosphere, based on the observed nearly constant radio light curves.

4. Visual & IR Light Curves

The lack of strong variability in the radio continuum emission from Miras is in sharp contrast to the extremely large amplitude (≈ 8 mag) of the visual light in stars such as χ Cygni and *o* Ceti. The problem of explaining how a star can periodically change visual light by factor of

1000 or more has been known for over 400 years. Pettit and Nicholson (1933) observed Miras and were able to separate the visual and infrared emission by placing an absorbing water cell in front of their bolometer detector. They noted that the effective temperatures deduced from their data could explain the infrared light curves, but failed by several magnitudes to explain the large amplitude of the visual light curves. They speculated that strong opacity in the visual, owing to the formation of molecules, might be a factor. However, little progress has been made on this long-standing problem until recently.

Reid & Goldston (2002) presented a simple explanation for the extremely large amplitudes of light curves of some oxygen-rich Mira variables. Metallic oxides, in particular TiO, form throughout the stellar atmosphere as the star cools when approaching minimum light. When this happens, the visual light can be almost completely absorbed at large radii, extending the visual photosphere to nearly twice its nominal size. At these large radii, temperatures can fall to ≈ 1400 K and essentially all of the star's radiation then emerges in the infrared. Since almost no optical light is emitted at these low temperatures, Mira variables can decrease their visual light by 8 or more magnitudes, almost "disappearing" to the human eye.

While calculating a visual light curve for a Mira variable using state-of-the-art dynamical models is a challenging enterprise, Reid & Goldston (2002) took a first step with simple models. Their models place a time variable blackbody star in an extended atmosphere, assumed to be in hydrostatic equilibrium. Assuming equilibrium chemistry for the formation of TiO (the dominant opacity source in the visible for O-rich Miras), they then performed wavelength-dependent radiative transfer calculations and produced model visual light curves. Figure 2 displays their model light curves for χ Cygni, a Mira whose visual light curve has an exceptionally large amplitude. Reid & Goldston found that, in order to fit the light curves, the temperature and density at minimum light at $\approx 2\ R_*$ must be ≈ 1400 K and $\sim 10^{-12}$ gm cm^{-3}, respectively.

5. Summary

Radio frequency observations of O-rich Mira variables, such as o Ceti, W Hya, R Leo, and χ Cygni, provide data which constrain the kinematics and physical conditions at $2 \rightarrow 5\ R_*$. Radio continuum emission comes predominantly from H^- free-free interactions. An optical depth of unity is observed at $\approx 2\ R_*$ and requires a temperature and density of $\approx$ 1500 K and $\sim 10^{-12}$ gm cm^{-3}, respectively. The "radio photospheres" of o Ceti and W Hya appear somewhat elongated.

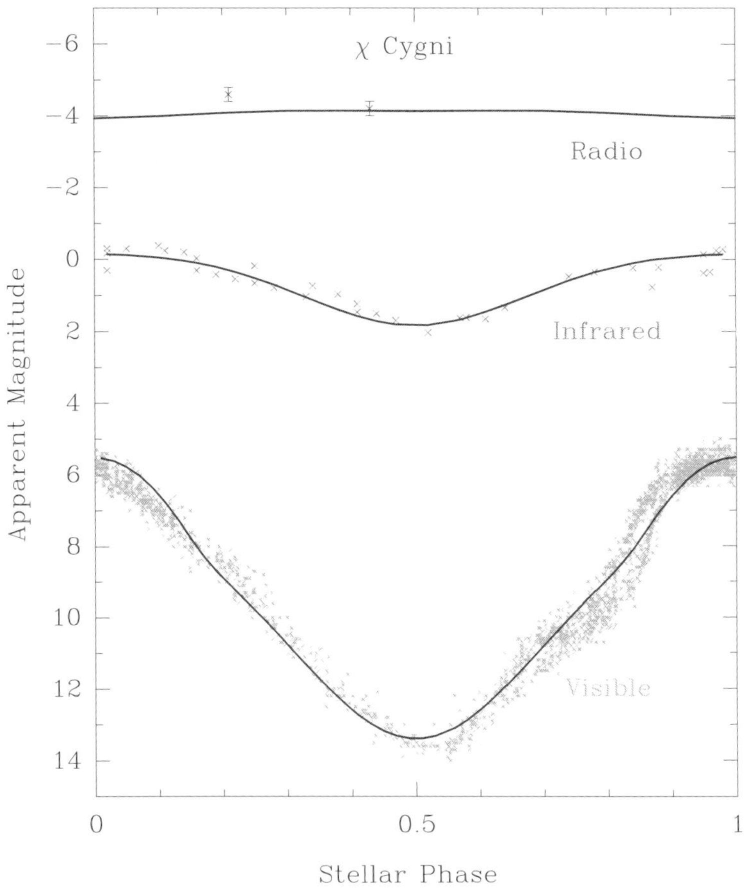

Figure 3. Observed and modeled light curves for χ Cygni from Reid & Goldston (2002). Model light curves (solid lines) were obtained by adjusting model parameters to best fit the optical and infrared light curves (crosses). The radio light curve was then calculated, following the procedures in Reid & Menten (1997a), without adjusting any model parameters.

SiO masers surround the "radio photosphere" between about 3 $\rightarrow$ 5 R_*. They often appear as partial rings and trace outwardly and inwardly moving material, possibly indicative of large scale convection. Temperatures and densities inferred from SiO masers are ≈ 1000 K and $\sim 10^{-14}$ gm cm^{-3}, respectively. Both the radio continuum light curves and the SiO maser proper motions suggest that, outside of 2 R_*, shocks are mostly damped and propagate at $\lesssim 5$ km s^{-1}.

The extremely large amplitudes of the visual light curves of O-rich Miras can be understood via a "double instability." First, the variable ionization of H and He below the stellar surface causes pulsations, leading to large changes in temperature over the stellar cycle. Second, the changes in temperature lead to the variable formation of molecular oxides, most importantly TiO, and to apparent sizes in strong absorption bands of ≈ 2 R_*. At minimum light at 2 R_* the temperature can fall so low that almost no visible light is emitted. Simple models which reproduce these light curves require temperatures and densities of ≈ 1400 K and $\sim 10^{-12}$ gm cm^{-3}, respectively, at radii of ≈ 2 R_* at minimum light. These conditions are similar to those inferred from the radio observations.

References

Celis, S. L. 1978, A&A, 63,53
Diamond, P. J. & Kemball, A. J. 1999, IAU Symp. 191: Asymptotic Giant Branch Stars, 191, 195
Lockett, P. & Elitzur, M. 1992, ApJ, 399, 704
Pettit, E. & Nicholson, S. B. 1933, ApJ, 78, 320
Reid, M. J. & Menten, K. M. 1990, ApJ, 360, L51
Reid, M. J. & Menten, K. M. 1997a, ApJ, 476, 327
Reid, M. J. & Menten, K. M. 1997b, Ap&SS, 251, 41
Reid, M. J. & Goldston, J. E, 2002, ApJ, 568, 931
Smak, J. I. 1966, ARAA, 4, 19

Questions:

M. Feast: If I remember correctly, Joe Smak discussed the effects of TiO absorption on the amplitudes of oxygen Miras in an early volume of Annual Reviews. It has generally been considered that the fact that the amplitudes of carbon Miras are mostly smaller than oxygen Miras is due to the lower effect of the variation of band strength in the carbon variables. It would be nice to see this treated more quantitatively in the general way you have done for the oxygen stars.

M. Reid: Smak (1966), as well as Pettit & Nicholson (1933), Celis (1978) and others, have suggested that TiO absorption probably contributed to the phenomenon of large amplitude Miras. However, I have not found *quantitative* work on this subject after extensively searching the literature. Regarding carbon-rich Miras, I plan to perform an analysis, similar to the one by Reid & Goldston (2002), but replacing TiO with C_2 and CN as the dominant source of opacity.

MAGNETIC FIELDS IN THE ENVELOPES OF LATE-TYPE STARS

Circular Polarization of H_2O Masers

W.H.T. Vlemmings, P.J. Diamond, H.J. van Langevelde
Sterrewacht Leiden
Niels Bohr weg 2, 2300 RA Leiden, The Netherlands
vlemming@strw.leidenuniv.nl

Abstract We present the first circular polarization measurements of circumstellar H_2O masers around a sample of late-type stars. These observations are used to obtain the magnetic field strength in the H_2O maser region with both an LTE and non-LTE analysis. We find fields from a few hundred milliGauss up to a few Gauss, indicating a solar-type r^{-2} dependence of the magnetic field on the distance. No linear polarization is detected to less than 1%.

Keywords: masers - polarization - stars: circumstellar matter - stars: magnetic fields - techniques: interferometric

Introduction

The role of magnetic fields in the late stages of stellar evolution is still unclear. Blackman et al. (2001) have shown that AGB stars could produce fields of several hundreds of Gauss. Such strong fields can play an important role in driving stellar winds and shaping the outflows.

Until recently, information on the magnetic field in the circumstellar envelopes was obtained by polarimetric observations of SiO masers at $\approx 2-4\ R_*$ from the central star, and OH maser at a distance of $1000-10000$ AU. The SiO observations indicated fields of 5-10 Gauss for Mira stars and up to 100 Gauss for supergiants (e.g. Barvainis et al., 1987). However, using a non-Zeeman interpretation of the observed circular polarization, the magnetic fields could be a factor 1000 less (Wiebe & Watson, 1998). OH maser observations indicated field strengths of 1-2 mG (e.g. Szymczak & Cohen, 1997).

Y. Nakada et al. (eds.), Mass-Losing Pulsating Stars and their Circumstellar Matter, 291–294.

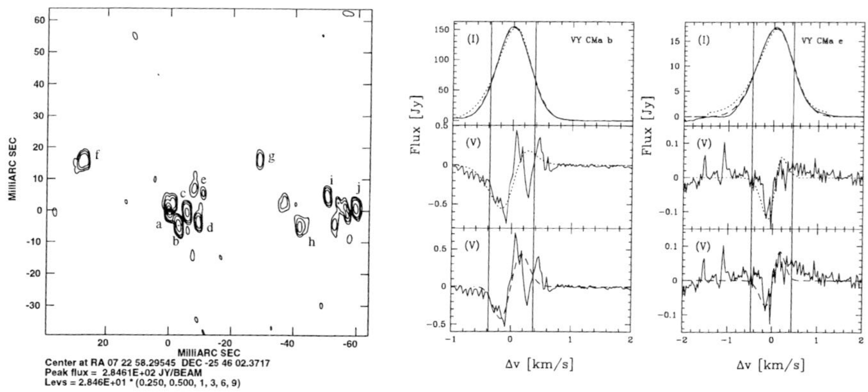

Figure 1. (left) Total intensity image of the H_2O maser features around VY CMa. (right) Total power (I) and V-spectra for selected maser features of VY CMa. The bottom panel shows the best fitting LTE model (dashed), the middle panel shows the best fitting non-LTE model (dotted).

Now we have been able to determine the magnetic field strengths in the intermediate region, at a few hundred AU, where the H_2O masers occur. Although H_2O is a non-paramagnetic molecule, it has been possible to observe the circular polarization on some of the strongest circumstellar H_2O maser features. We have used both the LTE analysis presented in Vlemmings et al. (2001) and non-LTE models based on the models presented in Neduloha & Watson (1990). A full description of the analysis methods, observations and results are presented in Vlemmings et al. (2002).

1. Observations

We have observed 4 late-type stars with the VLBA, the supergiants S Per, VY CMa and NML Cyg, and the Mira variable star U Her. To get the highest spectral resolution, required for the circular polarization measurements, the data were correlated twice. Once with modest spectral resolution (0.1 km/s), to get all 4 polarization combinations (RR, LL, RL and LR), and once with high resolution (0.027 km/s), with only RR and LL. The calibration was mainly performed on the modest spectral resolution data and the solutions were copied and applied to the high resolution data. This data set was then used to produce circular polarization and total intensity image cubes.

2. Results

We have examined the strongest H_2O maser features around the 4 stars observed. Circular polarization down to 0.2% of the total intensity was detected on $\approx$ 50% of the brightest maser features. An example of the features around VY CMa is shown in Fig.1. We rule out any systematic effects as a cause of the observed spectrum, because various percentages of circular polarization are observed as well as different directions. No linear polarization was detected above the limit of $\approx$ 0.5%.

The magnetic field strengths were determined with both the LTE and the non-LTE method. The LTE method predicts the circular polarization spectrum to be directly proportional to the derivative of the total power spectrum. We found that the observed spectra were narrower, which can only be explained with the non-LTE analysis. The non-LTE field strengths are $\approx$ 40% lower than those determined by the LTE method.

From the observations and analysis, we estimate the magnetic field strengths in the H_2O maser region to be $\approx$ 200 mG for S Per and VY CMa. The field around NML Cyg is $\approx$ 500 mG, while the Mira variable U Her shows a much higher field of $\approx$ 1.5 G.

3. Conclusions

Our results favor the non-LTE approximation and because we do not detect any linear polarization, a non-Zeeman interpretation is also highly unlikely. The lack of linear polarization can be easily explained in the non-LTE case, because linear polarization is only produced by strongly saturated masers. A line widths analysis indicates that the circumstellar H_2O masers are not saturated. Even for large angles between the line of sight along the maser and the direction of the magnetic field we do not expect any linear polarization. In the LTE analysis, the lack of linear polarization can only be explained by having the maser line of sight beam along the magnetic field lines.

We can compare the strength of the magnetic field in the H_2O maser region with the values obtained from SiO and OH maser polarization observations. This seems to indicate that the magnetic field strength values inferred from the SiO maser observations are indeed due to the normal Zeeman effect, although Elitzur (1996) has argued that the field strength can still be a factor 10 lower on both SiO and OH masers. Fig.2 shows the dependence of the magnetic field strength on distance from the star. Our observed values are plotted at the observed maximum extent of the H_2O maser region. Because our observations are most sensitive for the highest magnetic fields, we are actually probing the inner edges

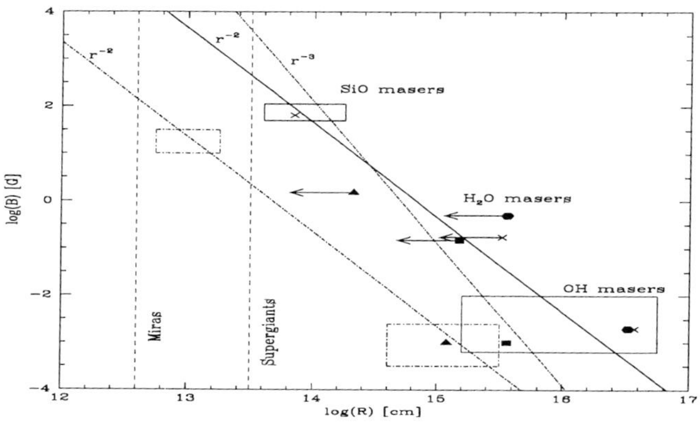

Figure 2. Magnetic field strength B, as function of distance R from the star. Dashed-dotted boxes are the SiO and OH maser estimates for Mira stars, solid boxes are those for supergiant stars. Symbols indicate observations; U Her is represented by triangles, S Per by the square, VY CMa by the crosses and NML Cyg by the hexagonal symbol. The dashed vertical lines are an estimate of the stellar radius.

of the H_2O maser shell. The arrows indicate the typical thickness of such a shell. These results indicates that the magnetic field strength is best represented by a solar-type dependence on distance (r^{-2}). The exact shape of the magnetic field strength cannot easily be determined from our observations. The SiO polarization maps indicate mostly radial field lines close to the star.

The magnetic pressure of the field in the H_2O maser region dominates the thermal pressure by a factor of 20. Using the solar-type field, extrapolated surface field strengths are of the order of 100 – 1000 Gauss, strong enough to drive and shape the outflows.

References

Barvainis, R., McIntosch, G., & Predmore, C.R. 1987, Nature, 329, 613
Blackman, E.G., Frank, A., Markiel, J.A., et al. 2001, Nature, 409, 585
Elitzur, M. 1996, ApJ, 457, 415
Nedoluha, G.E., & Watson, W.D. 1992, ApJ, 384, 185
Szymczak, M., & Cohen, R.J. 1997, MNRAS, 288, 945
Vlemmings, W., Diamond, P.J., & van Langevelde, H.J. 2001, A&A, 375, L1
Vlemmings, W., Diamond, P.J., & van Langevelde, H.J. 2002, A&A, accepted
Wiebe, D.S., & Watson, W.D. 1998, ApJ, 503, L71

PHASE DEPENDENCE OF 43 GHz SiO MASERS IN THE MIRA VARIABLE TX CAM

Jiyune Yi, R.S. Booth, J.E. Conway
Onsala Space Observatory, Chalmers University of Technology, 43992, Onsala, Sweden
jiyune@oso.chalmers.se

Abstract We have conducted simultaneous observations of $v = 1$ and $v = 2$ $J = 1 \rightarrow 0$ SiO masers at 43 GHz in the Mira variable TX Cam using the VLBA, over three different stellar phases. The spatial distribution of the two maser line emissions is considered to be a strong constraint on the modeling of circumstellar SiO maser.

We find that the distribution and the separation of the masing components in the $v = 1$ and $v = 2$ $J = 1 \rightarrow 0$ changes over different stellar phases. SiO maser rings are observed in both lines, however, the disruption of the maser ring is found when the SiO maser flux becomes weaker. The ring radii and the number of masing components also vary depending on the stellar phase. We argue that a conclusion in favor of one maser pumping scheme based on a single epoch observation could be misleading.

Keywords: Stars: individual: TX Camelopadalis, AGB and post-AGB, circumstellar matter, mass-loss ; Instrumentation: high angular resolution ; Masers

1. Introduction

Circumstellar SiO masers can serve as a unique tool to study the innermost part of circumstellar envelopes of AGB stars. SiO masers are believed to occur in the region close to a stellar photosphere inside the dust condensation region. VLBI mapping of $v = 1$ $J = 1 \rightarrow 0$ SiO masers toward Mira variables over the last several years has demonstrated that the masers arise within a few stellar radii of the parent star, forming a ring-like structure. High resolution observations of SiO masers can provide direct information on the physics of extended atmospheres

Y. Nakada et al. (eds.), Mass-Losing Pulsating Stars and their Circumstellar Matter, 295–298.

of evolved stars, where the stellar pulsation and shocks dominate and consequently complicate the physics of the region.

7mm observations of SiO masers in the $v = 1$ and $v = 2$ $J = 1 \rightarrow 0$ lines can play a significant role to build a plausible SiO maser model which can be coupled with a pulsating AGB stellar model. A standard radiative pump scheme has difficulty producing strong $J = 1 \rightarrow 0$ transition in both $v = 1$ and $v = 2$ at the same position in the extended atmosphere. To test the pumping mechanisms it is therefore of interest to determine the spatial distribution of the masers in each of these lines. We had performed simultaneous observations of these two lines using the VLBA in 1998; the results of our experiments (Jiyune Yi et al. 2002) leaded us to believe that a single epoch observation may not be able to give reliable answers on the pumping mechanism because of signs of stellar phase dependent properties.

We report our new multi-epoch, simultaneous observations of $v = 1$ and $v = 2$ $J = 1 \rightarrow 0$ SiO masers in TX Cam using the VLBA. We discuss the spatial distribution of masing components over different stellar phases with the resolution of sub-mas (milliarcsec). We examine the phase dependent characteristics of these masers.

2. Observation

We observed the 7 mm $v = 1$ and $v = 2$ SiO maser lines simultaneously toward TX Cam, using the VLBA. Three epochs of observations were made in May and September 2000, and January 2001, which are apart by its stellar phase of $\phi \sim 0.2$. The synthesized beam was less than 0.4 mas $\times$ 0.2 mas.

3. Results

Due to the limited space for this paper, we present only images of observations made in January 2001 and report preliminary results.

3.1 Observations in January 2001

The optical phase at this epoch is $\phi \sim 0.07$. Figure 1 shows SiO maser maps of $v = 1$ (left) and $v = 2$ (right) $J = 1 \rightarrow 0$ integrated over all velocity channels containing emission. A full ring shape distribution of maser clumps is found in both maps; the velocities of maser emissions range from ~ 2 to ~ 20 km s^{-1} in both lines. Many masing clumps appear in both lines and their velocity ranges correspond to each other. There are three clumps (marked A, B, & C in Figure 1) showing partial displacement with a linear size of about 1 to 2 mas between $v = 1$ and $v = 2$ $J = 1 \rightarrow 0$ line emissions. However, other common clumps

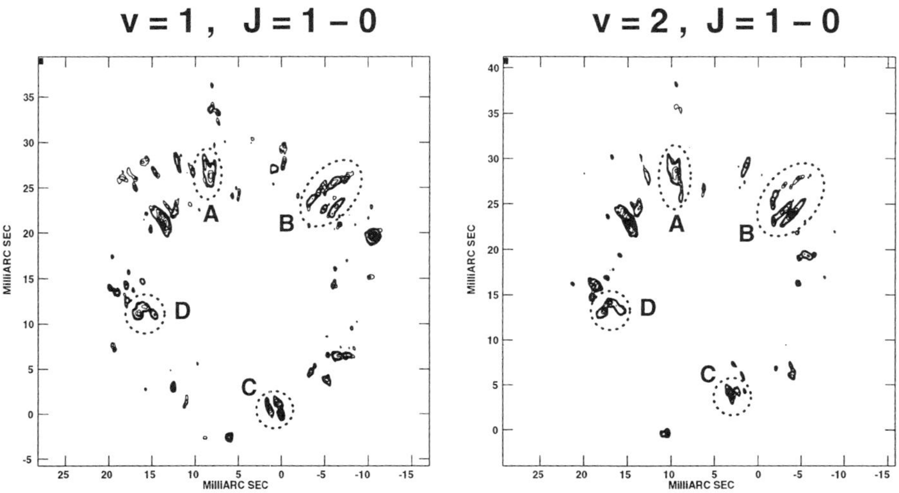

Figure 1. The VLBI maps of the $v = 1$ $J = 1 \rightarrow 0$ (left) and $v = 2$ $J = 1 \rightarrow 0$ (right) SiO maser emission toward TX Cam observed in January 2001.

observed in both lines show spatial coincidence. We will explore its implication further in the following discussion.

3.2 Observations in September and May 2000

The optical phase of TX Cam observed in September 2000 is $\phi \sim 0.85$. A ring-like maser distribution is also seen at this epoch, but emission is weaker and the numbers of masing clumps are fewer compared to those observed in January 2001. Some strong maser complexes (marked B, C, & D in Figure 1) which were seen in 2001 January have not yet developed at this epoch. The diameter of the maser ring size is approximately 4 mas smaller than the one observed in January 2001.

The corresponding stellar phase of observations in May 2000 is $\phi \sim 0.64$. Masers are not developed well at this epoch and due to fewer and weaker masing clumps, the maser rings shown in the later two epochs are not so clear at this epoch. The size of the maser distribution is smaller than those of two epochs observed later.

4. Discussion and Conclusion

Our multi-epoch, simultaneous observations of SiO masers in the $v = 1$ and $v = 2$ $J = 1 \rightarrow 0$ show the phase dependent properties of SiO masers.

I. The ring size become larger from May 2000 to January 2001. We found that the maser ring radii are ~ 2 mas larger in the maps of both lines observed in January 2001.

II. At the two stellar phases of $\phi \sim 0.85$ (Sep.2000) and $\phi \sim 0.07$ (Jan.2001), there is a significant overlap among the features in the $v = 1$ and $v = 2$ maps. Nevertheless, there is generally a small net inward displacement of the $v = 2$ ring. The displacement of $1 \sim 2$ mas is readily seen in some clumps between the two lines at the epoch of $\phi \sim 0.07$. It is difficult to quantify any displacement at the maser minimum since the rings are severely disrupted.

III. Among three epochs, SiO masers are minimum at its optical phase of $\phi \sim 0.64$ in May 2000 and the ring-like distribution of masers is not apparently seen at this epoch. The non-spherical maser distribution in both line were also found in our '98 observations (Jiyune Yi et al. 2002). Therefore, we confirm the prediction of maser ring disruption at the maser minimum phase.

The phase dependent properties of SiO masers which we found are predicted by the models of Gray & Humphreys (2000) and Humphreys et al (2002). The spatial coincidence of the two masers observed in a single epoch observation cannot be simply used in favor on one pumping model against another due to the maser phase dependence. Depending on the distance of the observed star or stellar phase, it may become difficult to determine the ring separation even with the VLBA resolution. Although we have confirmed maser phase dependent properties in general, there could be some discrepancies in details between the models and observational results, such as the size of ring separation differing by different Mira variables or different stellar phases. More discussion on the implications of our observational results are needed to be made further in the near future.

References

Gray, M.D., & Humphreys, E.M.L. 2000, New Astronomy, 5, 155

Humphreys, E.M.L., Gray, M.D., Yates, J.A., Field, D., Bowen, G.H., & Diamond, P.J. 2002, A&A, 386, 256

Yi, Jiyune, Booth, R.S., Conway, J.E., Winnberg, A., & Diamond, P. J., 2002, in IAU Symp 206, 266 Cosmic MASERs: From Protostars to Blackholes, ed. V. Migenes

MODELS OF MASERS IN AGB STARS

Getting the Big Picture

E.M.L. Humphreys

Onsala Space Observatory

S-43992 Onsala, Sweden

liz@oso.chalmers.se

M.D. Gray (UMIST), J.A. Yates (UCL), D. Field (Aarhus), G.H. Bowen (Iowa State), P.J. Diamond, A.M.S. Richards (Jodrell Bank)

Abstract Coupling maser models to dynamical stellar envelope models provides a new step towards the full understanding of the physics of mass-losing stars. Here, we describe how such numerical simulations of H_2O and SiO maser emission yield maser variability and the proper motions of maser clumps during a stellar pulsation cycle. These calculations place important constraints on stellar models and provide new insights on the conditions leading to, and predominantly affecting, stellar maser emission. This work also constitutes the foundations towards our final goal: the development of dynamical models of AGB stars which calculate accurately and self-consistently the time-dependent intensity and location of SiO, H_2O and OH maser emission, in addition to the inclusion of time-dependent chemistry and dust formation. Such models will provide a 'big picture' view of mass-losing stellar envelopes throughout their pulsation cycles.

Keywords: Masers, mass loss, radiative transfer, hydrodynamical models

Introduction

Masers are particularly useful for studying the mass-losing process in pulsating stars, since they are located in several different regions surrounding the star. Firstly, SiO masers are imaged using VLBI to exist in the extended atmosphere/inner circumstellar envelope (CE) of AGB and Red Giant stars (see Miyoshi et al. and Yi et al., this proceedings; hereafter this proc.). Their dynamical behaviour traces both the infall and outflow of gas close to the star. Secondly, water masers occur further

Y. Nakada et al. (eds.), Mass-Losing Pulsating Stars and their Circumstellar Matter, 299–302.

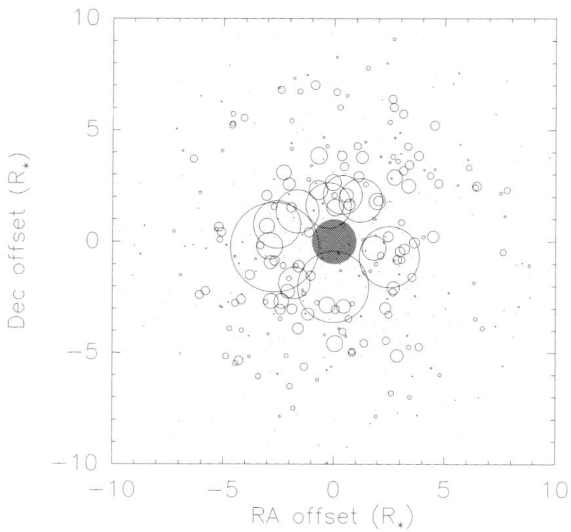

Figure 1. 22 GHz H_2O maser emission at a single stellar phase. Intensities of the masers are linearly proportional to the diameters of the black circles shown. Circles are centred on the maser projected coordinates. The grey disk represents the stellar disk of mean radius 1.1 AU.

out in the CE (Vlemmings et al., this proc.). In AGB stars of low mass loss rate ($\dot{M}$), 22 GHz water masers are found in the region in which the dust-driven acceleration of circumstellar gas is taking place. In OH/IR stars, of higher $\dot{M}$, they reside in the already-formed, outflowing stellar wind. In this case, radial amplification (i.e. along the radius vector of the star) is the norm, and the lineshapes resemble those of 1612 MHz OH masers. Lastly, OH masers reside in the stellar wind, some at radii of 1000 R_* or so. (see Etoka et al., this proc.)

1. Approach to modelling

The aim of our work is to couple hydrodynamical models of mass-losing stars with maser models, thereby reproducing the location and time-dependent behaviour of different maser species *and* predicting new observable features of the maser emission. The reader is referred to the references given at the end of this contribution for a detailed description of the modelling method.

2. Results: H_2O masers

22 GHz H_2O maser amplification is predominantly tangential in our model star of $\dot{M} = 1.8 \times 10^{-7}$ $M_\odot yr^{-1}$, as shown in Figure 1. Maser

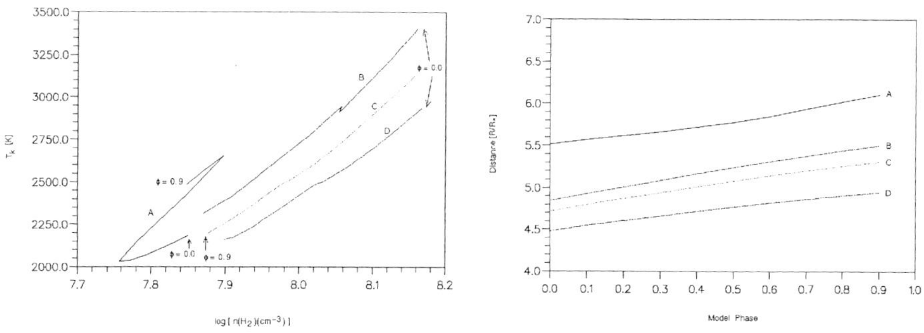

Figure 2. Temporal attributes: a) *(left)* Physical conditions experienced by four bright 22 GHz components throughout a stellar pulsation cycle of 332 days ($\equiv \Delta\phi =$ 1.0; b) *(right)* the motions of these components within the CE.

emission persists throughout the cycle, originating from the accelerating wind region of the CE model. The temperature – density space encountered by four H_2O maser components which are bright for much of the cycle is shown in Figure 2a. The corresponding motions of the components in the CE are shown in 2b. Motion of the 22 GHz H_2O components is *outflow only*, with a velocity of $\sim$5 kms^{-1} for this model star.

3. Results: SiO masers

Figure 3a shows the spatial distribution of SiO maser emission at a single epoch. Movies of the simulations, showing the tangentially-amplifying SiO maser ring dimming and brightening, contracting and expanding, during the stellar cycle are available on http://www.saturn.phy.umist.ac.uk:8000/~mdg/sio/movie.html. Turning again to the tracking of individual masing components throughout the stellar pulsation cycle, Figure 3b shows the density – temperature conditions experienced by four SiO components which are bright for much of the stellar pulsation cycle, with their *infall (4.2 kms^{-1}) and outflow (3.2 kms^{-1}) motions* shown in Figure 3c. Note that the components A – D are not the same as those shown in Figure 2.

4. Outlook

At present, these modelling methods are limited by (i) a lack of *oxygen-rich* dynamical stellar envelope models, noting that some models which do not include time-dependent dust formation processes have been

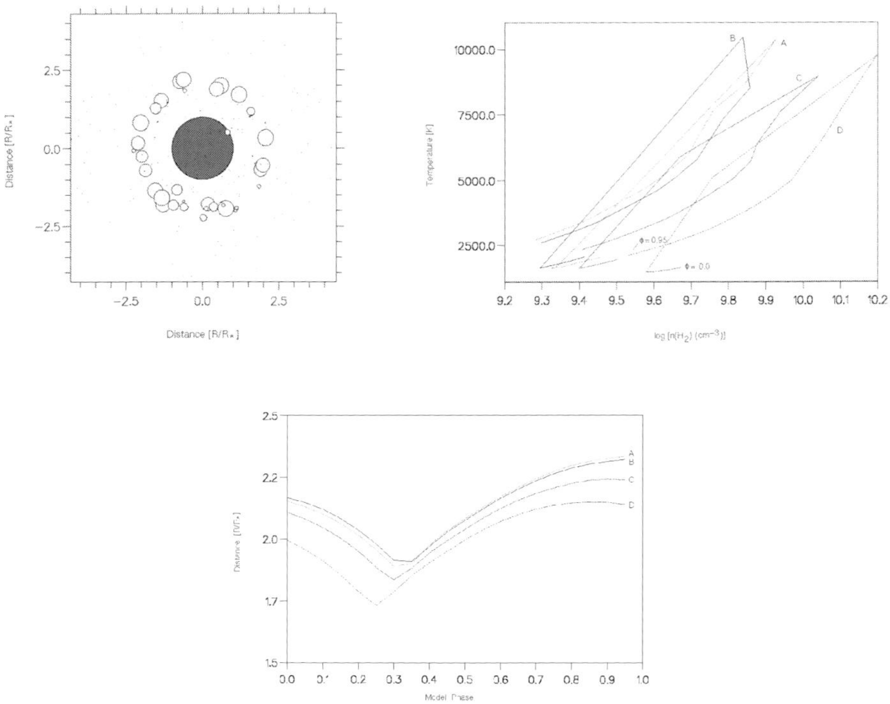

Figure 3. a) *(top left)* As for Fig. 1 except $v = 1$ $J = 1 - 0$ SiO masers (43 GHz); b) *(top right)* physical conditions encountered by four bright SiO components during the cycle; c) *(bottom)* accompanying motions within the CE.

produced by the M. Scholz group and (ii) the lack of polarised radiation transport in the SiO and H_2O maser models.

Acknowledgments

I am very grateful to the LOC for providing a travel bursary to the workshop and to the Swedish STINT foundation for funding this work.

References

Humphreys, E.M.L., Gray, M.D., Yates, J.A., et al. 1996, MNRAS, 282, 1359

Gray, M.D., & Humphreys, E.M.L., 2000, New Astronomy, 5, 155

Humphreys, E.M.L., Yates, J.A., Gray, M.D., Field, D., & Bowen, G. 2001, A&A, 379, 501

Humphreys, E.M.L., Gray, M.D., Yates, J.A., et al. 2002, A&A, 286, 256

VLBA OBSERVATIONS OF THE v=1 AND 2 SiO J=1-0 MASER EMISSIONS IN VY CMa

Makoto Miyoshi
National Astronomical Observatory Japan
2-21-1, Osawa, Mitaka Tokyo, Japan 181-8588
miyoshi@miz.nao.ac.jp

Abstract We performed VLBA observations of SiO masers in VY CMa. While some maser spots exist that have no counterpart in another line, the difference of distributions of the SiO J=1-0 maser emissions in v=1 and v=2 states is within 0.2 mas (= 0.3 AU at the distance of 1.5 kpc) in average.

Keywords: SiO masers(v=1,2,J=1-0), VLBA, VY CMa

1. Introduction

What is the pumping mechanism of SiO maser emissions? Though about 30 years have passed since the discovery of SiO masers (Kaifu et al. 1974, Snyder and Buhl 1974), the question is still a matter of controversy. The comparison of the spatial distributions of several SiO transitions in different states is the important key to settle the question. Miyoshi et al. (1994) performed the VLBI observations of SiO masers in VY CMa and W Hya with Kashima- Nobeyama Interferometer. They found the distributions of SiO J= 1-0 at the v= 1 and the v= 2 states are the same within 1 mas in average from synthesis imaging. The similarity is in favor of collisional pumpings (Lockett and Elitur 1992). Desmurs et al. (2000), using VLBA found clear difference between the distributions of the two maser lines in TX Cam and IRC +10011 , which gives favor to the radiative pumping of SiO masers (Bujarrabal and Nguyen-Q-Rieu(1981), Deguchi and Iguchi(1976), Western and Watson (1983)). Here I show the new result on SiO masers in VY CMa with VLBA.

Y. Nakada et al. (eds.), Mass-Losing Pulsating Stars and their Circumstellar Matter, 303–306.

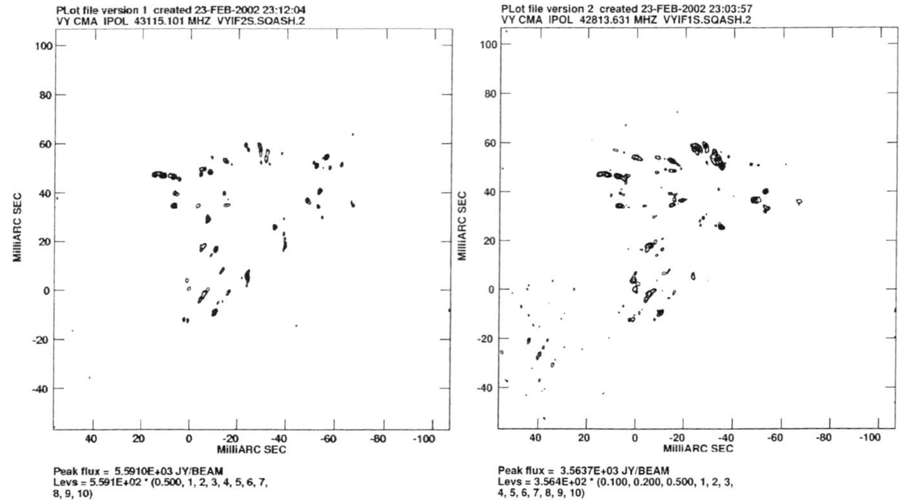

Figure 1. The left panel shows the spatial distribution of the J=1-0, v=1 SiO masers in VY CMa. The right shows that of J=1-0, v=2 SiO masers.

2. Observations and Result

VLBA observations of the SiO J=1-0 in v=1 and v=2 in VY CMa were performed in October 1998. The reduction was done using NRAO AIPS. From careful self-calibration sequences to the strong maser velocity channels, the residual phase fluctuations of visibilities became less than 2° , which corresponding to 40 μsec in delay. After the calibrations, synthesis imaging of every channel was performed using IMAGR with 50 μas spacings, 4096×4096 griding points. Fig 1 shows the totally integrated maps of all velocity channels. Though it is clear that some of maser spots have no counterpart in another line, the distribution as the whole is the same. To compare the distributions of two maps from VLBI, how to make the alignment of two maps is quite important point. Miyoshi et al.(1994) aligned the maps with assumption that $v_{LSR} = 22.17km/sec$ maser spots in the both lines were at the same point. We used the following manner. Calculating cross correlations of brightness distributions of the two maps in the same velocity channels, the lag that maximizes the correlations as a whole was applied to the alignment between the v=1 and v=2 maps.

Fig. 3 shows the intensity and distance relations, which is the same as with Fig 4 in Miyoshi et al. (1994). In here clean components in each channel map are treated as maser spots. Most of strong maser compo-

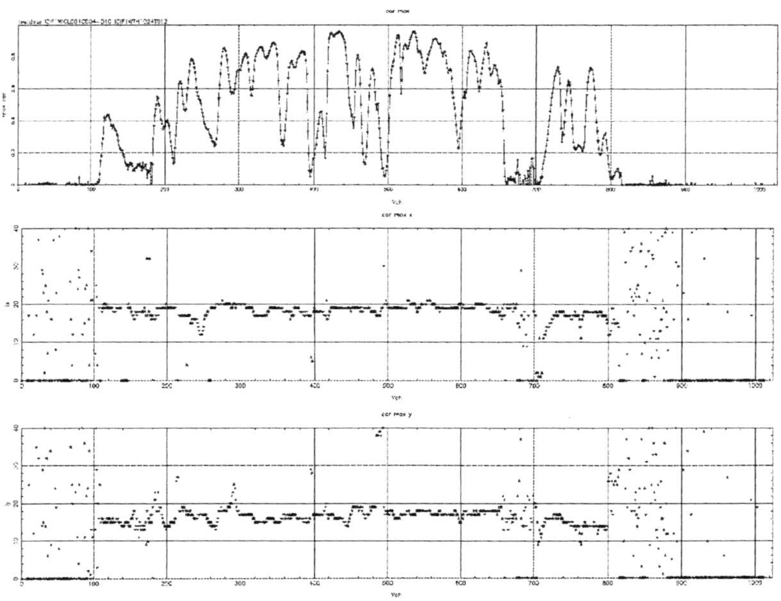

Figure 2. The cross correlations of channel maps between v=1 and v=2. The top panel shows the maximum cross correlations of each channel (the full scale is 100%). The middle shows the lags in $\Delta\alpha$ and the bottom shows those in $\Delta\delta$ that maximize the correlation of each channel (1-lag is 50 μas). X-axis is frequency channels (1024ch, full span 8MHz).

nents have the same velocity counterparts of the another line within 0.2 mas(=0.3au @1.5kpc). This tendency was also seen in Miyoshi et al. (1994), but become more notable in this new results.

3. Discussion

This result shows that most of the two SiO masers in VY CMa have the counter part each other within 0.2 mas in average. While some SiO maser spots have no counter part in another line.

Elitzur made a comment that the collisional pumping model can explain the quite similarity of the distributions of the two lines. Mochizuki et al. (2002) shows that different distributions of the two SiO masers in star forming region Orion KL. The pumping mechanism of SiO masers may not be common in different circumstances. Detailed theoretical calculations will be expected to make clear the pumping mechanism of SiO masers. Today we have sufficient spatial resolution to investigate SiO masers. Remaining observational difficulty is how to align two maps. It is quite important when we compare the brightness distributions of the

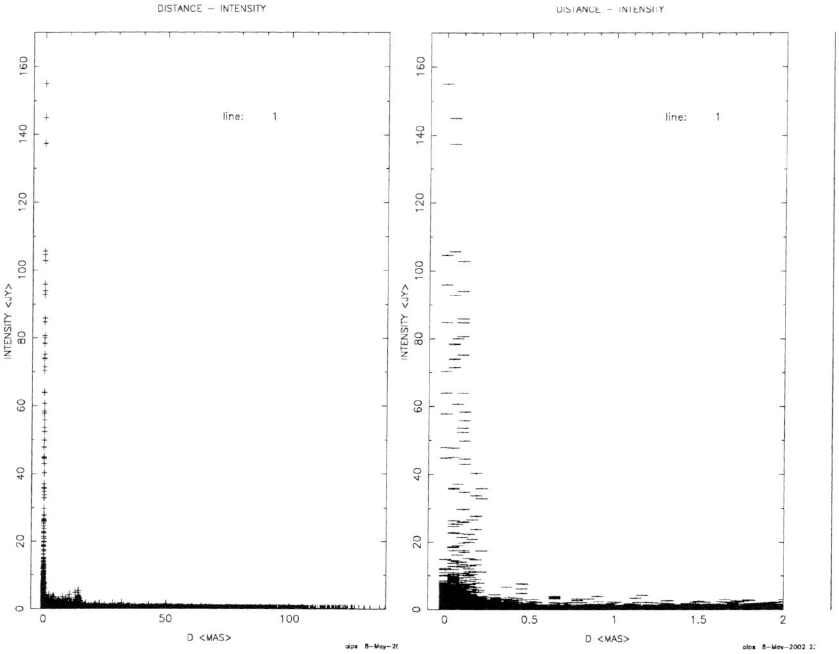

Figure 3. The relation between the intensity of v=1 spots and the angular distance to the nearest v=2 spot with the same velocity channel. Here we treat clean components as maser spots. These followed the same manner as in fig 4 in Miyoshi et al. (1994). The left panel shows the large scale up to D= 140 mas while the right panel shows the relation in less than 2 mas.

two images. Hachisuka et al. (2002) show that delta VLBI observations have the possibility to overcome the alignment problem.

References

Bujarrabal, V., & Nguyen, Q. R. 1981, A&A, 102, 65
Deguchi, S., & Iguchi, T. 1976, PASJ, 28, 307
Desmurs, J. F., Bujarrabal, V., Colomer, F., & Alcolea, J. 2000, A&A, 360, 189
Hachisuka, K. 2002, in this volume
Kaifu, N., Buhl, D., & Snyder, L. E. 1975, ApJ, 195, 359
Lockett, P., & Elitzur, M. 1992, ApJ, 399, 704
Miyoshi, M. et al. 1994, Nature, 371, 395
Mochiduki, N. 2002, in this volume
Snyder, L. E., & Buhl, D. 1975, ApJ, 197, 329
Western, L. R., & Watson, W. D. 1983, ApJ, 275, 195

OPTICAL AND MICROWAVE SPECTROSCOPY OF LONG-PERIOD VARIABLE STARS

G. M. Rudnitskij[1], M. I. Pashchenko[1], V. F. Esipov[1], and E. E. Lekht[1,2]
[1] *Sternberg Astronomical Institute, Moscow State University*
13 Universitetskij prospekt, Moscow, 119899 Russia
[2] *Instituto Nacional de Astrofísica, Óptica y Electrónica*
Luis Enrique Erro No. 1, Apdo Postal 51 y 216, 72840 Tonantzintla, Puebla, México
gmr@sai.msu.ru

Abstract The results of multiwavelength studies (optical spectroscopy, H_2O maser line, radio continuum) of a sample of long-period variable stars are presented. For some of the stars studied (R Aql, RR Aql, W Hya, U Ori) the H_2O maser emission has been detectable throughout the monitoring interval, for more than 20 years. A comparison of the H_2O maser monitoring of these stars with their visual light curves suggests a correlation of their maser and optical brightness. There is a phase lag $\Delta\varphi \sim 0.1 - 0.4P$ (P is the star's variability period) of the H_2O maser variations relative to the optical variability. There are also stars ('transient masers', R Leo, R Cas, U Aur) that have exhibited short-duration flares in the H_2O maser line with several years of no detectable emission between them. The H_2O maser flares were preceded by flares of the Hα emission line. The common cause of such sequence of events can be the consecutive impact of a shock wave on different layers of the stellar atmosphere and circumstellar envelope.

Keywords: Miras, spectroscopy, emission lines, masers

Since 1980 we have been performing radio spectral observations of the maser emission of late-type variable stars in the $\lambda = 1.35$ cm rotational line of the H_2O molecule (6_{16}–5_{23}, 22235.08 MHz) on the 22-meter radio telescope, Pushchino Radio Astronomy Observatory, Russian Academy of Sciences, with a helium-cooled FET amplifier ($T_{\rm noise} = 150$ K), 128-channel filter-bank spectrometer, resolution 7.5 kHz ($\Delta V_R = 0.1$ km/s). The 3σ sensitivity of the H_2O observations is about 10 Jy. The interval between the H_2O observational sessions is on the average 1–1.5 months. The sample includes M-type long-period variables (Miras and semireg-

Y. Nakada et al. (eds.), Mass-Losing Pulsating Stars and their Circumstellar Matter, 307–310.

ulars R Leo, R Aql, R Cas, U Ori, W Hya and others) as well as M-supergiants (VX Sgr, VY CMa, S Per).

In 1994 we began regular optical spectroscopic observations of the stars on the 125-cm telescope of the Crimean Observatory, Sternberg Astronomical Institute, with a diffraction spectrograph and CCD camera, $\Delta\lambda = 0.25$ Å/pixel (Esipov et al. 1999). Main attention has been paid to the Balmer emission lines (Hα and others).

In November 1995 34 stars were observed on the Australia Telescope Compact Array at $\lambda\lambda = 6$ and 3 cm for a systematic search of radio continuum emission. The program included o Cet, U Ori, U Her, R Aql and some southern-hemisphere Mira and semiregular variables. The result was negative with upper limits typically ~0.1–0.3 mJy (Chapman and Rudnitskij 2002).

The H_2O maser monitoring has allowed us to divide the stars into two groups.

1. Stable masers, emitting throughout 22 years (R Aql, RR Aql, RS Vir, S CrB, X Hya).

2. Transient masers, displaying strong flares intermitting with long quiescent intervals without detectable H_2O emission (R Leo, R Cas, U Aur).

For stable, permanently emitting stars we could trace a correlation between their visual light curves and H_2O maser variations. The H_2O line flux usually followed the visual brightness with a phase delay $\Delta\varphi$ of 0.1–0.4 stellar periods (P). For RS Vir ($P = 354^d$) the cross-correlation function between the visual light curve and H_2O maser variations shows maximum correlation for a delay τ of 4–5P (Lekht et al. 2001).

A special case is U Ori. This star was undetectable in the H_2O line in June 1980, but by October 1980 it flared and has remained since a 'permanent' (Group 1) maser, displaying, too, light-curve-correlated maser variations with some delay $\Delta\varphi$. The value of $\Delta\varphi$ was changing itself with a 'superperiod' of 9 years, i.e, almost 9 light cycles of $P = 372^d$ (Rudnitskij et al. 2000).

For three 'transient' (Group 2) masers (R Leo, R Cas, U Aur) the Hα flares were followed in 1–1.5 years by an H_2O maser flare (Esipov et al. 1999). In between R Leo showed also a SiO maser flare. Such sequence of the flares can be explained by a consecutive action of a moderate-strength shock ($v_{\rm sh} \sim 6 - 10$ km s^{-1}) on various layers of the stellar atmosphere and inner circumstellar envelope.

We connect the flaring events (observed once per several stellar light cycles) with the effect of a shock wave propagating through the stellar atmosphere and then through the inner layer of the circumstellar envelope, hosting the masers (Fig. 1, *left*). The cross-correlation function

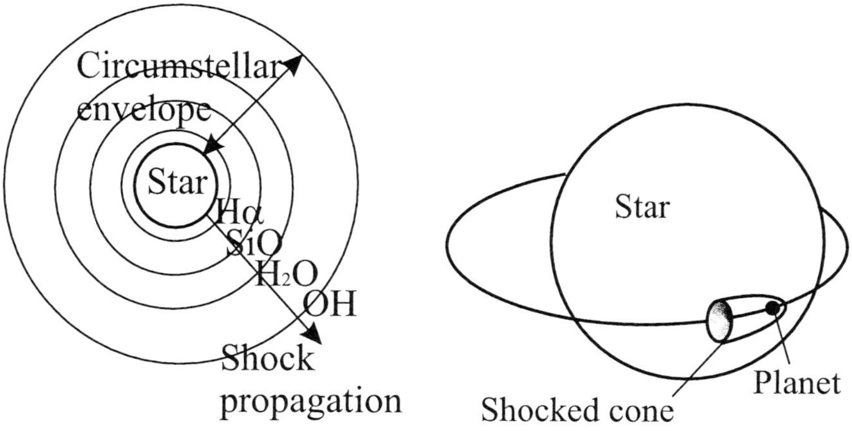

Figure 1. *Left:* Consecutive excitation by an outward propagating shock of the layers responsible for the Hα, SiO, H_2O and OH emission. *Right:* A shock induced by a planet revolving in a close orbit near the stellar photosphere.

for RS Vir, peaking at $\tau \sim 4 - 5P$, suggests the travel time of a shock from the stellar surface to the H_2O masering layer in the circumstellar envelope (at $r_{\rm m} \sim 10^{14}$ cm) of about this interval. Then we estimate the mean shock velocity as $v_{\rm sh} \sim r_{\rm m}/\tau \sim 7$ km s^{-1}, i.e., not a strong shock.

The rare occurrence of such episodes of the Hα emission followed by an H_2O maser flare can be explained by the following mechanisms.

1. Merging shocks (Wood 1979). Overtaking and merging of consecutive stellar-pulsation-driven shocks is possible. Merged shocks can produce stronger Balmer and maser emission once per several stellar cycles.

2. Appearing and disappearing quasi-stationary spherical layer (Hinkle et al. 1984) at a distance $r \sim 10^{14}$ cm from the stellar surface. This layer has physical conditions favourable for the H_2O maser generation, when pumped by a relatively slow shock with $v_{\rm sh}$ just about the value estimated for RS Vir, 7 km s^{-1} (Rudnitskij and Chuprikov 1990).

3. A low-mass companion in an elliptical orbit (Rudnitskij 2000; Struck et al. 2002; Fig. 1, *right*). This model does not imply a 'global' spherical shock, engendered by the stellar pulsation. Shocks in Miras may be intrinsically weak, as suggested above by the lack of associated radio continuum, though this continuum should accompany the observed strong Balmer line emission. An alternative may be a local hot spot, connected with the shocked tail behind a planetary (or brown dwarf) companion to the red giant. This spot can emit sufficiently strong optical hydrogen lines, remaining very weak in the radio continuum because of its small angular size. If the planet is revolving in an eccentric orbit

with a period of several years, the Hα–H_2O events are connected with the periastron passages of the planet. Details are given in Rudnitskij's poster (this volume).

Thus, our long-term observations have shown that some effects connected with Miras' light variations, 'superperiods' in the delay $\Delta\varphi$ and in the appearance of Balmer emission lines and molecular maser flares, can be understood only if monitored on a sufficiently long time interval, covering at least 10 to 20 light cycles of the stars.

References

Chapman, J.C., & Rudnitskij, G.M. 2002, PASA, submitted

Esipov, V.F., Pashchenko, M.I., Rudnitskij, G.M., & Fomin, S.V. 1999, Astron. Lett, 25, 672

Hinkle, K.H., Scharlach, W.W.G., & Hall, D.N.B. 1984, ApJS, 56, 1

Lekht, E.E., Mendoza-Torres J.E., Rudnitskij, G.M., & Tolmachev, A.M. 2001, A&A, 376, 928

Rudnitskij, G.M. 2000, in ASP Conf. Ser. 203, The Impact of Large-Scale Surveys on Pulsating Star Research, (San Francisco: ASP), 384

Rudnitskij, G.M., & Chuprikov, A.A. 1990, SvA, 34, 147

Rudnitskij, G.M., Lekht, E.E., Mendoza-Torres J.E., Pashchenko, M.I., & Berulis, I.I. 2000, A&AS, 146, 385

Struck, C., Cohanim, B.E., & Willson, L.A. 2002, ApJ, 572, L83

Wood, P.R. 1979, ApJ, 227, 220

Discussion

Sandra Etoka: A comment about the time scale in the occurrence of similar events at different wavelengths: superperiods and flare events in H_2O and OH, in particular found in both of our works on long-period monitoring of AGB stars, might be connected.

Georgij Rudnitskij: Indeed, in R Leo we have two examples of such chains of flare events: SiO$\rightarrow H_2O\rightarrow$OH in 1980s, H$\alpha\rightarrow$SiO$\rightarrow H_2O$ in 1990s (similar single chains in U Aur and R Cas). It would be interesting if R Leo flares also in OH within a couple of years, this would strengthen the evidence.

3D MODELING OF 1612 MHz OH MASERS

Monte Carlo modeling of the maser shells and the amplified stellar image

W.H.T. Vlemmings
Sterrewacht Leiden
Niels Bohr weg 2, 2300 RA Leiden, The Netherlands
vlemming@strw.leidenuniv.nl

Abstract We present the first results of our 3D Monte Carlo maser radiative transfer code, used to model the 1612 MHz OH maser shell and the amplification of emission from the stellar radio-photosphere.

Keywords: circumstellar masers - radiative transfer - methods: Monte Carlo

Introduction

In the circumstellar envelope of AGB stars, both the main line (1665 and 1667 MHz) and the satellite line (1612 MHz) OH maser transitions have been observed extensively. The 1612 MHz masers occur at $\approx 10^{16}$ cm from the central star.

The 1612 MHz maser line profile often shows a characteristic double peaked spectrum that is assumed to be the signature of an expanding shell (e.g. Reid et al. 1977). The exact shape however, is determined by local conditions in the maser shell and the level of saturation.

Recently it has been suggested that emission from the radio-sphere of the star will be amplified by the maser screen in front of the star. This amplification will cause a bright maser spots with several observable characteristics (Vlemmings, 2000). Strong evidence for this theory was presented recently when the most blue-shifted 1667 MHz maser spot of U Her was found to coincide with the Hipparcos optical position to within 15 mas (van Langevelde et al., 2000 & these proceedings)

Modeling the various observed shapes of the maser profile by, and examining the effect of a background star are the main goals of this work.

Y. Nakada et al. (eds.), Mass-Losing Pulsating Stars and their Circumstellar Matter, 311-312.

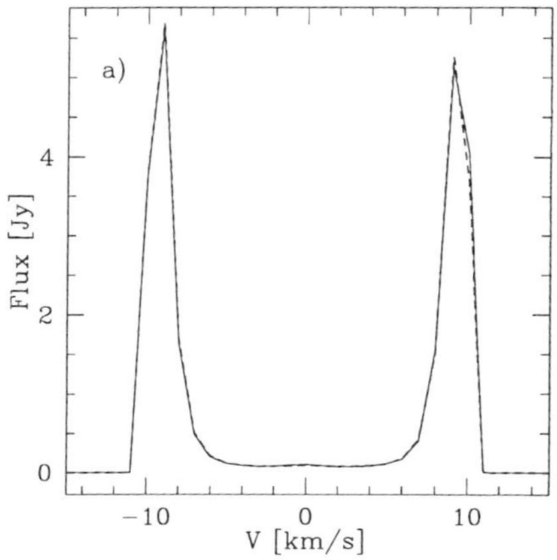

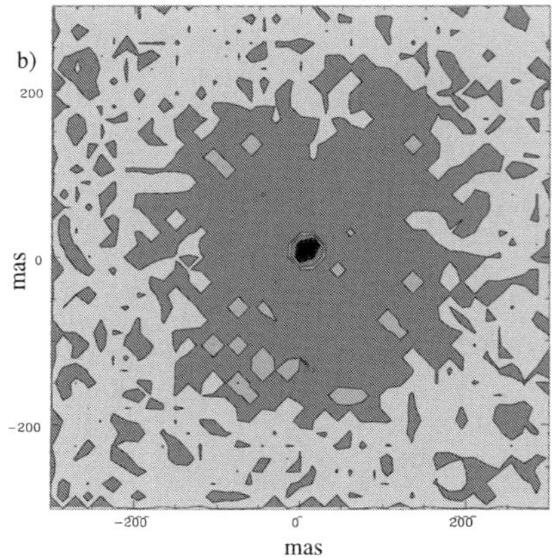

Figure 1. a) Modeled OH maser spectrum with (solid) and without (dashed) background star. b) Map of the most blue-shifted velocity channel for the model with a background star. Black is the highest intensity.

1. Modeling & Results

Maser emission is highly anisotropic as a result of beaming due to the maser geometry and the structure of velocity coherent paths through the medium. Additional anisotropy can be caused by density or pumping inhomogeneities and the contributions of the background source. This means that full 3-dimensional modeling is essential to accurately describe the maser.

Here we have used a fully 3D Monte Carlo code that solves the 1612 MHz OH maser in a way similar to the 2D method used by Spaans & van Langevelde (1992).

The first results are presented in Fig.1 for a typical OH shell with and without a background star. The expansion velocity is 10 km/s and the OH number density is 4 cm^{-3} throughout the maser shell. As can be seen, the effect of a background star on the OH maser spectrum is small, while the high resolution channel map clearly shows the high brightness stellar image.

References

Reid, M.J., Muhleman, D.O., Moran, J.M. et al., 1977, ApJ, 214, 60

Spaans, M., & van Langevelde, H.J. 1992, MNRAS, 258, 159

van Langevelde, H.J., Vlemmings, W., Diamond, P.J., et al. 2000, A&A, 357, 945

van Langevelde, H.J. et al., these proceedings

Vlemmings, W., & van Langevelde, H.J. 2000, in Proc. European VLBI Network Symp., 5, 189

EXPLORING STELLAR-PULSATION-DRIVEN SHOCK WAVES REVEALED IN THE KINEMATICS OF WATER MASERS AROUND RT VIR

Hiroshi Imai
Joint Institute for VLBI in Europe
Postbus 2, 7990 AA Dwingeloo, the Netherlands
imai@jive.nl

Abstract We report multi-epoch Very Long Baseline Array (VLBA) observations of water masers around the semiregular variable RT Virginis (RT Vir). The 3-D kinematics of 61 maser features describes a circumstellar envelope expanding roughly spherically with a velocity of ~10 km s^{-1}. Systematic radial-velocity drifts of masers were found with amplitudes of around 1 km s^{-1} yr^{-1}. From one maser feature, a quadratic position shift was discovered with an acceleration rate of 33 km s^{-1} yr^{-1}, implying a passage of a shock wave driven by the stellar pulsation. We estimated a distance to RT Virginis of ~220 pc on the basis of the statistical parallax and the model-fitting methods.

Keywords: Masers, semiregular variables, pulsation, RT Vir

The VLBA observations had been made at 5 epochs during 1998 May–August, with a separation of 3 weeks from epoch to another. The correlated data had spectral resolution of 0.056 km s^{-1}. An obtained synthesized beam was 0.4 mas and 1.0 mas in R.A. and decl. directions, respectively. About 60 maser features had been detected at every epoch and totally 61 maser features had been detected at least twice and measured their relative proper motions and radial-velocity drifts.

Fig. 1 shows the 3-D motions of the H_2O masers. While the H_2O masers exhibit a radial-velocity gradient in the east–west direction, the 3-D maser kinematics roughly fits a spherically-expanding flow model with a velocity of $\simeq$10 km s^{-1}. The velocity dispersion is almost equal in all directions within a factor of 1.7. Taking into account that evolved stars with relatively large mass-loss rates ($\dot{M}$ $>10^{-6}$ $M_{\odot}$ yr^{-1}), except supergiants, have envelopes with spherically symmetric morphology and

Y. Nakada et al. (eds.), Mass-Losing Pulsating Stars and their Circumstellar Matter, 313–314.

kinematics, the circumstellar envelope of RT Vir ($\dot{M} \simeq 3\times10^{-6}$ $M_{\odot}$ yr^{-1}, Yates et al. 1994) is likely in the transition when the mass-loss rate increases as the envelope evolves and the intrinsic bipolarity of the flow get being obscured by the growing envelope.

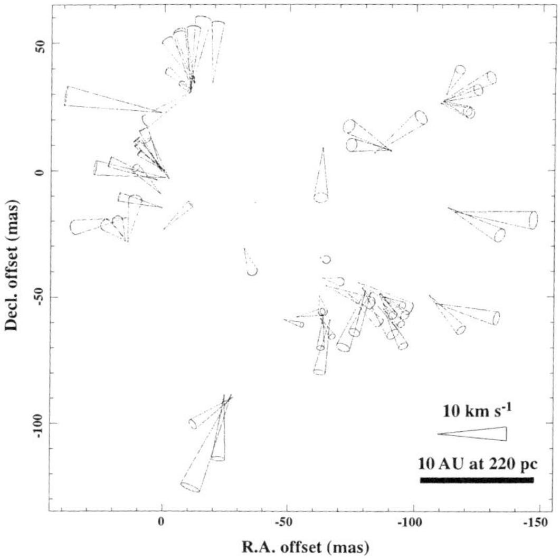

Figure 1 Velocity field of H_2O masers in RT Vir. A 3-D velocity vector of a feature is indicated by a cone.

We found an H_2O maser feature in RT Vir, which exhibits an apparently constant acceleration motion along a straight line. The acceleration of the feature is stronger on the sky plane ($\simeq$33 km s^{-1} yr^{-1}) than that along the line-of-sight ($\simeq$1 km s^{-1} yr^{-1}). Such a bias in an acceleration motion is expected from beaming effect of maser radiation. Large acceleration only along the sky plane is detectable because of necessity of a long coherent path of radial velocity in a maser cloud. The detected amplitude of the acceleration is roughly consistent with that expected by models of the pulsation-driven shock waves (Höfner et al. 1995).

With the kinematics of H_2O masers in RT Vir, the statistical parallax and the model-fitting methods assuming a radially expanding flow and uniformly random motions, respectively, gave distance values to RT Vir of 237$\pm$29 pc and 85$\pm$12 pc for the whole data of the proper motions. Excluding several features contracting towards the star, these methods gave distance values of 214$\pm$28 pc and 226$\pm$16 pc, respectively. These values are quite consistent with each other, and the distance value of $\sim$220 pc is adopted.

References

Höfner, S., Feuchtinger, M. U., & Dorfi, E. A. 1995, A&A, 297, 815
Yates, J. A., & Cohen, R. J. 1994, MNRAS, 270, 958

VLBI OBSERVATIONS OF WATER MASERS IN THE CIRCUMSTELLAR ENVELOPE OF IRC+60169

Hiroshi Sudou
Astronomical Institute, Tohoku University
Aramaki, Aoba, Sendai 980-8578, Japan
sudou@astr.tohoku.ac.jp

Abstract Water masers around an AGB star IRC+60169 have been observed at four epochs using the Japanese VLBI networks. The distribution of the red-shifted features exhibits a ring-like structure, the size of which is 30 mas and corresponds to the inner radius of the maser shell. This implies that dense gas around the star obscures red-shifted emission.

Keywords: masers, circumstellar matter

1. Water masers in IRC+60169

IRC+60169 has been classified as an IRC/AFGL object by using the IRAS two-color diagram (van der Veen & Habing 1988; Takaba et al. 1994). It shows a clear double peaked spectrum of water masers which is a typical signature of IRC/AFGL objects and OH/IR stars. The shape of the spectrum exhibited a violent time variation. In 1977 the red-shifted component was stronger than the blue-shifted one and in 1982 an intensity change between these components has occurred (Engels et al. 1988). This fact is thought to be related to the temporary variation in the mass-loss flow.

2. Blocking model

The water masers of IRC+60169 were observed using the Japanese domestic VLBI Networks on 1992 June 3, 1997 March 30, 1997 May 17, and 1998 April 4.

In order to obtain the averaged distribution among several years, we have made the superposition of the distribution of the maser components of IRC+60169 with respect to the brightest component. Figure 1a shows

Y. Nakada et al. (eds.), Mass-Losing Pulsating Stars and their Circumstellar Matter, 315–316.

that the blue-shifted component appeared to be in a limited region, such as blue-shifted components around Mira variables. On the other hand, in figure 1b, the red-shifted components appeared to be in the ring-like region around the blue-shifted component. This morphology can be explained by the blocking effect proposed by Takaba et al. (1994).

The size of the obscuration is roughly estimated to be ~30 mas (21 AU), by using the inner diameter of the distribution of the red-shifted component. This fact implies that the blocking material is dense gas with a density of Hydrogen molecules higher than 10^{10} cm^{-3} and quenches water maser emission (Cooke & Elitzur 1985). This dense-gas region corresponds to the region where SiO masers occur or to "the radio photosphere"(Reid & Menten 1997).

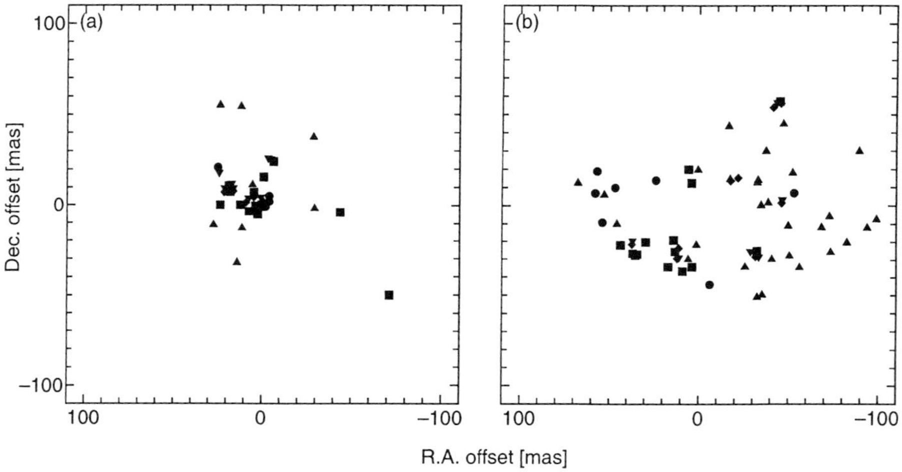

Figure 1. A superposition of the maser feature distributions between 1990 and 1998 of (a) the blue-shifted components and (b) the red-shifted components. The filled circles, triangles, inverse triangles, lozenges, and squares show the VLA data of Colomer et al. (2000), the 1st, 2nd, 3rd, and 4th epochs of the present work, respectively

References

Colomer, F., Reid, M.J., Menten, K.M. & Bujarrabal, V. 2000, A&A, 355, 979
Cooke, B. & Elitzur, M. 1985, ApJ, 295, 175
Engels, D., Schmid-Burgk, J. & Walmsley, C.M. 1988, A&A, 191, 283
Reid, M.J. & Menten, K.M. 1997, ApJ, 476, 327
Takaba, H., Ukita, N., Miyaji, T & Miyoshi, M. 1994, PASJ, 46, 629
van der Veen, W.E.C.J. & Habing, H.J. 1988, A&A, 194, 125

FIRST OH POLARIMETRIC IMAGES OF NML CYG

Sandra Etoka, Philip Diamond
Jodrell Bank observatory, University of Manchester, UK
setoka@jb.man.ac.uk

Abstract MERLIN interferometric observations of the supergiant NML Cyg were obtained in October 1993 in full polarisation, at 1612 and 1665 MHz. They reveal a complex distribution of the vectors of polarisation. At 1612 MHz, the majority of the components lies along a NW-SE geometrical axis along which a velocity gradient is clearly evident. The vectors of polarisation associated with these maser spots are roughly aligned on that same NW-SE axis. The distribution of the external and fainter components is spherical and the associated vectors of polarisation have a tangential distribution. At 1665 MHz, the maser spot distribution reveals an incomplete spherical shell for which the associated vectors of polarisation also show a tangential distribution. This dichotomy strongly suggests the existence of two distinct OH shells.

Introduction

The supergiant NML Cyg, classified as a supergiant OH/IR star by Bowers (1981), shows an asymmetric maser distribution both in H_20 and OH lines (Masheder, Booth & Davies 1974, Richards, Yates & Cohen 1996). Especially, the inner OH regions are expanding faster than the outer. It is suggested that an increase in the stellar luminosity by a factor of 2 or 3 over a 500-year period could result in the observed acceleration (Diamond, Norris & Booth 1984).

Maser emission characteristics

Masheder, Booth & Davies (1974) single-baseline maps of NML Cyg obtained in 1971 show a NW-SE elongated OH maser structure at 1612 MHz covering about 3 arcsec. The observations presented here, performed some 20 years later with better sensitivity and resolution, show that the whole 1612 MHz maser emission extension is about 5 arcsec while that of the 1665 MHz maser emission is only about 2.5 arcsec. The 1665 MHz maser emission shows a spherical distribution (cf. Fig. 1, where the area of the symbols are proportional to the maser spot in-

Y. Nakada et al. (eds.), Mass-Losing Pulsating Stars and their Circumstellar Matter, 317–318.

tensities and the length of the vectors to the linear polarisation). The majority of the maser spots at 1612 MHz lies along a NW-SE axis along which the associated vectors of polarisation are roughly aligned. Also, a velocity grandient is clearly evident along that axis (cf. Fig. 2). Away from that axis, in a more spherical distribution and a slightly greater total extent than the axisymmetrical maser spot distribution, are found very faint maser spots. These maser spots have a stronger polarisation and show a tangential orientation of their associated vectors of polarisation (cf Figs. 2).

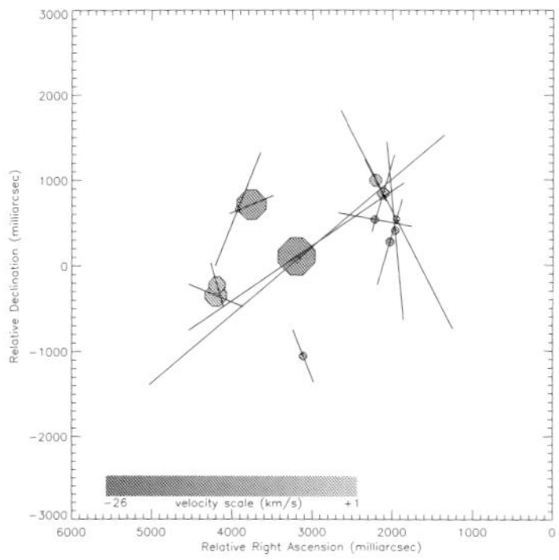

Figure 1. 1665 MHz maser spot distribution of NML Cyg.

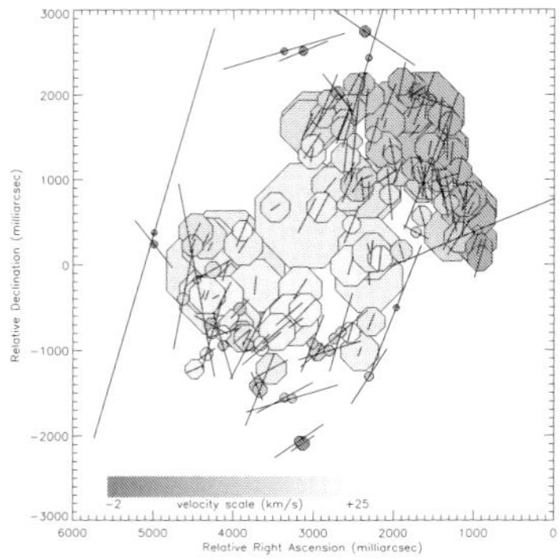

Figure 2. same as Fig. 1 for the 1612 MHz red-shifted emission.

Discussion

The interferometric observations presented here were the first done in full polarisation at these frequencies and with adequate resolution. They have revealed the presence of a quite complex distribution of the vectors of polarisation at the location of the OH maser emission. We suspect that the OH material observed belongs to 2 distinct shells. The hint of the existence of 2 distinct shells was already apparent in the velocity vs radial distance diagram of Diamond, Norris & Booth (1984, their Fig.4). The polarimetric information helps to distinguish which shell the maser spots belong to : the external components, showing a tangential distribution of their vectors of polarisation, most probably come from an older spherical shell, as the one observed in "standard OH/IR stars", while the components with the NW-SE orientation of their vectors of polarisation come from the most recent and probably bipolar outflow.

References

Bowers, P.F. 1981, AJ, 86, 1930
Diamond, P.J., Norris R.P., & Booth R.S. 1984, MNRAS, 207, 611
Masheder, M.R.W., Booth R.S., & Davies R.D. 1974, MNRAS, 166, 561
Richards, A.M.S., Yates J.A., & Cohen R.J. 1996, MNRAS, 282, 665

DETECTION OF A POSSIBLE BIPOLAR OUTFLOW AROUND THE SEMI-REGULAR VARIABLE STAR R CRATERIS BY THE JAPANESE VLBI NETWORK (J-NET)

José K. Ishitsuka
Department of Earth Science and Astronomy, University of Tokyo, Komaba, Tokyo 153-8902
pepe@chianti.c.u-tokyo.ac.jp

Abstract

We have monitored water vapor masers around the semi-regular variable star R Crateris (R Crt) with the Japanese VLBI Network (J-Net) at the 22 GHz band at four epochs with intervals of one month. With a typical positional accuracy of 0.5 mas in the R.A. and 0.6 mas in the declination directions, and with a velocity resolution of 0.105 km s^{-1}, we were able to measure twelve relative proper motions and radial-velocity drifts of maser features. The 3-D kinematics of the features indicates a bipolar expanding flow. The major axis of the asymmetric flow was estimated to be at P.A. = 136°. The existence of a bipolar outflow suggests that an evolved star had already formed a bipolar outflow at the early phase of the AGB star.

Keywords: AGB Stars, R Crateris, Water Vapor Masers

Introduction

The gas dynamics of circumstellar envelopes (CSEs) during the latest phase of the asymptotic giant branch (AGB) stars is one matter of debate to elucidate the physical mechanisms of energetic mass loss of evolved stars. The AGB stars evolve to proto-planetary nebulae (PPNs), and planetary nebulae (PNs). Recent HST infrared images (e.g. Sahai et al. (1998)) have shown asymmetric or bipolar outflows around the central object of PNs. On the other hand, the CSEs of AGB stars are generally

Y. Nakada et al. (eds.), Mass-Losing Pulsating Stars and their Circumstellar Matter, 319–320.

spherically symmetric (Manchado et al. (2000)). The most intriguing question is when the change of morphology occurs between the AGB and PN phases.

Maser action is predicted in the inner part of the CSE. Measurements of the relative proper motion of compact maser features with high spatial resolution can be used to investigate the structure of CSEs and to trace those asymmetries or bipolarities.

1. R Crateris a Low Mass-Loss Rate Star

R Crt is a semi-regular variable star of SRb type in spectral classification M7. The mass-loss rate of R Crt was estimated to be 1.0×10^{-6} $M_{\odot}$ yr^{-1} (Kahane, Jura (1994)) from CO observation.

2. J-Net Observations Results

We investigated the 3-D kinematics of water masers around the semi-regular variable star R Crt using data from multi-epoch observations with VLBI at short intervals see Ishitsuka et al. (2001). Some important results are as follows:

- We measured the three-dimensional velocity field, which indicates a possible bipolar outflow. A variance-covariance analysis applied to the velocity field supports the bipolarity of the outflow.

- The existence of bipolar outflow suggests that a Mira variable had already formed a bipolar outflow.

3. Asymmetries in CSE

From Ishitsuka et al. (2001) in table 5, low mass-loss rate stars seems to have asymmetric CSEs. Are low velocity gradients and low mass-loss rate indicators of asymmetries in circumstellar envelopes?

References

Ishitsuka, J. K., Imai, H., Omodaka, T., Ueno, M., Kameya, O., Sasao, T., Morimoto, M., Miyaji, T., Nakajima, J., & Watanabe, T. 2001, PASJ, 53, 1231

Johnston, K. J., Spencer, J. H., & Bowers, P. F. 1985, ApJ, 290, 660

Kahane, C., & Jura, M. 1994, A&A, 290, 183

Manchado, A., Villaver, E., Stanghellini, L., & Guerrero, M. A. 2000 in ASP Conf. Ser. 199, Asymmetrical Planetary Nebulae II: From Origins to Microstructures, ed. Kastner J. H., Soker N., & Rappaport S. (San Francisco: ASP), 17

Sahai, R., Hines, D. C., Kastner, J. H., Weintraub, D. A., Trauger, J. T, Rieke, M. J, Thompson, R. I., & Schneider, G. 1998, ApJ, 492, L163

A COMPARISON OF THE SiO MASERS IN ORION KL WITH THOSE OF LATE TYPE STARS

Nanako Mochizuki
The Institute of Space and Astronautical Science
motiduki@vsop.isas.ac.jp

Makoto Miyoshi
National Astronomical Observatory, Japan
miyoshi@miz.nao.ac.jp

Abstract Using the VLBA, we observed the SiO J=1-0 masers lines at the v = 1 and v = 2 states in the Orion KL. The spatial distribution of the Orion KL shows no regularity in two lines, though supergiant VY CMa shows good correlation. The environment where the SiO masers generate is different between late type stars and star forming regions.

Keywords: SiO masers, VLBA, Orion-KL

Introduction

The comparison of the spatial distributions of several SiO transitions in different states is a clue to solve the question what is pumping SiO maser emissions. The comparison of SiO J=1–0 at v=1 and v=2 states were made in some late-type stars (e.g. Miyoshi et al. 1994). We performed the comparison of two states of SiO masers in the nearest massive star forming region Orion KL (450pc; Genzel & Stutzki 1989).

The Orion KL is a rare star-forming region which has SiO maser sources and has been well studied in relation of SiO masers and central activity of this region (source I; Menten & Reid 1995). The spectrum shows double peak (Doeleman et al. 1999) and red-shifted components were situated in northwest, blue-shifted components situated in south-east (Morita et al. 1992).

Y. Nakada et al. (eds.), Mass-Losing Pulsating Stars and their Circumstellar Matter, 321–322.

Observation and Results

We observed the vibration states v=1 (ν=42.820539 GHz) and v=2 (ν = 43.122027 GHz), J=1–0 transitions of SiO masers in Orion KL region for 20 minutes in 24 Oct. 1998 using ten antennas of VLBA. The visibilities have a spectral resolution of 50 m s^{-1} for 8 MHz bandwidth of 2 IFs. The reduction was done using NRAO AIPS. After the self-calibration for V_{LSR} = 15.5km s^{-1} channel for each line IF, synthesis images of every channel were performed. SiO masers extended about 200 mas.

Discussion

To compare the distribution of two transition, we calculated cross correlations of brightness distributions of two maps in the same velocity channels. Figure 1 is the relation of the intensity of v=1 spots and the angular distance to the nearest v=2 spots with the same velocity channel. In spite of the v=1 and v= 2 transition shows similar distribution in the super-giant VY CMa (Miyoshi 2002), the distribution in the Orion KL shows no regularity. The differences of the correlation of two transitions in Orion KL and VY CMa suggest that the pumping mechanism of the SiO maser in star-forming region is different from that of late type stars, after all. Around the strong masers, the distributions of v=1 and v=2 were quite different. The difference in two transition lines of SiO maser in Orion KL suggests they traces different physical and chemical conditions. The mechanism of SiO maser in Orion-KL might be radiative pumping.

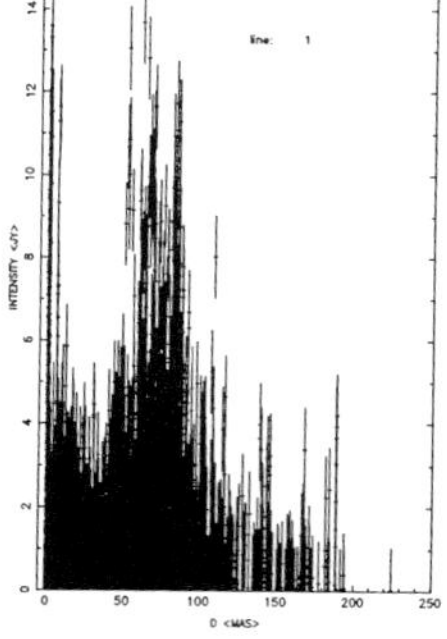

Figure 1. Comparison of the v=1 and v=2 states line of Orion KL. The intensity and distance relations.

References

Doeleman, S.S., Lonsdale, C.J., Pelkey, S. 1999, ApJ, 510, L55

Genzel, R. & Stutzki, J. 1989, ARA&A, 27, 41

Menten, K. & Reid,M. 1995, ApJ, 445, L157

Miyoshi,M., et.al. 1994, Nature, 371, 395

Miyoshi,M. 2002, *in this volume*

Morita, K.I., Hasegawa, T., Ukita, N., Okumura, S.K. & Ishiguro, M. 1992, PASJ, 44, 373

SILICON ISOTOPIC ABUNDANCES IN PULSATING SiO MASER STARS

B.W. Jiang
National Astronomical Observatories, Chinese Academy of Sciences
jiang@ns.bao.ac.cn

J.J. Shen
National Astronomical Observatories, Chinese Academy of Sciences
Department of Astronomy, Beijing Normal University
sjj@yac.bao.ac.cn

J. Sun
Department of Astronomy, Beijing Normal University
sunjin@bnu.edu.cn

Abstract The abundances of ^{28}Si and ^{29}Si in pulsating SiO maser stars are calculated by fitting the SiO maser emissivity from the model to the observed integrated SiO maser intensity. For the first example star R Cas, the abundances of ^{28}Si and ^{29}Si are 1.0e-4 and 3.4e-6 respectively, which corresponds to a ratio of 29, higher than the terrestrial. The calculation to another Mira variable NV Aur draws the ratio of 32, similar tendency as R Cas.

Keywords: abundance, maser

The abundances of silicon and its stable isotopes ^{29}Si and ^{30}Si have been obtained based on infrared and radio spectroscopic observation. Lambert et al.(1987) and Tsuji et al.(1994) analyzed the spectra near 4μm, i.e. the first overtone ($\Delta v = 2$) bands of SiO radical, of some late-type stars to measure the abundance ratio between ^{28}Si and ^{29}Si or ^{30}Si. But such analysis of near-infrared spectrum suffers from the uncertainty from serious line blending. Direct line intensity ratios of ^{28}SiO, ^{29}SiO and ^{29}SiO lines in the radio bands have also been used to infer the isotopic abundance ratios (see e.g. Ukita & Kaifu (1987), Wolff (1980)), although which may be too much simplified the real situation.

Y. Nakada et al. (eds.), Mass-Losing Pulsating Stars and their Circumstellar Matter, 323–325.

We intend to calculate the silicon isotopic abundances by fully solving a radiative transfer equation to fit the observational result of the SiO maser lines at 43GHz i.e. the J=1-0 rotational transitions. It is hoped the high S/N maser lines and careful modeling will bring about an accurate determination of the silicon abundance.

The basic frame of this model is imported from Zubko & Elitzur's code MOLPOP (private communication), which solves the statistical level population equations for a line-emitting molecular environment by using the escape probability approach to describe the interaction of the molecules with the radiation. Under the assumptions of spherical symmetry, large velocity gradient and collisions with only H_2, we calculate accurately the dilution factor of the stellar radiation field without simplification in order to match the fact that SiO maser emission comes close to the stellar atmosphere (Miyoshi et al. 2002). To couple with the location of SiO maser emission in the stellar atmosphere where the column density drops rather fast than in the steady wind region, the density structure is taken as r^{-3}.

Table 1 Values of the parameters to fit the observed SiO maser intensity of R Cas

Distance	160 pc
Radius	3.8e13 cm
T_{eff}	2500 K
T_{gas}	1500 K
n_{H_2}	3.5e10 cm^{-3}
d(lnV)/d(lnR)	0.8
V_0	0.6 km s^{-1}
n_{SiO}/n_{H_2}	1.0e-4

Figure 1 SiO maser photon emissivity of R Cas from the values of the parameters given in the left table

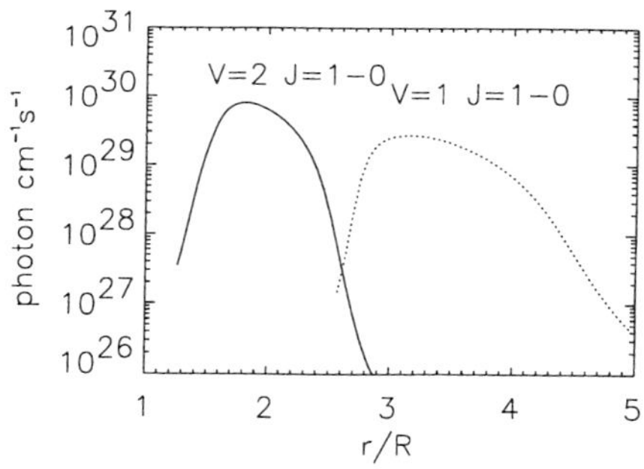

R Cas is chosen as the first object to calculate the abundance of silicon and its isotopes because the available abundant data on this object can put constraints on most of the stellar parameters. The observed ^{28}SiO and ^{29}SiO maser intensities are taken from Jiang et al. (1996), which correspond to 1.01e43 photons/s, 1.79e43 photons/s and 1.35e41 photons/s for ^{28}SiO J=1-0 V=1, V=2 and ^{29}SiO V=0 respectively. The abundance of ^{28}SiO is 1.0e-4 and the abundance of ^{29}SiO is 3.4e-6, and their ratio is 29. However, because the model can hardly produce the maser at V=0 without line overlap, we add the line overlap effect and change a little the velocity gradient to fit the observed ^{29}SiO V=0 J=1-0 emission.

For another Mira variable NV Aur, by adopting a distance of 410pc and a diameter of 7.0e13 cm, we derived the ^{28}SiO abundance ~7.2e-5 and ^{29}SiO abundance ~2.2e-6, therefore the ratio ^{28}Si/^{29}Si is 32. Similar to R Cas, the abundance ratio ^{28}Si/^{29}Si is higher than the terrestrial. Though there are two stars calculated at present and we can not conclude such tendency is true, more calculations to about 10 object shall be done in the future and reveal the truth of the Si isotopic ratio in SiO maser stars.

References

Jiang, B., et al. 1996, ApJS, 106, 463
Jiang, B. 2002, ApJ, 556, L37
Lambert, D., et al. 1987, Ap&SS, 133, 369
Miyoshi, M. 2002, this proceeding
Tsuji, T., et al. 1994, A&A, 289, 469
Ukita, N., & Kaifu, N. 1988, IAUC, 108, 51
Wolff, R. 1980, ApJ, 242, 1005

V

DISTRIBUTION OF MASER SOURCES IN THE GALAXY

INFRARED PROPERTIES AND GALACTIC DISTRIBUTION OF OH/IR STARS

H.J.Habing
Leiden Observatory
habing@strw.leidenuniv.nl

Abstract We discuss 766 masers detected in two surveys in the 1612 MHz OH maser line, one made with the Australian Compact Array and the other with the (U.S.) Very Large Array. The surveys covered the sky in l from $-45°$ to $+45°$, in b from $-3°$ to $+3°$ and over the radial velocity range from -300 km s^{-1} to +300 km s^{-1}. We identified the masers with sources found in published near- and mid-IR surveys and find that the large majority of the doubly peaked masers correspond to AGB stars, but that there exists a small group with a similar maser line profile but different infrared colors; probably they are very young post-AGB objects. We confirm with new evidence that OH/IR stars with larger outflow velocities are more massive and younger than those with a smaller outflow velocity. We find an asymmetry in the longitude distribution of the OH/IR stars in the inner Milky Way Galaxy and explain this by the assumption of the existence of a galactic bar.

Keywords: masers, galactic structure, AGB stars, post-AGB stars

1. Introduction

The 1612 MHz maser associated with AGB stars is strong and can be detected with existing radio telescopes in a major part of our Milky Way Galaxy (= MWG). This offers the possibility to study the distribution of a class of AGB stars in the interior parts of the MWG, from where no visual light can reach us. Several such surveys have been made in the past. For some the detection limit was much worse than now can be achieved. Others were based on IRAS data and this means that sources close to the galactic plane and in the inner Galaxy have been missed: there the IRAS survey was strongly affected by source confusion.

We have carried out two surveys for 1612 MHz sources, one in the Southern sky, using the Australian Compact Telescope Array (ATCA),

Y. Nakada et al. (eds.), Mass-Losing Pulsating Stars and their Circumstellar Matter, 329–337.

and the other in the Northern sky, using the Very Large Array (VLA) of the (U.S.) National Radio Astronomy Observatory. We covered the longitude range between $l = \pm 45^\circ$ and between latitudes $b = \pm 3^\circ$ and in velocity roughly between ± 300 km s^{-1}. The project was started by Huib Jan van Langevelde and myself and then taken over by Maartje Sevenster for her doctorate thesis in Leiden. Thereafter she continued the study at Mount Stromlo. Several others became involved; we mention especially Jessica Chapman, Herwig Dejonghe and Neil Killeen. All data are freely available at http://www.strw.leidenuniv.nl/ sevenste/

2. Two surveys

A summary of major properties of both surveys is given in table 1. The surveys were made with synthesis arrays, which were pointed at pre-selected grid points. Analysis of the responses by all possible interferometer combinations then yields a map of the radiation within the primary beam of each telescope and with a resolution determined by the length of the array. The sensitivity is highest at the center of the beam and decreases going outward. The detection limit is not uniform but it is very well known in each point of the field that we imaged; this is a considerable advantage in the analyses of the completeness of the surveys. The analysis shows that the VLA survey is 99% complete for sources above 2 Jy. For the ATCA survey the completeness is 95% for sources with peak flux density above 375 mJy. These detection limits may be compared to the flux densities of the strongest sources near the galactic center (Lindqvist et al., 1992): 3 Jy. We conclude that beyond 10 kpc from the Sun we begin to miss the detection of the weaker OH/IR stars; our survey will be incomplete from thereon.

Table 1. Specification of the two surveys

	ATCA	VLA
survey instrument	ATCA	VLA
longitude	$-45^\circ \leq l \leq +10^\circ$	$5^\circ \leq l \leq +45^\circ$
latitude	$-3 \leq b \leq +3^\circ$	$-3^\circ \leq b \leq +3^\circ$
# pointings	539	965
velocity range	-330, 410km s^{-1}	-306, +306 km s^{-1}
velocity res.	1.5 km s^{-1}	2.3 km s^{-1}
noise	$<$ 40mJy	32 mJy
completeness	$>$ 95% for $>$ 375 mJy	$>$ 99% for $>$ 500 mJy

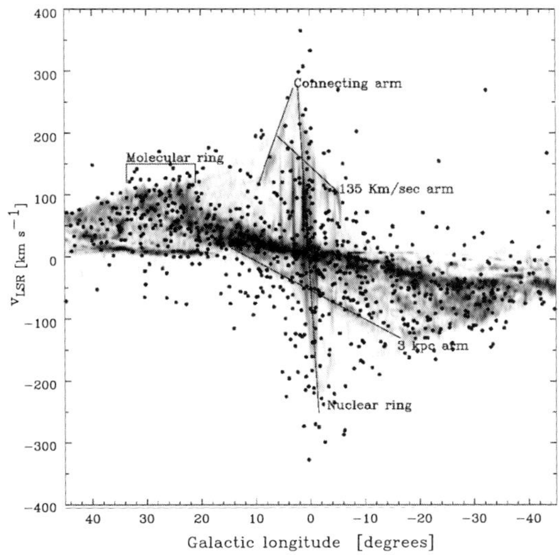

Figure 1. The distribution of the 1612 MHz OH masers from both surveys in galactic longitude and radial velocity. In grey the interstellar CO emission (Dame et al., 2001) and labels from (Fux, 1999). Courtesy Maria Messineo.

In our analysis of the observations we looked for emission peaks in the line profiles. Most of the detections consist of two emission peaks with the shape characteristic for OH/IR stars but a number of single peaks have been found as well. We are not at all certain about the nature of these single peak sources. Some will be OH/IR stars with one peak below the detection threshold, but others may have a different origin, e.g. in a star forming region. We remind the reader that V_*, the average of the two peak velocities corresponds to the radial velocity of the star and that the velocity difference of the two peaks, ΔV, is twice the expansion velocity, $V_{\rm exp}$, of the envelope around the AGB star that is the heart of the maser.

In the southern disk area $(-45^\circ \leq l \leq -10^\circ)$ we detected 166 double peak sources and 32 single sources. In the northern disk we detected 276 double peak sources and 10 single peaks. In the bulge area, $-10^\circ \leq l \leq +10^\circ$ we found 248 double peak sources and 32 single peaks.

Figure 1 shows what is perhaps the most important result of our two surveys: the distribution of all 766 stars in the (l, V_*)-plane. This diagram contains information on the kinematics of AGB stars in the inner MWG. Major features of the CO distribution are the same in the distribution of the OH-stars, but it is also obvious that the OH/IR stars have a wider distribution. This is to be expected: whereas the gas must flow in closed orbits the stars do not have to. This freedom allows the stars to migrate out of the contours of the gas motions.

3. Infrared Identifications

Several previous studies have lead to the conclusion that each 1612 MHz maser star can be identified with a stellar source with a spectral energy distribution peaking between 2 and 20 μm. In a recent publication (Ortiz et al., 2002) a cross identification was made between sources detected in the ISOGAL galactic plane survey at 8 and 15 μm (Omont et al., 2002, A&A accepted) with the OH/IR maser stars detected at 1612 MHz. *All* maser stars inside an ISOGAL field had an infrared counterpart.

The 766 OH/IR maser stars sources from the ATCA and VLA surveys have been cross identified with point sources from the infrared surveys IRAS, MSX and 2MASS (Sevenster, 2002a; Sevenster, 2002b) with the following results:
IRAS: An all sky survey with flux measurements at 12, 25, 60 and 100 μm. We expect that all 766 maser sources have a strong infrared counterpart, but we identified only 587 IRAS counter parts and even for those there frequently is no good flux measurement at 12, 25 and 60 μm.We missed 179 sources because of incompleteness of the IRAS data base due to source confusion at low galactic longitude and close to the galactic center.
MSX: A galactic plane survey at 4, 8, 12, 15, 21 μm. We found counterparts for 494 OH/IR stars.
2MASS: An all–sky survey in the JHK-bands. Not all data have been taken or they have not yet been made public. We identified 194 of our stars in this data base.

In the following text we use "colors". A color "$[a - b]$" is defined as $2.5\,log\,(S_b/S_a)$, where S_a is the flux density at wavelength λ_a. R_{21} and R_{32} are colours where "1" indicates the 12 μm IRAS photometric band, "2" the 25 μm band and "3" the 60 μm band. Figure 2 shows the distribution of the sources identified with IRAS sources in a two color diagram. The different symbols will become clear in figure 3. Of importance here is the fact that they are all 1612 maser sources. The curve is, from left to right, the locus of stars with an envelope of increasing optical thickness. Most masers are close to the curve and are AGB stars. Those that deviate from the curve are probably all post-AGB objects. Later we will discuss two samples of all objects with a well determined outflow velocity, $V_{\rm exp}$; the sources that in figure 2 are located to the left of the curve will form the sample "LI" and those to the right of the curve

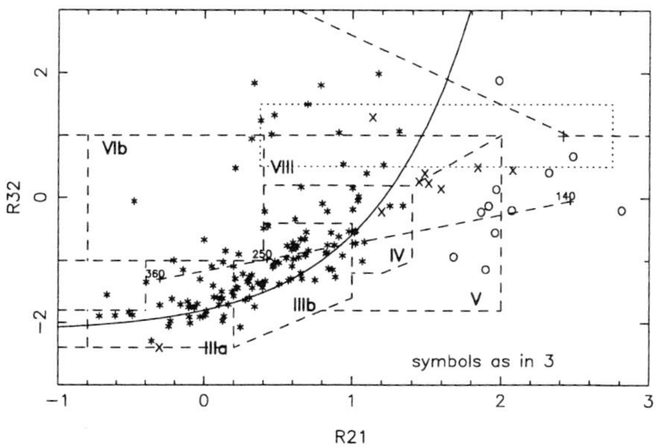

Figure 2. IRAS sources identified with 1612 MHz maser sources have been placed in a two color diagram (van der Veen and Habing, 1998). See the text for a discussion. From (Sevenster, 2002a).

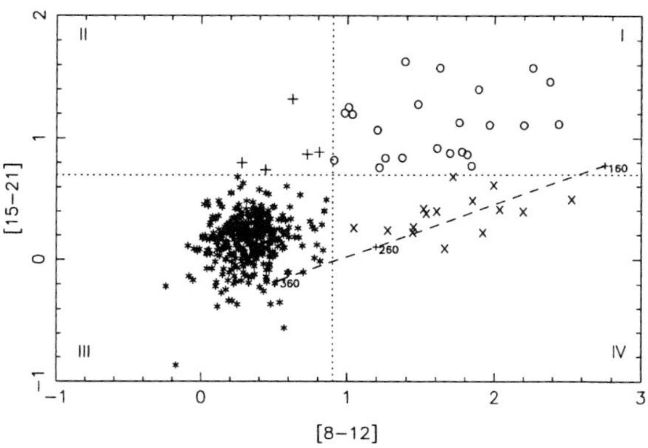

Figure 3. The distribution on the MSX ([8- 12], [15- 21]) plane of 1612 MHz masers identified as point sources in the MSX catalog. The dotted lines separate the diagram into quadrants. The sources are represented by different symbols in each of the quadrants (Sevenster, 2002a).

will be called "RI".

The analysis of the sample identified with MSX sources offers news compared to what was known from the comparison with the IRAS data base. In figure 3 we show the MSX colours [8-12] and [15-21] of the maser sources identified with MSX point sources. The distribution is remarkable; two dotted lines divide the diagram in four parts. In the lower left quadrant most sources are found; they concentrate around the point (0.3, 0.3). The other three quadrants contain a minority of points with no specific concentration. The stars are represented by a symbol according to the quadrant in which they are present. The same symbols have been used in figure 2. Looking at that figure again it is clear that the asterisks represent mass-losing AGB stars but that the other symbols indicate non-AGB stars– we suggest that they are all post-AGB stars.

Another diagram with unexpected properties is shown in figure 4. It contains the MSX [8-12] colors plotted against $V_{\rm exp}$, the expansion velocity of the circumstellar envelope. The surprise in this diagram is

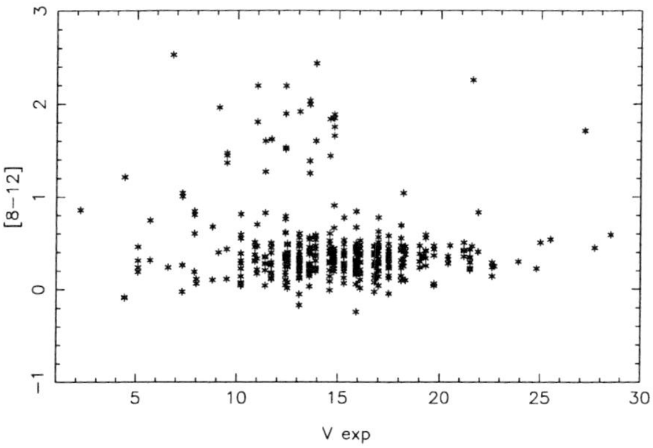

Figure 4. A distribution of 1612 MHz masers identified with MSX sources. The horizontal axis contains V_{exp} and the vertical axis the MSX color [8-12] (Sevenster, 2002a).

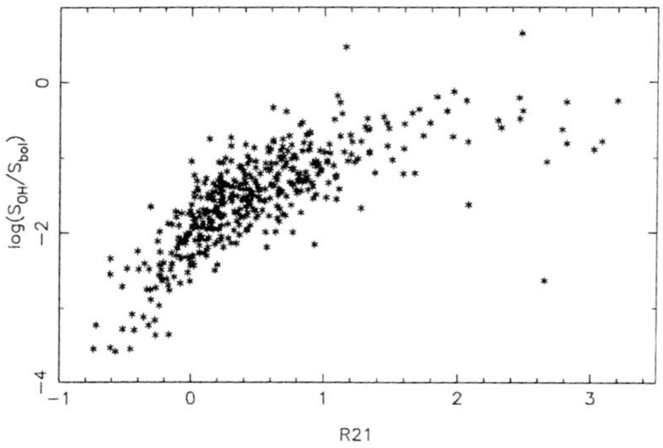

Figure 5. The logarithm of the ratio between the OH maser line flux and the bolometric flux as a function of the color R_{21}.

the set of points at $V_{\mathrm{exp}} < 15\,\mathrm{km\,s^{-1}}$ and at [8-12] > 1.0. The high [8-12] value makes it likely that these are post-AGB objects. This does not show, however, in the 1612 MHz line profile because that has the regular form of the AGB-stars. It leads to the conclusion that the regular 1612 MHz profiles does not *always* belong to an AGB star but occasionally may belong to a post-AGB star.

There is even more to conclude. In the second paper (Sevenster, 2002b) the LI and RI samples (defined above; cf. figure 2) are discussed. Their location in diagram 3 suggests that these sources will be post-AGB stars. The LI sources have a significantly higher outflow velocity V_{exp} than the RI sources and it is concluded that the LI sources have a higher main-sequence mass. Sevenster suggests 4 $M_{\odot}$ for the LI sources and 1.7 $M_{\odot}$ for the RI sources.

Finally, a long awaited result is shown in figure 5: it shows a ratio r as a function of R_{21}, where r is the ratio between the maser flux, S_{OH}, and the bolometric flux, S_{bol}. The diagram shows a clear correlation between R_{21} and r. The spread around the average is at least partially caused by the variability of each source and the fact that S_{OH} and S_{bol}

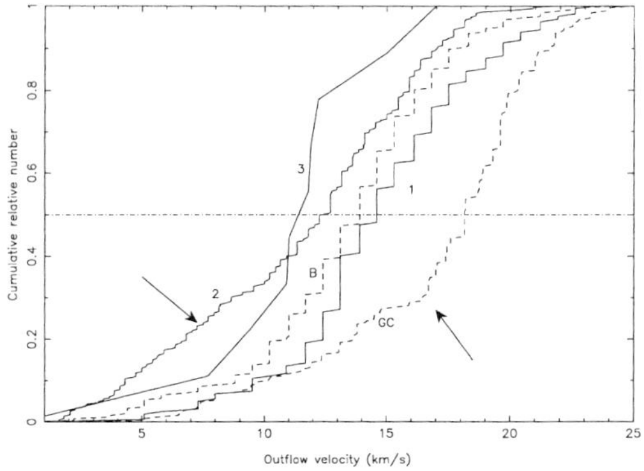

Figure 6. Cumulative histogram. Horizontally the outflow velocity, V_{exp} and vertically the fraction of all the masers with lower outflow velocities V_{exp}.

have been measured at different times.

4. Distribution in the Milky Way Galaxy

As I said at the beginning of my talk, the surveys were set up with the hope to obtain insight in the galactic distribution of AGB stars with thick envelopes and in their kinematics. The results have been published some time ago (Sevenster et al., 1999; Sevenster, 1999; Sevenster et al. 2000). I will now comment on two specific results.

The first result is contained in figure 6. It shows the latitude distributions of several samples of maser sources. Sample GC consists of maser stars close to the galactic centre (Sjouwerman, 1998); sample 1 contains the maser sources found in our ATCA survey at longitudes between $-10°$ and $-45°$; sample B the sources from the same survey, but located in the bulge ($-10° < l < +10°$). Sample 2 contains 1612 MHz masers detected in Arecibo (Chengalur et al., 1993) and sample 3 consists of an unfortunately very small sample of 1612 MHz stars in the outer MWG (Blommaert et al., 1993). The construction of the Arecibo telescope prevents it to look at the galactic centre and the sample of sources is quite local. It then follows that the various samples represent stars at increasing distance from the galactic centre, in the order GC, B, 1, 2, 3.

The diagram is cumulative: it gives the fraction of sources with V_{exp} below the value on the abscissa. A horizontal dash/dot line determines the median value (cumulative number = 0.5). This point is reached first by the most distant sample ("3") followed by 2, B, 1,and GC, which is a (slightly imperfect) ordering of increasing galactocentric distance. The ordering may be the consequence of two facts, probably effective simultaneously: a decrease in metallicity and a decrease in main-sequence

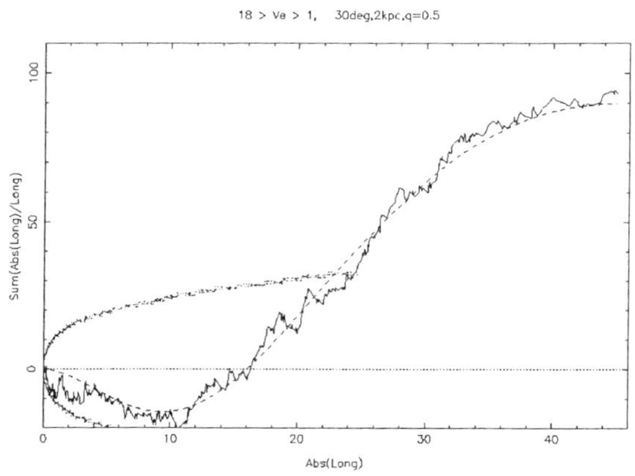

Figure 7. Vertically the number of sources with a positive longitude minus those with a negative longitude below the absolute longitude given on the horizontal axis.

mass of the AGB stars when moving outward from the galactic center. It stresses the point that V_{exp} is a parameter that measures a combination of initial stellar mass and metallicity and thus the age of the sample for which it has been determined.

Finally a few words about the bar in the inner MWG. There is a remarkable asymmetry in the longitude distribution of the 1612 MHz stars: there are systematically more with positive longitudes than with negative longitudes, as shown in figure 7 (Sevenster, 1999). On the horizontal axis the absolute longitude, $\mid l \mid$, is displayed and on the vertical axis *Sum*, the number of OH/IR stars at positive longitude between 0 and l minus the number of stars at longitudes between 0 and $-l$; mathematically expressed:

$$Sum = \Sigma_i \ (l_i/ \mid l_i \mid).$$

If the sources are distributed symmetrically with respect to $l = 0°$ then *Sum* should be zero. *Sum* is however a stochastic variable; the expectation value is 0, but actual realisations may have a value different from 0. The probability that this happens can be easily calculated from the binomial theorem. It then follows that the probability that *Sum* has a value outside of the parabola-like curve in figure 7 is less than 3%. We conclude that the distribution of the maser sources is indeed asymmetric with respect to $l = \pm 0°$: there are more sources on the positive side than on the side of negative longitudes. How to interpret this result? As we have argued above the ATCA and VLA surveys become incomplete beyond the galactic center. If there is a bar with closest distance to us at positive longitudes then we see the sources in the bar at positive longitudes, but the sources in the bar at negative longitudes are too faint to detect. The curve drawn through the observed value corresponds to a simple model for a bar.

5. Conclusions

1 We have detected 766 OH masers in the inner Milky Way Galaxy

2 Of those with a regular line profile, i.e. with two maser peaks, most are AGB stars but there is a small minority of post-AGB stars among them.

3 AGB stars with smaller outflow velocities have a broader distribution in b and must be of lower main-sequence mass than those with a higher $V_{\rm exp}$.

4 Counting the stars from $l = 0°$ one finds systematically more stars at positive l than at negative l. This can be explained by the existence of a galactic bar.

Acknowledgments

We are grateful to National Radio Astronomy Observatory and to ATCA for granting us so much observing time.

References

Blommaert, J. A. D. L., van der Veen, W. E. C. J., & Habing, H. J. 1993, A&A, 267, 39

Chengalur, J. N., Lewis, B. M., Eder, J., & Terzian, Y. 1993, ApJS, 89, 189

Dame, T. M., Hartmann, D., & Thaddeus, P. 2001, ApJ, 547, 792

Fux, R. 1999, A&A, 345, 787

Lindqvist, M., Habing, H. J., &Winnberg, A. 1992, A&A, 259, 118

Ortiz, R., Blommaert, J. A. D. L., Copet, E., Ganesh, S., Habing, H. J., Messineo, M., Omont, A., Schultheis, M. & Schuller, F. 2002, A&A, 388, 279

Sevenster, M. N., Saha, P., Valls-Gabaud, D. & Fux, R. 1999, MNRAS, 307, 584

Sevenster, M. N. 1999, MNRAS, 310, 629

Sevenster, M. N., Dejonghe, H., Van Caelenberg, K., & Habing, H. J. 2000, A&A, 355, 537

Sevenster, M. N. 2002, AJ, 123, 2772

Sevenster, M. N. 2002, AJ, 123, 2788

Sjouwerman, L. O., van Langevelde, H. J., Winnberg, A., & Habing, H. J. 1998, A&AS, 128, 35

van der Veen, W. E. C. J., & Habing, H. J. 1988, A&A, 194, 125

CIRCUMSTELLAR MASERS IN THE GALACTIC CENTER

Loránt Sjouwerman
National Radio Astronomy Observatory (NRAO), PO.Box 0, Socorro NM 87801, U.S.A.
lsjouwerman@aoc.nrao.edu

Abstract Blind and targeted surveys for maser emission (OH, H_2O, and SiO) in the central degree ($\leq$ 150 pc) of the Galactic center will be discussed, with the emphasis on masers in OH/IR stars. Using the masers, some recent clues about the kinematics and evolution of the central part of the Milky Way galaxy will be given. An outlook toward measuring proper motions and individual stellar orbits in the nucleus of the Galaxy with, for example, the new VERA array will be presented.

Keywords: Masers, Galactic center, AGB stars, Circumstellar matter, Astrometry

Introduction

Of the different types of masers found in the Galactic center (GC), here I will try to give a summary of the masers found in the circumstellar environment (CSE) of late-type stars, mostly Asymptotic Giant Branch (AGB) stars of the OH/IR type. I will not further mention maser emission related to massive star formation, nor mention maser emission generated by shocks between molecular clouds.

First, I will shortly recall the results of studies of OH/IR stars in the Galaxy that are relevant to this contribution in Sect. 1. In Sect. 2, a compilation of different surveys for maser emission in the GC is given. Section 3 describes three different samples of OH/IR stars in the GC, of which one population points to a starburst episode of the nucleus of our Galaxy some few Gyr ago. Finally, in Sect. 4, I will look forward to prospects for astrometry with AGB stars in the GC, which makes it possible to obtain e.g. a geometric distance to the GC and reveal the shape of stellar orbits and the central gravitational potential.

1. Circumstellar masers

After the mission of the Infrared Astronomical Satellite (IRAS) in the mid 1980-ies, it became clear that the IRAS color-color diagram was

Y. Nakada et al. (eds.), Mass-Losing Pulsating Stars and their Circumstellar Matter, 339–346.

an efficient tool to find circumstellar masers, in particular OH/IR stars (Olnon et al. 1984; van der Veen & Habing 1988). The dust particles in the circumstellar environments (envelopes, and/or shells) obscure the central star, and re-radiate the stellar emission in the infrared (IR). The radiation pressure on the dust, and the collisions with the gas molecules, causes the star to loose it mass and form this expanding circumstellar environment, with expansion velocities ($v_{\rm exp}$) ranging between a few, to about 30 km s^{-1}, depending on physical properties of the star and its environment (such as the dust-to-gas ratio δ, which I will use as synonym for metallicity). The circumstellar environments of the variable, oxygen-rich stars may provide the physical conditions to form and sustain maser emission from, in particular, SiO, H_2O, and OH molecules.

Most of these circumstellar masers are found in AGB stars, low and intermediate-mass stars with a degenerate C/O core that have evolved away from the main sequence toward the asymptotic giant branch of the color-magnitude or Hertzsprung-Russel diagram. These masers are also found in massive supergiant stars that have non-degenerate cores, but their number density is very small compared to the AGB stars.

By targeting IR sources with IRAS colors characteristic of CSE's, many double peaked 1612 MHz OH masers, the so-called OH/IR stars, have been found over the Galaxy (e.g. te Lintel Hekkert et al. 1991). The OH/IR stars are AGB stars with the highest mass-loss rates, and with denser obscuring shells compared to optically visible Mira-type and Red Giant Branch stars. The 1612 MHz OH maser originates in the outer part of the CSE, typically 1000 AU away from the star, where one observes a blue shifted and a red shifted maser from the expanding shell. The centroid of the blue and red shifted OH maser velocity reveals the line-of-sight velocity of the central star; half the difference defines the shell expansion velocity, $v_{\rm exp}$. As the OH masers are generally strong, OH/IR stars can be found all over the Galaxy and even in the Large Magellanic Cloud (LMC; e.g. Baud et al. 1979; Wood et al. 1992).

In the studies involving physical properties and Galactic distribution it is evident that there is a statistical correlation. Of course there is a blend of different OH/IR stars 'polluting' the samples, so the correlations should be considered in statistical sense and not be applied to individual objects. However, in general the more luminous, more massive, higher mass-loss rate, higher $v_{\rm exp}$, higher metallicity (δ, dust-to-gas ratio), and younger objects with a smaller stellar velocity dispersion are found in the Galactic plane, following Galactic rotation more closely. In short, there is a statistical (anti-)correlation of $v_{\rm exp}$ with age, in the sense that low luminosity, lower mass, low mass-loss rate Mira-type variables correlate with lower $v_{\rm exp}$ sources, and high luminosity, higher mass, high

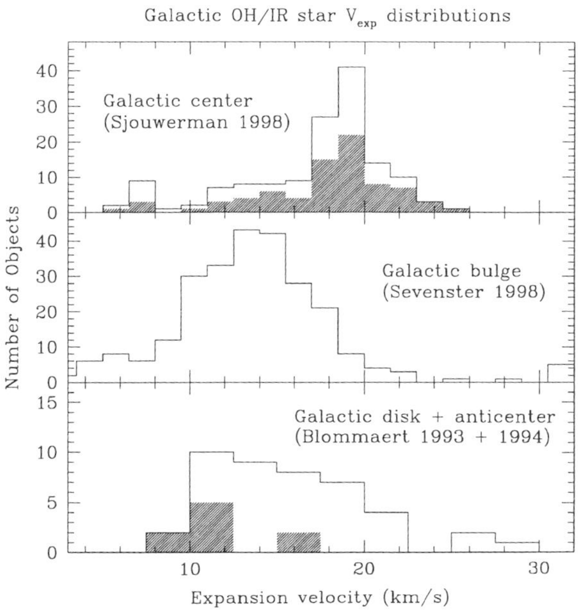

Figure 1 Distribution of v_{exp} for different samples. The bottom panel shows the combined distribution for OH/IR stars in the Galactic disk and anti-center (shaded), the center panel shows the distribution for the Galactic bulge. They are characteristic overall wide distributions due to a mixture of stars with e.g. different luminosities, ages and metallicity (δ). The Galactic center distribution (top) is different in the sense that *next to a wide* distribution, it also shows a *peaked* distribution - see Sect. 3 for an interpretation.

mass-loss rate visibly obscured OH/IR stars correlate with higher v_{exp} objects. However, it is also found that in the Galactic anti-center, and in the LMC, high mass OH/IR stars have low v_{exp}, and low mass Galactic bulge stars have higher mass-loss rates than low mass field stars.

Clearly, not only the bolometric luminosity (L_*), but also the metallicity (δ) of the stellar material influences the individual values of v_{exp}. It can be summarized as (Habing et al. 1994; Elitzur, these proceedings)

$$v_{\mathrm{exp}}^2 \propto \delta \cdot \sqrt{\mathrm{L}_*}$$

This explains why the correlation of v_{exp} with age only holds in statistical sense, as the observed samples generally consist of a mixture of stars with different chemical abundances and different mass and luminosities. A histogram of the distribution of v_{exp} for these samples is usually broadened due to this effect (Fig. 1).

A good review of current knowledge of AGB stars and their circumstellar environment is from Habing (1996). Elsewhere in these proceedings, sub-topics on Galactic distribution issues of AGB stars e.g. the period-luminosity relation (Groenewegen, Whitelock), astrometry (van Langevelde) and stellar kinematics and Galactic dynamics (Feast, Habing, Messineo, Winnberg) can be found.

2. Masing AGB stars in the GC

The onset of AGB studies in the GC started when Baud et al. (1979) and Olnon et al. (1981) surveyed for OH/IR stars toward the GC. However, the largest increase in newly found OH/IR stars came with a survey by Lindqvist et al. (1992b; also Winnberg et al. 1985), who used an interferometer to deal with source confusion as noted by Habing et al. (1983). Since then several other surveys in the 1612 MHz OH line have been performed, with different telescopes, sensitivity limits and sky and velocity coverages (van Langevelde et al. 1992a; Sevenster et al. 1997; Sjouwerman et al. 1998b). Note that all these OH surveys toward the GC were blind; IRAS was confused in the GC and the IRAS color-color diagram could not readily be used to find the OH/IR sources.

Because of this, the IR colors, luminosities and periods for the GC OH/IR stars were determined by classical photometry, which was difficult because of crowding, extinction, and the variable nature of the sources (e.g. Jones et al. 1994; Blommaert et al. 1998). A blind 2.2 μm IR monitoring survey by Glass et al. (2001) for long-period variables, and the color-color diagrams of the ISOGAL and MSX satellites revealed other AGB stars, i.e. evolved stars without OH masers.

As AGB stars are also known to exhibit H_2O and SiO masers, blind and targeted IR (and OH) surveys were initiated. Many H_2O masers were found, but they were not as numerous as the OH masers and fairly often also connected to star forming regions instead of AGB stars (Guesten & Downes 1983; Lindqvist et al. 1990; Taylor et al. 1993; Levine 1995; and also Nakada et al. 1993; Menten et al. 1997; Sjouwerman et al. 2002). The conclusion was that the H_2O masers in AGB stars were too few and also too variable to warrant a detailed study.

Much more successful and, as I'll show later, much more useful were the blind and targeted searches for SiO masers in AGB stars, especially close to Sgr A*, the black hole in the dynamical center of the Galaxy (e.g. Lindqvist et al. 1987, 1991; Izumiura et al. 1998; Nakada et al. 1993; Jiang et al. 1995; Menten et al. 1997; Deguchi et al. 2000, 2002; Miyazaki et al. 2001; Imai et al. 2002; Sjouwerman et al. 2002). Compared to the H_2O masers in AGB stars, the 43 GHz SiO masers are less variable. Also, the SiO masers are located very close to the star, within a few $\mathrm{km\,s^{-1}}$ of the stellar line-of-sight velocity and at this high frequency their image will be less broadened by interstellar scattering than for the low frequency OH masers. This makes the 43 GHz SiO masers ideal dynamical probes for the central part of the Galaxy. Pilot projects to detect 86 GHz with the 15m SEST (Winnberg 1987; Sjouwerman 1998, pers. comm.) were not successful because of sensitivity. Recently however, Messineo et al. (2002) successfully used the 30m IRAM telescope on an IR selected sample to obtain a detection rate in the 86 GHz SiO line of $\sim$60 %. Almost all of the $\sim$40 strongest 86 GHz SiO masers also show 43 GHz SiO masers, and because VLBI at 43 GHz is easier than at

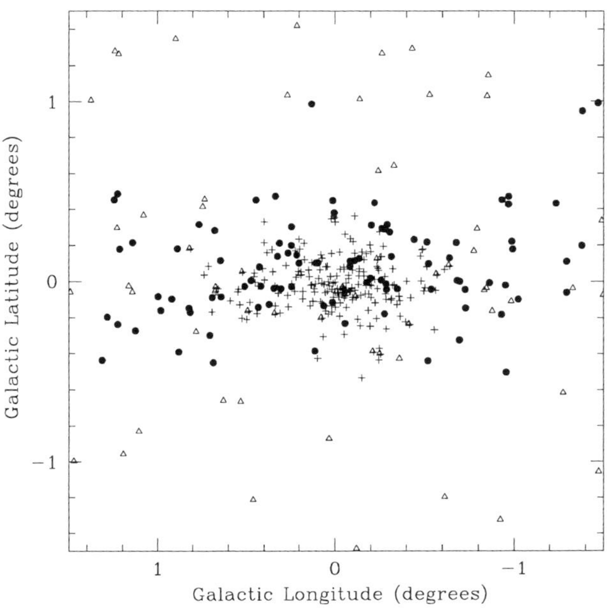

Figure 2 OH (crosses, triangles) and 86 GHz SiO (dots) maser detections in the GC. The 43 GHz SiO maser detections can be found in Sjouwerman et al, these proc. The OH masers were found in deep, blind, and limited sky coverage surveys (crosses: Lindqvist et al. 1992b; Sjouwerman et al. 1998b) - hence partly the concentration to the center (also number density) - and a shallow all-sky survey (triangles: Sevenster et al. 1997). The IR targeted 86 GHz SiO masers (Messineo et al. 2002, also these proc.) fill the area between the two OH surveys.

86 GHz, the 43 GHz SiO masers will be very important for dynamical studies (Sect. 5), e.g. with VERA (Sjouwerman et al., these proc.).

Most of our current knowledge of the AGB stars in the GC (Fig. 2) comes from the earlier OH maser studies in combination with the classical IR photometry and period determinations (also in the OH line; van Langevelde et al. 1993). The H_2O masers are too variable, and not many have been found to allow a detailed study. However, at the moment the 43 (and 86) GHz SiO masers are promising probes to further extend our knowledge about AGB stars in the GC (periods, luminosity, composition, evolution and structure of the GC, etc.).

3. Populations in the GC

Figure 1 displays the $v_{\rm exp}$ distribution for the OH/IR stars in the GC (top panel). Compared to the distribution for the Galactic bulge and Galactic disk plus anticenter, the distribution for the GC is much more peaked. All distributions show a broad histogram that can be explained by a general mixture of stars with different composition and luminosities. In the GC case, not only is the mean $v_{\rm exp}$ high, but in view of the overall distribution there is a clear excess of stars with almost the same $v_{\rm exp}$ (17 $< v_{\rm exp} < 20$ km s^{-1}). As the overall luminosities of the GC stars do not differ much from the OH/IR star luminosities in the bulge (van der Veen & Habing 1990; Blommaert et al. 1998; Wood et al. 1998), the relation

$v_{\rm exp}^2 \propto \delta \cdot \sqrt{L_*}$ gives away the overall high mean metallicity in the center of about 1.8 times the mean metallicity in the bulge.

In the GC, there is a clear distinction in mean Galactic rotation between low $v_{\rm exp}$ sources (no net rotation) and the high $v_{\rm exp}$ sources (strong rotation). The distinction appears at $v_{\rm exp} \approx$ 16.5-17 km s^{-1}, at the rise of the peak in the $v_{\rm exp}$ distribution, in contrast to the 18 km s^{-1} used on statistical grounds by Lindqvist et al. (1992a). The sources that make up the excess in the $v_{\rm exp}$ distribution, are strong Galactic rotators, as the high $v_{\rm exp}$ sources beyond the fall in the peak in the distribution (at 20 km s^{-1}). To investigate the statistical difference between the low and high $v_{\rm exp}$ groups, this means that *in addition*, one has to investigate the 17 $< v_{\rm exp} <$ 20 km s^{-1} sources separately (and not divide them up almost equally in the two earlier groups of $v_{\rm exp} <$ 18 km s^{-1}, and $v_{\rm exp}$ $>$ 18 km s^{-1}). In Sjouwerman et al. (1999) it is shown that the low and high $v_{\rm exp}$ groups have a wide range of physical properties, whereas the properties for the 17 $< v_{\rm exp} <$ 20 km s^{-1} group are also more peaked, leading to the suggestion that most of these stars have the *same* physical properties. Because the AGB phase is short compared to the total lifetime, most of the stars in this group should therefore have the same age - and have been formed at the same time. It is difficult to estimate an exact age because of many uncertainties, but between about 1 and 4 Gyr ago, the Galactic nucleus would have resembled a starburst galaxy.

Note that the general conclusion of Lindqvist et al. (1992a) is not altered, but next to a relatively old population of low $v_{\rm exp}$ stars that show no net Galactic rotation and a relatively young population of high $v_{\rm exp}$ stars that follow Galactic rotation closely, there also is a diminishing population of stars that follow Galactic rotation, has $v_{\rm exp}$ between 17 and 20 km s^{-1}, was formed in an isolated era about 2 Gyr ago, and of which the observed high metallicity OH/IR stars (of M = 1.5-2 $M_\odot$), are the most massive remaining representatives (Sjouwerman et al. 1999).

4. Astrometry in the GC

The best demonstration of astrometry with masers in the GC so far is by Menten et al. (1997). In an effort to find an IR position for Sgr A*, which is a radio source for which no unambiguous IR counterpart has been found, they identified several H_2O and SiO masers with IR point sources, thus linking the IR to the accurate radio reference frame, placing an upper limit to the (non-detected) IR emission from Sgr A*.

Because of interstellar scattering, and in lesser extent the intrinsic size of the masing region, the 1612 MHz OH masers are not suited for astrometry in the GC (van Langevelde et al. 1992b; Frail et al. 1994).

The intrinsic, as well as the scatter sizes for the 43 GHz masers measure about 1 mas at the GC, which means that in principle relative positions of the SiO masers, and thus the underlying stars, with respect to Sgr A* can be made with this precision. In the IR, the best accuracy is 50 mas (e.g. Ghez et al. 1998), which makes the 43 GHz masers (next to the IR work of Ghez and co-workers) very useful to measure the individual stellar orbits around the black hole, Sgr A*, also because the line-of-sight velocity of the star will be known instantaneously (in contrast to spectroscopic IR measurements, which are difficult to perform).

A star in a circular orbit around a 2.6×10^6 $M_\odot$ black hole would show a circular velocity of 106 $km\,s^{-1}$ over the orbital radius in pc. For a distance of the GC of 8 kpc and a typical stellar velocity of 100 $km\,s^{-1}$, this would mean a proper motion of 2.6 mas yr^{-1}. This is measurable with VLBI instruments like the VLBA and VERA. In the VLBA case the feasibility of obtaining mas accurate positions with respect to Sgr A* was (Sjouwerman et al. 1998a) and will (Reid et al., in preparation) be demonstrated. However, because of the rapid phase fluctuations at 43 GHz, VLBI at these frequencies requires the phase to be tracked from a nearby calibrator (e.g. Sgr A* itself!) to the targeted SiO maser, for example with the VLBA in a 20/20 sec calibrator/target cycle (Sjouwerman et al. 1998a) and wasting observing time. The dual beam VERA does not have this disadvantage, because one beam can be pointed at a maser while the other can be observing simultaneously while being pointed at a calibrator source 0.5-2 degrees away.

With VLBI it is possible to measure individual stellar proper motions using the 43 GHz SiO masers as accurate to about 50 $km\,s^{-1}$ within 5 years, and possibly individual stellar orbits within the next decade. The ultimate test will be to compare the proper motions and orbits for the sources which could be measured both in the 43 GHz SiO maser as well as in the IR.

References

Baud, B., Habing, H.J., Matthews, H.E., & Winnberg, A. 1979, A&AS, 35, 179
Blommaert, J.A.D.L., van der Veen, W.E.C.J., & Habing, H.J. 1993, A&A, 267, 39
Blommaert, J.A.D.L., van Langevelde, H.J., & Michiels, W.F.P. 1994, A&A, 287, 479
Blommaert, J.A.D.L., van der Veen, W.E.C.J., van Langevelde, H.J., Habing, H.J., & Sjouwerman, L.O. 1998, A&A, 329, 991
Deguchi, S., Fujii, T., Izumiura, H., Kameya, O., et al. 2000, ApJS, 128, 571
Deguchi, S., Fujii, T., Miyoshi, M., & Nakashima, J. 2002, PASJ, 54, 61
Frail, D.A., Diamond, P.J., Cordes, J.M., & van Langevelde, H.J. 1994, ApJ, 427, L43
Ghez, A.M., Klein, B.L., Morris, M., & Becklin, E.E. 1998, ApJ, 509, 678
Glass, I.S., Matsumoto, S., Carter, B.S., & Sekiguchi, K. 2001, MNRAS, 321, 77
Guesten, R., & Downes, D. 1983, A&A, 117, 343

Habing, H.J. 1996, A&ARv, 7, 97
Habing, H.J., Olnon, F.M., Winnberg, A., et al. 1983, A&A, 128, 230
Habing, H.J., Tignon, J., & Tielens, A.G.G.M. 1994, A&A, 286, 523
Imai, H., Deguchi, S., Fujii, T., Glass, I.S., Ita, Y., Izumiura, H., Kameya, O., Miyazaki, A., Nakada, Y., & Nakashima, J. 2002, PASJ, 54, L19
Izumiura, H., Deguchi, S., & Fujii, T. 1998, ApJ, 494, L89
Jiang, B.W., Deguchi, S., Izumiura, H., Nakada, Y., Yamamura, I. 1995, PASJ, 47, 815
Jones, T.J., McGregor, P.J., Gehrz, R.D., & Lawrence, G.F. 1994, AJ, 107, 1111
Menten, K.M., Reid, M.J., Eckart, A., & Genzel, R. 1997, ApJ, 475, L111
Messineo, M., Habing, H.J., Sjouwerman, L.O., Omont, A., & Menten, K.M. 2002, A&A, in print
Miyazaki, A., Deguchi, S., Tsuboi, M., et al. 2001, PASJ, 53, 501 + 669
Levine, D.A. 1995, PhD thesis, University of California, Los Angeles, U.S.A.
Lindqvist, M., Winnberg, A., Johansson, L.E.B., & Ukita, N. 1987, A&A, 172, L3
Lindqvist, M., Winnberg, A., & Forster, J.R. 1990, A&A, 229, 165
Lindqvist, M., Winnberg, A., Johansson, L.E.B., & Ukita, N. 1991, A&A, 250, 431
Lindqvist, M., Habing, H.J., & Winnberg, A. 1992a, A&A, 259, 118
Lindqvist, M., Winnberg, A., Habing, H.J., & Matthews, H.E. 1992b, A&AS, 92, 43
Nakada, Y., Onaka, T., Yamamura, I., et al. 1993, PASJ, 45, 179
Olnon, F.M., Walterbos, R.A.M., Habing, H.J., Matthews, H.E., Winnberg, A., Brzezinska, H., & Baud, B. 1981, ApJ, 245, L103
Olnon, F.M., Habing, H.J., Baud, B., Pottasch, S.R., de Jong, T., & Harris, S. 1984, ApJ, 278, L41
Sevenster, M.N., Chapman, J.M., Habing, H.J., Killeen, N.E.B., & Lindqvist, M. 1997, A&AS, 122, 79
Sjouwerman, L.O., van Langevelde, H.J., & Diamond, P.J. 1998a, A&A, 339, 897
Sjouwerman, L.O., van Langevelde, H.J., Winnberg, A., & Habing, H.J. 1998b, A&AS, 128, 35
Sjouwerman, L.O., Habing, H.J., Lindqvist, M., van Langevelde, H.J., & Winnberg, A. 1999, in "The Central Parsecs of the Galaxy", eds. H.Falcke et al., ASP Conf. Ser. 186, p.379
Sjouwerman, L.O., Lindqvist, M., van Langevelde, H.J., & Diamond, P. 2002, A&A, in print
Taylor, G.B., Morris, M., & Schulman, E. 1993, AJ, 106, 1978
Te Lintel Hekkert, P., Caswell, J.L., Habing, H.J., Haynes, R.F., Haynes, R.F., & Norris, R.P. 1991, A&AS, 90, 327
van der Veen, W.E.C.J., & Habing, H.J. 1988, A&A, 194, 125
van der Veen, W.E.C.J., & Habing, H.J. 1990, A&A, 231, 404
van Langevelde, H.J., Brown, A.G.A., Lindqvist, M., Habing, H.J., & de Zeeuw, P.T. 1992a, A&A, 261, L17
van Langevelde, H.J., Frail, D.A., Cordes, J.M., Diamond, P.J. 1992b, ApJ, 396, 686
van Langevelde, H.J., Janssens, A.M., Goss, W.M., Habing, H.J., & Winnberg, A. 1993, A&AS, 101, 109
Winnberg, A., Baud, B., Matthews, H.E., Habing, H.J., & Olnon, F.M. 1985, ApJ, 291, L45
Wood, P.R., Whiteoak, J.B., Hughes, S.M.G., Bessell, M.S., Gardner, F.F., & Hyland, A.R. 1992, ApJ, 397, 552
Wood, P.R., Habing, H.J., & McGregor, P.J. 1998, A&A, 336, 925

SiO MASER SOURCES NEAR THE GALACTIC CENTER

Shuji Deguchi
Nobeyama Radio Observatory
National Astronomical Observatory
deguchi@nro.nao.ac.jp

Abstract I summarize results of the SiO maser surveys of the Galactic center region, which have been made in the past 3 years at Nobeyama. About 600 color selected IRAS Sources in the inner ($< 3°$) and outer ($< 10°$) bulge, the 300 large amplitude variables in the nuclear disk ($< 12'$), and the stars in the Galactic-center star cluster ($< 1'$), have been observed with Nobeyama 45-m telescope. Especially in the Galactic center, mapping and monitoring observations have been made. The kinematic of stars very near to the galactic center (within a few pc) and a flare of SiO masers of IRS 10 are discussed. The flare of IRS 10 seems to repeat in every two years.

Keywords: AGB stars; the Galactic center; Masers

1. Introduction

SiO masers are useful to learn kinematics of stars in the Galaxy. They have been found in mass-losing AGB stars in the Galactic bulge and disk, where interstellar extinction often hinder optical radial-velocity measurements. The molecule, SiO, has several rotational transitions at radio frequencies. Two rotational transitions in the vibrationally excited states, $J = 1$–0, $v = 1$ and 2 at 43.122 and 42.821 GHz, respectively, emit strong maser radiation; intensities of these two lines are nearly equal in many stars. With the 45-m radio telescope at Nobeyama, these two lines can be observable simultaneously with the S40 receiver of the bandwidth of 0.4 GHz. The pumping mechanism of SiO has been studied well (see Bujarrabal 1994). Radial velocities of SiO masers very often fall at the middle velocity of OH 1612 MHz double peaks, where the two OH peaks were emitted at approaching and receding parts of the

Y. Nakada et al. (eds.), Mass-Losing Pulsating Stars and their Circumstellar Matter, 347–354.

expanding envelope of mass losing stars (Jewell et al. 1991). Thus, the SiO radial velocity coincides with the stellar velocity.

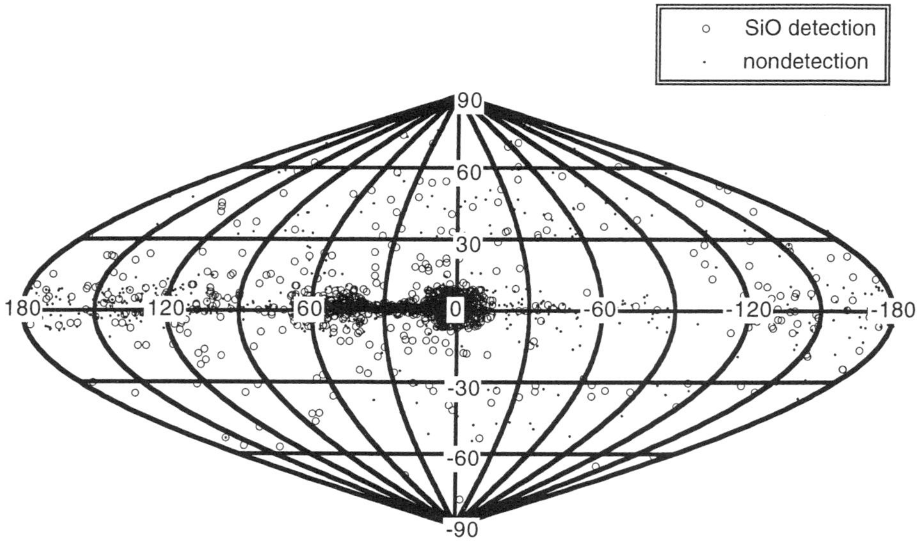

Figure 1. Plot of SiO maser sources in the galactic coordinates. Some sources in the southern sky were observed at Mopra observatory.

We have surveyed about 2500 Galactic IRAS/MSX sources all over the sky and detected about 1300 sources till to the present. Figure 1 shows a sky map of the surveyed sources in the Galactic coordinates. The past SiO maser surveys which were made at Nobeyama are listed in Table 1. The SiO maser database is now available at *http://www.nro.nao.ac.jp/SiO_database.*

Because of the high source density, the IRAS Point Source Catalog is quite incomplete within one degree from the Galactic center. Therefore, blind surveys in this region had been made in OH and SiO maser lines (Sjouwerman et al. 1998; Miyazaki et al. 2001). This situation, however, has now changed owing to new ground-based surveys as 2MASS and DENNIS, and owing to space-based mid-infrared surveys as MSX and ISOGAL. The ground-based mid-infrared array cameras with a resolution of better than 1 arcsecond are now available. A pointed survey in SiO has been made towards the Large Amplitude Variables (LAVs; Glass et al. 2001) within 12′ of the Galactic center (Imai et al. 2002).

Table 1. Past Major SiO maser surveys at Nobeyama.

Period	P.I.	l range	b range	papers
1992–94	H. Izumiura	$\|l\| < 15°$	$\|b\| > 3°$	Izumiura et al. 1995
1994–95	B.W. Jiang	$90 < l < 270°$	$\|b\| < 3°$	Jiang et al. 1996, 1999
1997–99	H. Izumiura	$-10 < l < 40°$	$\|b\| < 3°$	Deguchi et al. 2000ab
1999–02	J. Nakashima	$40 < l < 70°$	$\|b\| < 5°$	to be published
2000	A. Miyazaki	$\|\Delta RA\| < 7'$	$\|\Delta Dec\| < 13'$	Miyazaki et al. 2001
2000	S. Deguchi	$\|\Delta l\| < 2'$	$\|\Delta b\| < 2'$	Deguchi et al. 2002
2001–02	H. Imai	$\|\Delta l\| < 12'$	$\|\Delta b\| < 12'$	Imai et al. 2002

The SiO maser sources within a few parsec (one arc minute) from the Galactic center (Sgr A*, the dynamical center of the Galaxy), are quite interesting from the following view points. They are useful to align between near-infrared and radio coordinate frames, enabling the position of Sgr A* to be pinpointed with an accuracy of better than 0.1 arcsecond on near-infrared images (Menten et al. 1997). Detections of accelerating motions of stars near the Galactic center fixed the mass of the central object at about $3 \times 10^6 M_\odot$ (Ghez et al. 2000).

Using the 45-m telescope at Nobeyama, Izumiura et al. (1997) detected 14 SiO maser sources around the Galactic center. This was a set of pointed observations toward the galactic center and two one-beam offset positions, SiO source positions were not determined with accuracy better than the beam size. In order to remedy the positional uncertainties, we made the 30″ grid mapping and additional 20″ triangle mapping toward the galactic center, giving the positions of SiO sources with an accuracy of 5–10″ (Deguchi et al. 2002). Because the SiO maser intensity is time variable, we have monitored the SiO masers near the Galactic center since 1999. We found an SiO maser flare in the −27 km s^{-1} component (IRS 10 EE) during March–June 2000 and March–May 2002. We report these results and possible periodicity of two years for maser flares of IRS 10 in this paper.

2. Galactic Center Star Cluster

The star cluster at the galactic center consists of mixed stellar population (Krabbe et al. 1995). It involves a number of late-type stars that are potential candidates for OH/SiO masers. Deep surveys in OH 1612 MHz, H_2O 22 GHz, and SiO 43 GHz masers have been made (Sjouwerman et al. 1998; Menten et al. 1997). We have made medium deep surveys of the galactic center in SiO in 1997, 1999, and 2000, with the 45-m telescope at Nobeyama. The detections are listed in Izumiura et al.

(1998), and Deguchi et al. (2002). Including all the SiO sources in the past several years (excluding the overlapped ones), we detected 21 SiO sources in 200″ × 100″ area of the galactic center. We now know 8 SiO maser sources within 20″ radius of the Galactic center. One interesting source is ~ -336 km s^{-1} sources, which is probably located closely to the Galactic center.

The identifications of these sources to OH 1612 MHz and Glass et al.'s objects were made. However, because of the positional uncertainties, the identifications of the -336, -61, and $+85$ km s^{-1} components were not yet made to any known objects successfully; these SiO sources have no OH 1612 MHz counterparts.

Figure 2 shows the longitude-velocity diagram of the galactic center SiO sources and nuclear disk sources (LVPs) within 12′ from the galactic center. It is curious that two sources near the galactic center have large negative radial velocities of ~ -300 km s^{-1}. It is possible that these are bulge sources with highly eccentric orbits (van Langevelde et al. 1992) and are not bounded to the galactic-center massive object (the central blackhole). It was a puzzle that these high velocity sources appeared only at negative velocity sides. However, we now know several cases of large positive velocities, which are indicated in figure 1 (+244 and +156 km s^{-1}).

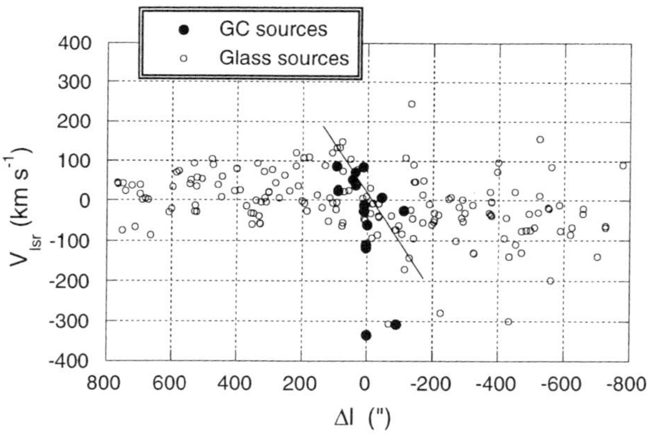

Figure 2. Longitude-velocity diagram for the Galactic-center sources within 120″ (filled circles) and within 12′ (unfilled circles; Glass et al's sources).

The velocity dispersion (from the average linear fit) of the 15 sources within 120″ is 108 km s^{-1} (54 km s^{-1} excluding the extreme two sources). This value gives the mass inside 100″ (4.1 pc) as $\sim 1 \times 10^7 M_\odot$ from the virial theorem. This value is consistent with the mass of the Galactic center region obtained by Krabbe et al. (1995), showing the velocity dispersion of $\sim$ 154 km s^{-1} for late-type stars within 12$''$ of the galactic center.

3. Flare of SiO masers in IRS 10EE

Intensities of SiO masers lines change significantly on a time scale of one year. In March 2000, we noticed that the SiO masers of the -27 km s^{-1} component (IRS 10 EE) had flared up to T_a $\sim$0.5 K ($\sim$1.5 Jy), which was strongest among the SiO masers near the Galactic center. The SiO maser intensity of this source was $\sim$0.3 Jy in 1997 and 1999. The flare lasted for more than two months. Near-infrared monitoring observations of this source over 4 years (Wood et al. 1998; assigned LWHM 65) gave a period of 736 days for the light variation. Extrapolating the light curve given in Wood et al., we estimated the infrared maximum of this source came near the end of 1999.

It is interesting to know whether the maser flare is associated with the pulsating activity of this star or any other temporal variation in the envelope. If it is associated with the stellar pulsation, the flare will repeat in every two years. We monitored this source between February and May in the year of 2002 and found that it brightened again. Figure 2 shows the line profiles of SiO masers in March 2002. This observation clearly indicates the periodicity of about 2 years of the SiO intensity variation, which is consistent with the near-IR periodicity.

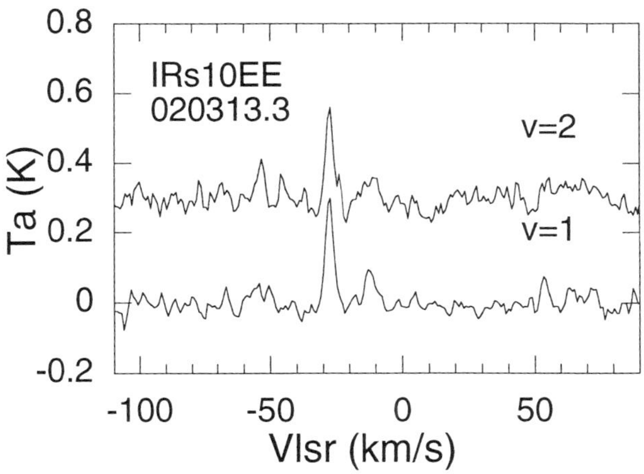

Figure 3. Spectra of SiO maser lines (J = 1–0, v = 1 and 2) toward IRS 10EE (the -27 km s^{-1} component) taken on March 13, 2002. Weaker emissions from other sources are also recognizable.

Intensities of SiO lines of IRS 10EE, $T_a \simeq 0.5$ K, in May 2000 were comparable with the intensity of the SiO $J = 1$–0 $v = 0$ maser of Sgr B2 MD5 (a young star). If they are scaled to the distance of 500 pc, they are comparable to the Orion SiO masers. Therefore, the intrinsic luminosity of SiO masers of IRS 10EE at the flare time was comparable with those of Orion or Sgr B2 SiO sources in the star-forming regions.

Figure 4 shows the time variation of intensities of the IRS 10EE SiO masers. It clearly show the two-year periodicity, though the observations in 1998 was missing.

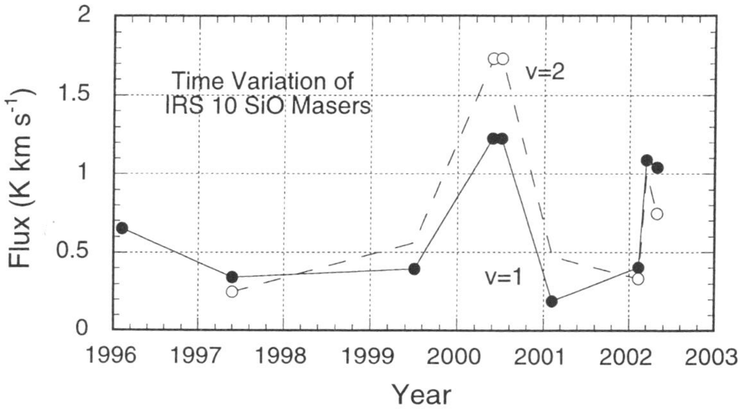

Figure 4. Time variation of the line flux of SiO masers in IRS 10EE. We did not observe in 1998.

The galactic center SiO masers are quite weak in general (~ 0.3 Jy). Only the source IRS 10 became very strong in intensity. The -120 km s^{-1} source, IRS 7, which is another SiO source closer to the Galactic center, was not detected for the last few years with the NRO 45-m telescope except in 1997 (Izumiura et al. 1998). The 12 μm SUBARU COMICS image with a spatial resolution better than $0.5''$ (Okada et al. 2002) showed that the bright stars around the Galactic center are IRS 1, IRS 3, IRS 10, IRS5, and IRS7. This image also showed a complex ring of dust grains surrounds the galactic center along the circumnuclear disk. Near the IRS 10, the 12 μm feature seems to be bent toward IRS 10.

It is quite important to measure the proper motions of maser stars relative to Sgr A* with VLBA. However, a major obstacle to do this was the weak intensities of the SiO maser lines of the galactic center objects. If we perform VLBI experiments during the flare of IRS 10 SiO masers, which occurs periodically in every two years, we can measure the positions of IRS 10 EE accurately with high signal-to-noise ratio. Thus to predict the SiO maser flare time accurately is important.

Accelerating motions of stars, if directly measured, gives the mass of the central object (see Ghez et al. 2000). It can be evaluated as GM_G/R^2, where M_G and R are the mass of the central object and the distance from the central object. In the case of IRS 10, which is about $8''$ apart from Sgr A*, the accelerating motion is of the order of 0.5 μas per year, which is too small to be measured for the present day VLBI technique. Because the accelerating motion is proportional to R^{-2}, the experiment will become more realistic if we find any maser sources within $1''$ from Sgr A*.

4. Summary

We have made mapping and monitoring observations of SiO maser sources near the Galactic center and have detected 21 SiO sources. Among these, 5 sources were identified to the previously known OH 1612 MHz sources, 4 to the large amplitude variables observed at near-infrared wavelengths, and 2 to previously detected SiO sources with accurate known positions with VLA. Four year monitoring observations of SiO masers in IRS 10EE established the two year periodicity of SiO maser flares.

Acknowledgments

The author wishes to thank Dr. S. K. Gorny for making the SiO maser database. He also thank members of the SiO maser Survey team, Drs. T. Fujii, I. S. Glass, Y. Ita, H. Imai, H. Izumiura, O. Kameya, A. Miyazaki, M. Miyoshi, Y. Nakada, J. Nakashima, and N. Ukita for this long-term research. This research was partly supported by Scientific Research Grant (C2) 12640243 of Japan Society for Promotion of Sciences.

References

Bujarrabal, V. 1994, A&A, 285, 953

Deguchi, S., Fujii, T., Izumiura, H., Kameya, O., Nakada, Y., Nakashima, J., Ootsubo, T., & Ukita, N. 2000a, ApJS, 128, 571

Deguchi, S., Fujii, T., Izumiura, H., Kameya, O., Nakada, Y., & Nakashima, J. 2000b, ApJS, 130, 351

Deguchi, S., Fujii, T., Miyoshi, M., & Nakashima, J. 2002, PASJ, 54, 61

Ghez, A. M., Morris, M., Becklin, E. E., Tanner, A., & Kremenek, T. 2000, Nature, 407, 349

Glass, I. S., Matsumoto, S., Carter, B. S ., & Sekiguchi, K. 2001, MN, 321, 77

Imai, H., Deguchi, S., Fujii, T., Glass, I., Ita, Y., Izumiura, H., Kameya, O., Miyazaki, A., Nakada, Y., & Nakashima, J. 2002, PASJ, 54, L19
Izumiura, H., Deguchi, S., & Fujii, T., 1998, ApJ, 494, L89
Izumiura, H., Deguchi, S., Hashimoto, O., Nakada, Y., Onaka, T., Ono, T., Ukita, N., & Yamamura, I. 1995, ApJ, 453, 837
Jewell, P. R., Snyder, L. E., Walmsley, C. M., Wilson, T. L., & Gensheimer, P. D. 1991, A&A, 242, 211
Jiang, B. W., Deguchi, S., & Ramesh, B. 1999, PASJ, 51, 95
Jiang, B. W., Deguchi, S., Yamamura, I., Nakada, Y., Cho, S. H., & Yamagata, T. 1996, AJ, 111, 231
Krabbe, A., Genzel, R., Eckart, A., Najarro, F., Lutz, D., Cameron, M., Krocker, H., Taccorni-Garman, L.E., et al. 1995, ApJ, 447, L101
Menten, K.M., Reid, M.J., Eckart, A., & Genzel, R. 1997, ApJ, 475, L111
Miyazaki, A., Deguchi, S., Tsuboi, M., Kasuga, T., Takano, S. 2001, PASJ, 53, 699
Nakashima, J. 2002, Ph.D. Thesis, Graduate University for Advanced Studies.
Okada, Y., Onaka, T., Miyata, T., Sakou, S., Takubo, S., Honda, M., Yamamshita, T., Okamoto, Y., et al. 2002, Poster paper presented at the annual spring meeting of Japan Astron. Soc.
Sjouwerman, I.O., van Langevelde, H.J., Winnberg, A., & Habing, H.J., 1998, A&A,S, 128, 35
van Langevelde, H.J., Brown, A. G. A., Lindqvist, M., Habing, H.J., & de Zeeuw, P.T. 1992, A&A, 261, L17
Wood, P.R., Habing, H.J., & McGregor, P.I., 1998, A&A, 336, 925

Questions/Comments

- question by H. van Langevelde: These stars on large negative velocities are the same as previously detected in OH ? We seem to detect them only at negative velocities.
  Answer: -340 km s^{-1} component has no OH counterpart. Because of the small number, it is hard to say whether the concentration to the negative-velocity side is real, and/or any maser-beaming effect.

- comment by M. Reid: K. Menten and I have been observing the SiO masers within 12$''$ of Sgr A* with the VLA and VLBA in order to measure proper motions. We now have propoer motions for 6 stars, including IRC 10 EE, which will be published soon.

- comment by I. Sjouwerman: I have some 43 GHz flux densities for IRS 10 EE around the maximum that you missed in the monitoring in 1998–1999.

INFRARED STUDIES OF GALACTIC CENTRE OH/IR STARS

Joris Blommaert
Institut voor Sterrenkunde, K.U. Leuven
Celestijnenlaan 200B, B-3001 Leuven, Belgium
joris.blommaert@ster.kuleuven.ac.be

Abstract An overview of infrared studies of OH/IR stars in the Galactic Centre is presented. From modelling of the observed Spectral Energy Distributions we find that the mass loss rates range from 10^{-6} to $10^{-4} M_{\odot}$/yr. The averaged $M_{\rm bol}$ ranges from -5.5 to -4 mag. The OH/IR stars fall predominantly below the Mira PL-relation. There is a large overlap in luminosities between the kinematically different groups of OH/IR stars. The metallicities however are likely to be different.

Keywords: AGB stars, OH/IR stars, Galactic Centre, luminosities, mass loss, PL relation, stellar populations

Introduction

The sample of Galactic Centre (GC) OH/IR stars is interesting in two aspects. First, it allows to study a group of AGB stars with high mass-loss rates for which the distance is more or less known. Second, by studying the kinematics and the luminosities (and thus ages) of these stars we can learn more about the history of this enigmatic part of our Galaxy, a study which has always been hampered strongly in the visible because of the high and patchy interstellar extinction towards the GC.

In this paper I give an overview of the different infrared projects that aimed at studying the nature of the GC OH/IR stars.

1. OH surveys - kinematic populations

An overview of surveys to detect the OH 1612 MHz maser line from late type stars in the Galaxy is presented by Habing at this conference. These surveys were performed to study the kinematics of stars in different parts of our Galaxy like the disk and bulge. A first deep

Y. Nakada et al. (eds.), Mass-Losing Pulsating Stars and their Circumstellar Matter, 355–362.

survey of the GC, using the VLA was made by Lindqvist et al. (1992a). Work by Sjouwerman et al. (1998) extended the survey to lower OH luminosity sources, but this did not reveal another population of AGB sources. After separating the sample into stars with low (< 18 km/s) and high (> 18 km/s) circumstellar shell expansion velocities, Lindqvist et al. (1992b) found that the kinematical and the spatial distributions of these two subsamples are quite different. Stars with low expansion velocities behave as if they are members of the slowly rotating spheroid, with an isotropic velocity distribution. The subsample of stars with higher outflow velocities is distributed in a flatter system. It was suggested that the subsample with high outflow velocities has a formation history different from the low expansion velocities one; it might be a later addition to the GC, e.g. via a merger.

The outflow velocity of the circumstellar shell is related to the luminosity of the star and to the properties of the dust (see also Elitzur, these proceedings).

2. Infrared Observations

Several projects aimed at observing the OH detected sources in the infrared, where most of the stellar energy is expected to be radiated. In the nineties of last century, several groundbased projects were performed: Jones et al. (1994), Blommaert et al. (1998), Wood et al. (1998). Jones et al. observed 15 sources in the near-infrared and monitored the stars in the L'- band. Blommaert et al. observed just over 30 targets in the 1-12 μm wavelength range and Wood et al. monitored 102 fields of 1 square arc-minute centred on OH/IR stars in the K-band. Periods and amplitudes were determined for 80 sources, together with single epoch data from the J, H and L-band. With the "Infrared Space Observatory" Blommaert et al. (2002) observed 26 stars in the 2-60μm wavelength range. The ISOGAL survey used the ISOCAM instrument to map about 16 square degrees in the Galactic Plane and Bulge (Omont et al. 2002). The ISOGAL catalogue was searched for the GC OH/IR stars by Ortiz et al. (2002).

Data from these projects are combined in the following sections to give an overview of our understanding of the OH/IR stars in the GC.

3. Physical properties

3.1 Characterization

Ortiz et al. (2002) searched for counterparts in the ISOGAL catalogue of the OH detected sources. The mid-infrared colour-magnitude diagram

(Fig. 1) confirms the expectation that the OH sources correspond to mass-losing AGB stars. In comparison with other Long Period Variable stars detected in the same area the OH/IR stars are redder, and are thus suffering from higher mass loss rates ($> 10^{-7} - 10^{-6} M_{\odot}$/year).

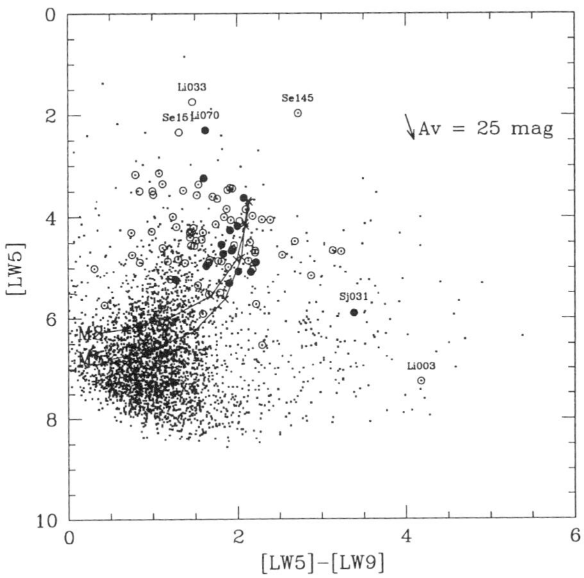

Figure 1. ISOGAL 15 μm vs 7-15 μm colour-magnitude diagram (taken from Ortiz et al. 2002). Filled circles denote the OH/IR stars for which simultaneous 2-filter measurements exist. The open circles have 7 and 15 μm measurements at different epochs. The wider spread in colour for the latter can be explained by the variability of the sources. For comparison ISOGAL sources in a sub field are shown as small dots. These are expected to be predominantly AGB stars with lower mass loss rates than the OH/IR stars. The curves indicate tracks of M5 and M8 giants with $5000 L_{\odot}$ luminosity and mass loss rates ranging from 1×10^{-8} up to $1 \times 10^{-5} M_{\odot}$/year.

3.2 Period - Luminosity

Wood et al. (1998) conducted a monitoring-program of the OH sources in the K-band. Most of the stars have periods in the range of 400 to 800 days, but longer periods do exist. Fortunately their monitoring-program overlapped in time with the ISO-observations by ISOGAL and Blommaert et al. (2002). In this way, we were able to derive amplitudes at 15 μm. Fig. 2 shows the relation between the difference in K-magnitude and the 15 μm magnitudes taken at the same epochs. Assuming that the variations at both wavelengths occur simultaneously, we find that the amplitude at 15 μm is about 1 magnitude (halve the value

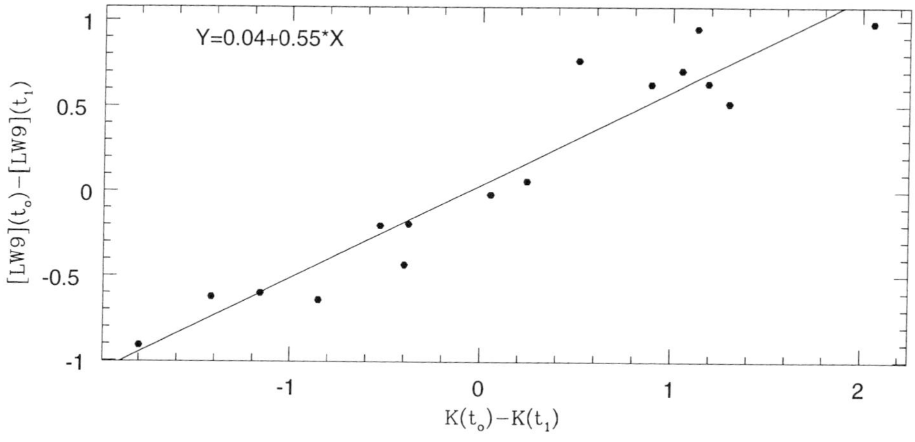

Figure 2. Comparison of the 15 μm magnitudes with the K-magnitudes taken at the same epoch.

at 2.2 μm). Assuming that the amplitude scales linearly between these 2 wavelengths and remains the same at 25 μm as at 15 μm we correct our data to an average luminosity. The averaged M_{bol} ranges from -5.5 to -4 mag.

With the periods determined by Wood et al. (1998) we can now place the OH/IR stars in the Period-Luminosity diagram. Fig. 3 shows the OH/IR stars of Blommaert et al. (2002), together with the P-L relation determined by Glass et al. (1995) for LPVs in the SGR I bulge window. For the comparison the luminosities are calculated assuming a distance of 8.7 kpc. Predominantly the stars fall below the PL relation, and would even do more so if a distance of 8 kpc was assumed. This is in agreement with the earlier work by Blommaert et al. (1998) and Wood et al. (1998) and the findings of Ortiz et al. (2002). The interpretation given in Blommaert et al. (1998) for the fact that Miras do follow a relatively narrow sequence in the P-L diagram whereas OH/IR stars don't, is that the GC OH/IR stars have evolved off the Mira PL relation towards longer periods, because of a significant decrease in the stellar mass due to the high mass losses. At the same time, the luminosity which is related to the core mass hardly changes. Looking at this diagram one should also realise that this phase is only a small fraction in time in comparison with the lower mass-loss rate phase. As pointed out by Wood et al. (1998) the number of Mira-type LPVs in the GC area is about 10 times higher.

Figure 3. Period-luminosity diagram for OH/IR stars observed with ISO (Blommaert et al. 2002). The period-luminosty relation is taken from LPVs in the SGR I window, a distance of 8.7 kpc was used.

3.3 Mass loss

The Spectral Energy Distributions obtained in Blommaert et al. (2002) are fitted with models from the radiative transfer code by Groenewegen (1992). Silicate dust was used with $\chi_{60} = 70\,\mathrm{cm}^2\mathrm{g}^{-1}$ and a gas to dust ratio of 200. A M10 giant with $T_{\rm eff} = 2500$ K was used as the central star. The expansion velocity of the circumstellar shell was taken from the OH spectrum. By modifying mass loss rate and the luminosity, a good fit to the observed SED, was searched for. An example is given in Fig. 4. The individual total mass loss rates range from 10^{-6} to $10^{-4} M_\odot$/yr in our sample.

Some characteristics:

- mass loss rates increase for increasing amplitudes (like also shown in Whitelock et al. 1991)
- mass loss rates versus period follow the same trend as in Vassiliadis & Wood (1993)
- $\dot{M} c V_{\rm exp}/L_* > 1$ occurs (contrary to the assumption used in Vassiliadis & Wood (1993) and in agreement with discussions in for instance Habing et al. (1994))
- No relation between mass loss rate and luminosity

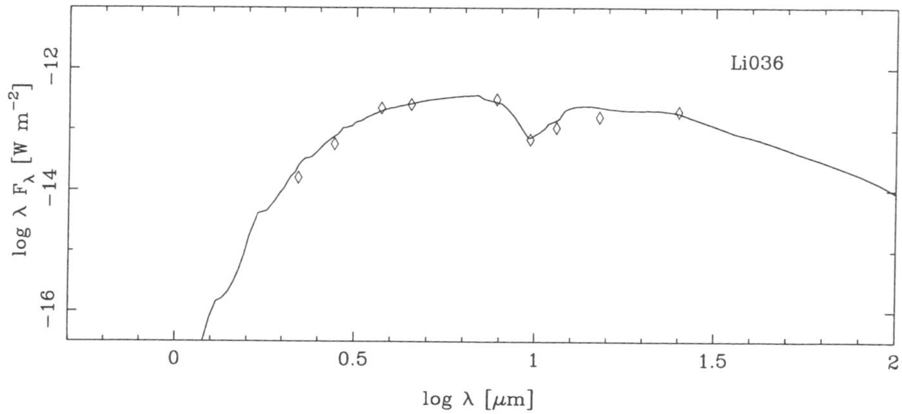

Figure 4. An example of a model fit to the ISO data (Blommaert et al. 2002).

4. OH/IR star populations

The existence of these stars in the GC is already a point of debate: does the existence prove the presence of a young population in the GC, or may the GC population be old? As was mentioned in the introduction, the study of the kinematics of the OH sample led Lindvist et al. (1992b) to the conclusion that 2 OH/IR star populations exist in the GC and that these can be separated on basis of the outflow velocities.

If stars with high expansion velocities are younger than the other group, one expects to find higher luminosities for the high expansion velocity group. Blommaert et al. (1998) found that the high expansion velocities contain higher luminosity stars but that there was also a large overlap in the luminosity distributions of the 2 groups. Ortiz et al. (2002) also compared the measured luminosities with a higher number of stars and did not find any evidence for a distinction on basis of luminosities between the 2 groups. The disadvantage of the latter study is the fact that the luminosities have not been corrected for variability and that the luminosities spread over a larger range. Nevertheless, it seems that the differences between the 2 groups, if they exist, are not very strong.

In Fig. 5, I plot 3 (l-v)-diagrams for OH/IR stars with different luminosities and/or expansion velocities, using the *average* luminosities from Wood et al. (1998) and Blommaert et al. (1998, 2002). The first plot shows the (l-v) diagram for sources with $L_* > 10,000 L_\odot$. These luminosities correspond to ages of approximately 1 Gyr or less and initial masses above 2 $M_\odot$ (Vassiliadis and Wood, 1993). The majority of

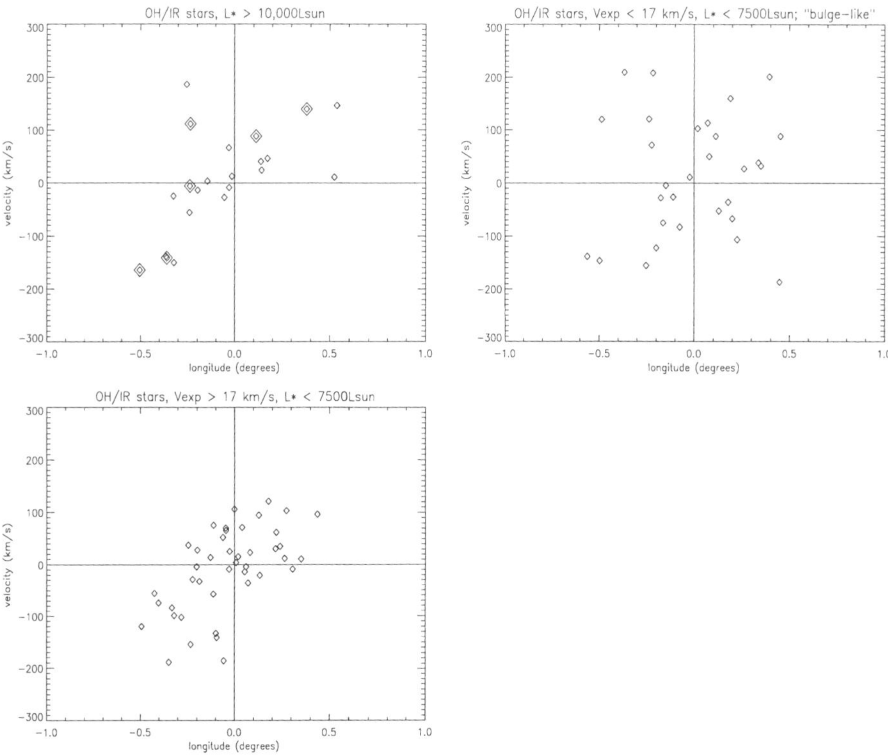

Figure 5. (l,v) diagrams for 3 sub-groups of the Linqvist et al. (1992a) sample.

these objects follow a line in the diagram which can be interpreted as a rotation curve through the centre.

The second plot contains sources that correspond to the bulk of bulge OH/IR stars (van der Veen and Habing 1990): $L_* < 7500 L_\odot$. The expansion velocity was limited to 17 km/ s which is about the highest outflow velocity in their paper. Clearly, we now find a sample with a very high dispersion in velocities which has a very different kinematical behaviour than the first group.

The last l-v diagram contains sources with similar luminosities as the "bulge"-like stars but with higher expansion velocities. As can be expected from Lindqvist's analysis these stars have a more disk-like kinematics. Nevertheless their luminosities are not different from the previous group. The higher expansion velocity are to be explained by differences in the dust properties (Habing et al. 1994, Elitzur & Ivezić 2001 and Elitzur, these proceedings) which may be related to a difference in metallicity of the stars. For instance, OH/IR stars in regions of lower metallicity like the outer Galaxy (Blommaert et al. 1993) and the LMC (Wood et al. 1992) show relatively low V_{exp}.

What are the ages connected for these latter groups? Comparing with the models of Vassiliadis and Wood (1993) we find an initial mass of around 1.5 $M_{\odot}$ and an age of a few Gyrs. The ages and initial masses for these luminosities are however quite uncertain. Van der Veen and Habing (1990) estimate ages ranging from a few Gyrs to well over 10 Gyrs for luminosities around $5000L_{\odot}$. A limitation in the estimates of the ages of the stars is the fact that in the Vassiliadis and Wood models the highest metallicity used, is solar whereas the metallicities here could be higher (Wood et al. 1998). Further discussions on the ages of the OH/IR stars can be found in Sevenster (1999) and Sjouwerman et al. (1999). The next step has to come from a determination of the metallicities of the OH/IR stars. In this respect it is worthwhile mentioning that near-infrared spectroscopy is being obtained on ISOGAL sources in the GC, including OH/IR stars (Schultheis et al., in preparation and Messineo et al., in preparation). Also the enlargement of the AGB samples can be envisaged through the SiO maser survey on GC ISOGAL sources (Messineo, these proceedings).

References

Blommaert, J.A.D.L., van der Veen, W.E.C.J, & Habing, H.J. 1993, A&A, 267,39

Blommaert, J.A.D.L., van der Veen, W.E.C.J, Habing, H.J., van Langevelde, H., & Sjouwerman, L. 1998, A&A, 329, 991

Elitzur, M., & Ivezić, Z. 2001, MNRAS, 327, 403

Glass, I.S., Whitelock, P.A., Catchpole, R.M., & Feast, M.W. 1995, MNRAS, 273, 383

Groenewegen, M.A.T. 1992, PhD thesis, Univ. of Amsterdam

Habing, H.J., Tignon, J., Tielens, & A.G.G.M. 1994, A&A, 286, 523

Jones, T.J., McGregor, P.J., Gehrz, R.D., & Lawrence, G.F. 1994, AJ, 107, 1111

Lindqvist, M., Winnberg, A., Habing, H.J., & Matthews, H.E. 1992a, A&ASS, 92, 43

Lindqvist, M., Habing, H.J., & Winnberg, A. 1992b, A&A, 259, 118

Ortiz, R., Blommaert, J.A.D.L., Copet, E., Ganesh, S., Habing, H. J., Messineo, M., Omont, A., Schultheis, M., & Schuller, F. 2002, A&A, 388, 279

Sevenster, M. 1999, MNRAS, 310, 629

Sjouwerman, L.O., van Langevelde, H.J., Winnberg, A., & Habing, H.J. 1998, A&AS, 128, 35

Sjouwerman, L.O., Habing, H.J., Lindqvist, M., van Langevelde, H.J., & Winnberg, A. 1999, in ASP Conf. Ser. 186, The Central Parsecs of the Galaxy, eds. H. Falcke, A. Cotera, W.J. Duschl, F. Melia, & M.J. Rieke (San Francisco ASP), 379

Vassiliadis, E., & Wood, P.R. 1993, ApJ, 413, 641

van der Veen, W.E.C.J, & Habing, H.J. 1990, A&A, 231, 404

Whitelock, P.A., Feast, M.W., & Catchpole, R. 1991, MNRAS, 248, 276

Wood, P., Whiteoak, J.B., Hughes, S.M.G., et al. 1992, ApJ, 397, 552

Wood, P., Habing, H.J., & McGregor, P.J. 1998, A&A, 336, 925

86 GHz SiO MASER SURVEY OF LATE-TYPE STARS IN THE INNER GALAXY

Maria Messineo, Harm Habing
Leiden Observatory, the Netherlands
messineo@strw.leidenuniv.nl

Lorant Sjouwerman
National Radio Astronomy Observatory, Socorro, U.S.A.

Alain Omont
Institut d'Astrophysique de Paris, France

Karl Menten
Max-Planck-Institut für Radioastronomie, Germany

Abstract We present observations of 86 GHz (J=2-1, v=1) SiO maser emission in the Inner Galaxy. Using the IRAM 30m telescope, we targeted late-type stars selected from the ISOGAL and MSX catalogues, in order to measure radial velocities of a large number of stars. Our 255 new measurements double the number of maser line-of-sight velocities known toward the inner Galaxy. Combining our measurements with those from other surveys, in particular OH maser surveys, will allow a kinematic analysis of the stellar orbits in the inner Galaxy. Analyzing the colours and luminosities of OH and SiO maser emitters also yields new clues on the conditions under which different masers may occur.

Keywords: stars: late-type – masers – Galaxy: kinematics and dynamics

Introduction

Maser emission from circumstellar envelopes of mass-loosing late-type stars is often strong enough to be detected throughout the Galaxy. Masers can indicate the line-of-sight velocities of the respective stars to within a few km s^{-1}. Frequently, detected maser lines are from OH

Y. Nakada et al. (eds.), Mass-Losing Pulsating Stars and their Circumstellar Matter, 363–366.

at 1.6 GHz, H_2O at 22 GHz, and SiO at 43 and 86 GHz (Habing 1996). Until recently, only a few hundred stellar line-of-sight velocities, mainly from OH/IR stars, were known toward the inner Galaxy ($30° < l < -30°$ and $|b| < 1$). This number is too small to allow for a good multicomponent analysis of the Galactic structure and dynamics. Obtaining more stellar line-of-sight velocities is therefore an urgent need. We are therefore conducting a search for 86 GHz SiO ($v = 1, J = 2 \rightarrow 1$) maser line emission (Messineo et al. 2002) towards late-type stars in the inner Galaxy. This survey is conducted with the IRAM 30m telescope. Many of our 86 GHz masers have been observed successfully also at 43 GHz (Sjouwerman et al., this volume).

1. Source selection

SiO maser emission is detected from several transitions towards oxygen-rich Asymptotic Giant Branch (AGB) stars. However, it occurs more frequently towards M-type Mira stars than towards other AGB stars. Furthermore, the relative strengths of different SiO maser lines are observed to vary with AGB type (Nyman et al. 1993 and references therein), indicating that the SiO maser properties depend on the stellar mass loss rate. The ratio of the SiO maser flux at 43 and at 86 GHz ranges between 2 and 10 in OH/IR stars, but is ~0.4 in Mira stars and in supergiants. Thus, in particular for Mira-like stars, the 86 GHz ($v = 1, J = 2 \rightarrow 1$) SiO maser line is a good handle to measure stellar line-of-sight velocities, complementing previous OH surveys. Mira stars also have the advantage of being more numerous than OH/IR stars.

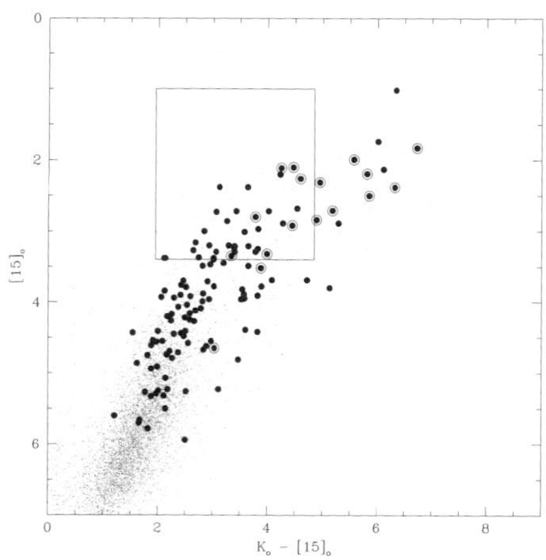

Figure 1 $[15]_o$ versus (K_o – $[15]_o$) ISOGAL-DENIS colour-magnitude diagram (extinction corrected). LPV stars from Glass et al. 2001 are shown as big dots of which the encircled ones mark known OH/IR stars. For comparison, the small dots indicate all the ISOGAL sources within 5 degrees from the Galactic centre. The box defines the region of our sample.

The late-type stars to be searched for maser emission were selected from the combined ISOGAL-DENIS catalogue (Omont et al. 1999) and

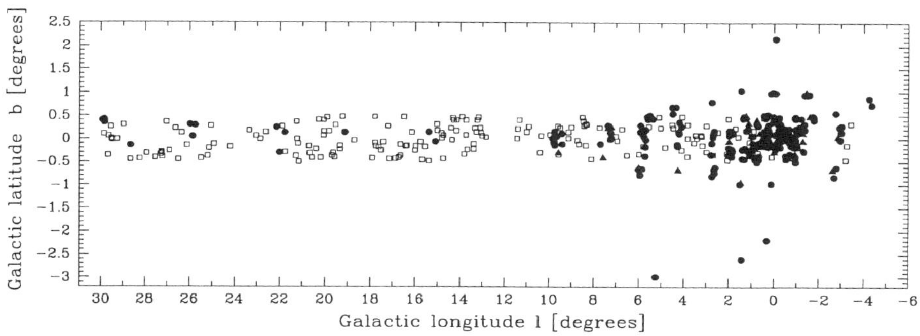

Figure 2. Location of the observed sources, irrespective of detection and non-detections, in Galactic coordinates. Filled circles denote sources present in both catalogues, open squares denote MSX sources and filled triangles ISOGAL sources.

from the MSX catalogue (Egan et al. 1999) on the basis of colour and magnitude. Analyzing the location of the LPV stars of Glass et al. (2001) in the different ISOGAL-DENIS colour-magnitude diagrams, we found that in the $(K_o - [15]_o)$ versus $[15]_o$ colour-magnitude diagram the LPVs without OH masers separate well from the OH/IR stars (Fig. 1). The OH/IR stars, having 10^2 to 10^3 times higher mass-loss rates than Mira stars, are the brightest objects at 15 μm and have $(K_o - [15]_o)$ colour redder than 4 mag. Therefore, we selected sources from the ISOGAL catalogue by their $[15]_o$ magnitude, and their $(K_o-[15]_o)$ and $([7]_o-[15]_o)$ colours, in a transition region between Mira and OH/IR stars (see Fig. 1). We excluded the brightest and bluest sources, which are most likely foreground stars, and the reddest sources, which likely are either high mass-loss OH/IR stars or young stellar objects. We further excluded sources with $[15]_o > 3.4$, since they are –given the general correlation of SiO maser emission and IR luminosity (Bujarrabal et al. 1987)– likely to show SiO maser emission fainter than our detection limit of 0.2 Jy. For the MSX sources similar selection criteria were applied, and in addition variability information was taken into account.

To avoid duplicating the OH maser line-of-sight measurements, all known OH maser sources were excluded. The search was limited to the Galactic plane between $l = -4^\circ$ and $l = +30^\circ$ and $|b| < 1^\circ$ (Fig. 2); the lower limit in longitude is imposed by the northern latitude of the IRAM 30-m telescope.

2. Results

We observed 441 positions, and detected SiO ($v = 1, J = 2 \rightarrow 1$) maser lines towards 268 of them, a detection rate of 61%. Because many of these stars we expect to be variable, this detection rate is a lower limit to the actual percentage of 86 GHz SiO maser emitters.

The SiO maser detection rate slightly increases with the mIR flux density at 7 and 15 μm. The detection rate is 71 % for the bright mIR sources at magnitude [15] $\sim$ 1.8 ($\sim$ 3.8 Jy), and decreases to 53 % for the fainter mIR sources, with [15] $\sim$ 3.2 ($\sim$ 1 Jy).

We observed 15 LPVs found by Glass et al. (2001) and detected SiO maser emission from 11 of them (73 %). The fraction of OH-emitting LPVs passing our colour-magnitude selection is much smaller (23 %).

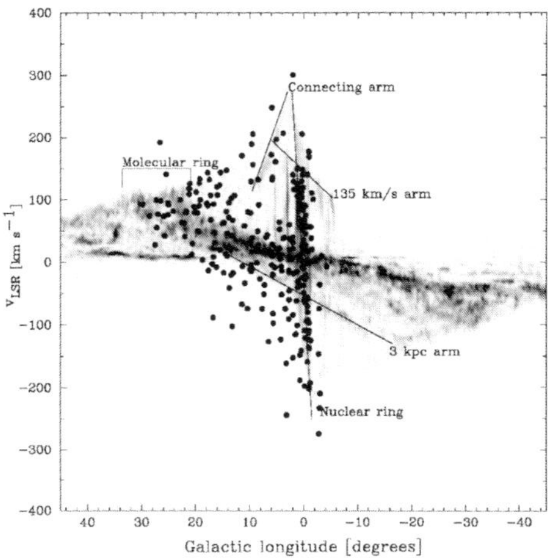

Figure 3 Stellar (l, v) diagram overlayed on the grayscale CO (l, v) diagram from Dame et al. (2001). The SiO 86 GHz masers are shown as dots. Gas features are labelled as in Fig. 1 of Fux (1999).

The line-of-sight velocities of the 86 GHz SiO maser sources range from -274 to 300 km s^{-1}, consistent with previous stellar maser measurements and with the ^{12}CO gas velocities (Fig. 3). An appreciable number of SiO maser sources are far from the CO emission, at negative velocities between $20° > l > 0°$, a range forbidden for pure circular rotation. At longitude 0° the stellar distribution follows the high velocity gas component of the **nuclear disk**; similar results are found by Imai (this volume).

References

Bujarrabal, V., Planesas, P., & del Romero, A. 1987, A&A, 175, 164
Dame, T. M., Hartmann, D., & Thaddeus, P. 2001, ApJ, 547, 792
Egan, M. P., Price, S. D., Moshir, M. M., et al. 1999, AFRL-VS-TR-1999, 1522
Fux, R. 1999, A&A, 345, 787
Glass, I. S., Matsumoto, S., Carter, B. S., & Sekiguchi, K. 2001, MNRAS, 321, 77
Habing, H. J. 1996, A&AR, 7, 97
Messineo, M., Habing, H., Sjouwerman, L., Omont, A., & Menten, K. 2002, A&A in press
Nyman, L.-A., Hall, P. J., & Le Bertre, T. 1993, A&A, 280, 551 miyoshi.tex
Omont, A., Ganesh, S., Alard, C., et al. 1999, A&A, 348, 755

HOW DOES THE GALACTIC BULGE ROTATE?

Anders Winnberg
Onsala Space Observatory
S - 439 92 Onsala, Sweden
anders@oso.chalmers.se

Abstract The bulge of our Galaxy can be studied, unaffected by interstellar absorption, using maser amplified radio lines from the envelopes of AGB stars. A large number of such stars showing OH, H_2O and SiO emission is now known in the Galactic Bulge. The positions and line-of-sight velocities of these stars have been measured with high precision. In the present work, which is on-going, I analyse a sample of AGB stars with OH maser emission (OH/IR stars) taken from several surveys. I demonstrate strong evidence for a differential rotation of the Bulge with a mean rotational velocity which is more or less constant and independent of galactocentric distance.

Keywords: Galactic Bulge, kinematics, rotation, AGB stars, OH/IR stars

Introduction

In the analysis of the kinematics of stars in the Galactic Bulge the so-called $(l - v)$ diagram (line-of-sight velocity plotted versus Galactic longitude) has played an important role (cf. McGinn et al., 1989; Lindqvist et al., 1992b; Durand et al., 1998). In all cases a linear regression line has been fitted to the data points in the diagram and this line has been interpreted as representing the mean rotation ('mean streaming motion') of the stellar sample. However, such a rotation corresponds to that of a solid body whereas the sky distribution of the stars is compatible with a volume density distribution proportional to r^{-2} (isothermal sphere). One should remember that the tracer stars are passing through evolutionary stages which most stars are believed to pass and that they therefore represent well the bulk of stars.

Is there more information present in $(l - v)$ diagrams than is visible at a first glance? One hint is obvious when comparing $(l - v)$ diagrams of different samples with different extent in the Bulge area. This can be seen in the papers referred to above: the red giants in McGinn et al.

Y. Nakada et al. (eds.), Mass-Losing Pulsating Stars and their Circumstellar Matter, 367–371.

(1989) show a mean rotation of 1930 km s^{-1} deg^{-1}, the OH/IR stars in Lindqvist et al. (1992b) show 178 km s^{-1} deg^{-1}, and the planetary nebulae in Durand et al. (1998) show 9.9 km s^{-1} deg^{-1}.

I want to demonstrate that there is indeed more information in $(l - v)$ diagrams than one can see at a first glance and that there is strong evidence for differential rotation of the Bulge.

I assume rotational symmetry because I do not see any strong evidence for triaxiality in the data. Perhaps future observations, including proper motion measurements, will be able to demonstrate clearly the bar nature of the Bulge in terms of stellar kinematics.

1. Survey Data

In the present work I use only published data of OH/IR stars. The data are published in the following papers:

1 te Lintel Hekkert et al. (1991): From this all-sky survey of IRAS point sources I limit myself to $|l| \leq 15°$, $|b| \leq 15°$.

2 Sevenster et al. (1997)

3 Habing et al. (1983): Despite the fact that only 34 OH/IR stars were found in this survey I include it in my analysis because several stars were not found in subsequent surveys due to intensity variations.

4 Lindqvist et al. (1992a)

5 Sjouwerman et al. (1998)

These five surveys have very different sky coverages, sensitivities and observing modes. A combination of all of them therefore is questionable. However, I think that as long as I am dealing only with the positions and line-of-sight velocities of the stars I should be on the safe side, provided that the surface densities follow the same general trend with galactocentric distance. I have calculated the surface densities and find that they fall off with distance as r^{-1} (corresponding to a volume density fall-off as r^{-2}) and possibly somewhat steeper in the outer regions.

Because of this I have concluded that it is permissible to treat the stars from all the five surveys as if it were one big survey. The total number of OH/IR stars in this combined survey is 546.

2. Novel Analysis

I have tried a more sophisticated kinematical analysis than those applied so far, by dividing the sample of OH/IR stars into a series of elliptical annuli. I have formed the $(l - v)$ diagram for the stars in each annulus and calculated the corresponding regression line. Then I have considered the slopes of these regression lines as a function of projected distance to the Galactic centre. The idea behind this is that in an r^{-2} density distribution most of the stars are located close to the tangential directions, except perhaps in the outermost regions.

At this stage I have not attempted any corrections for foreground stars in the Disk. This could possibly be done later by using the IR properties of the stars, if they are known. In any case I suspect that at least for the inner annuli such a correction would be rather small due to the small areas and the scarceness of OH/IR stars in the Disk.

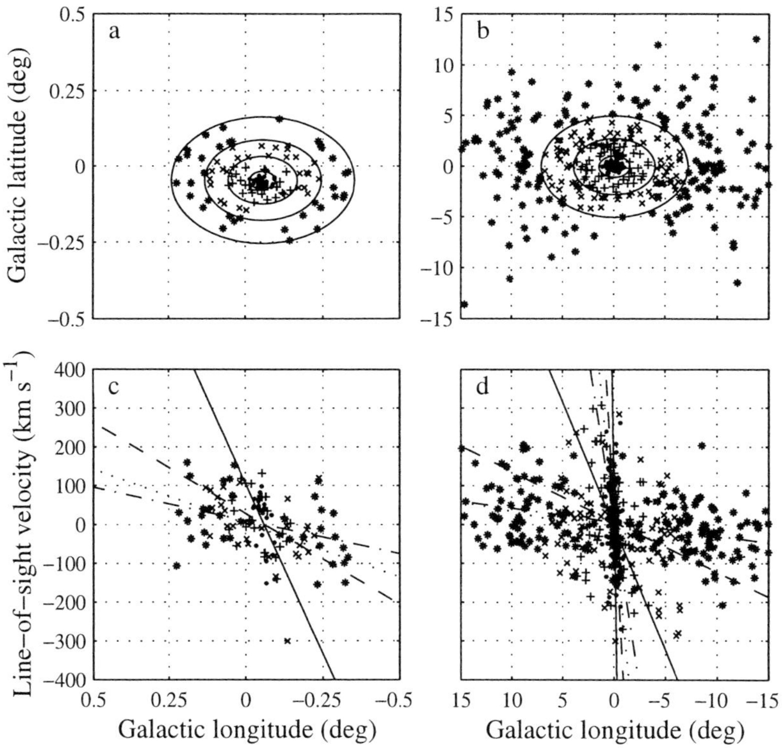

Figure 1. The sky distributions and $(l - v)$ diagrams of all the OH/IR stars in the five surveys combined together (a and c: the central $1^\circ \times 1^\circ$ area; b and d: the full $30^\circ \times 30^\circ$ area). The stars have been divided into concentric elliptical annuli with axial ratios of 0.7 and centred on the Galactic centre. The stars are represented by symbols which cycle through the scheme: points, plus signs, crosses, and asterisks. The way in which the corresponding regression lines in the $(l-v)$ diagrams are drawn cycle through the following scheme: solid, dashed, dotted, and dot-dashed.

The division into elliptical annuli and the regression analysis of the $(l - v)$ diagrams are demonstrated in Figure 1. I have chosen the sizes of the ellipses more or less arbitrarily but with the consideration in mind that the slope of the regression lines change rather rapidly close to the centre. However, I have experimented with several different division schemes and they have all given the same result.

There is a general trend that in spite of very large statistical errors the slopes of the regression lines decrease with increasing annulus sizes. In Figure 2 these slopes are plotted against the semi major axes of their corresponding elliptical annuli together with their 1-σ statistical errors. The weighted regression line for these points is drawn as a solid line.

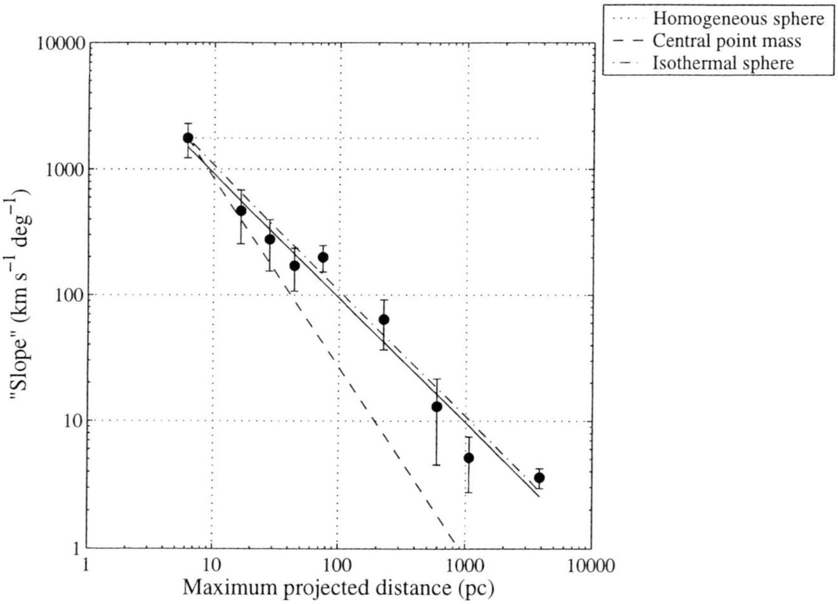

Figure 2. The "slopes" (derivatives) of the regression lines in Figure 1 as a function of the semi major axes of their corresponding elliptical annuli converted into linear distance. The 1-σ error bars are shown and the weighted regression line is drawn (solid). Three model lines are presented: homogeneous sphere (dotted), central point mass (dashed), and isothermal sphere (dot-dashed).

One can apply a very simple-minded interpretation of the diagram in Figure 2 by considering some extreme cases and assuming circular orbits. In this case the line-of-sight velocity V_{los} is a simple function of the circular velocity V_c, the galactocentric distance r and the Galactic longitude l:

$$V_{los} = V_c \frac{R_0}{r} \sin l \tag{1}$$

where R_0 is the distance to the Galactic centre. When derivating this one gets an expression for small l which is related to the 'slope of slopes' (the slope of the regression line in Fig. 2) for circular orbits:

$$\frac{dV_{los}}{dl} \propto \frac{V_c}{r} \tag{2}$$

For a homogeneous sphere $V_c \propto r$ and therefore the 'slope of slopes' is constant (dotted line in Fig. 2). This obviously is incompatible with the data. For the opposite extreme, namely a central point mass, $V_c \propto r^{-1/2}$ and therefore the 'slope of slopes' is $\propto r^{-3/2}$ (dashed line). Despite the large error bars this model is likely incompatible as well. For the intermediate equilibrium model, the isothermal sphere, $V_c =$ const and the 'slope of slopes' is $\propto r^{-1}$ (dot-dashed line), very close to the derived value.

3. Conclusion

I conclude from this work that the Galactic Bulge rotates as one would expect, i.e. like an isothermal oblate spheroid with a more or less constant rotational velocity. It is fairly easy to estimate the value of this rotational velocity from the available data and it is about 70 ± 20 km s^{-1} out to 1 kpc from the Galactic centre. Due to non-circular orbits of the stars there is a large spread from the mean rotation which I estimate to 80 ± 20 km s^{-1}.

Acknowledgments

This work is supported by The Swedish Research Council.

References

Durand, S., Acker, A., & Zijlstra, A. 1998, A&AS, 132, 13

Habing, H.J., Olnon, F.M., Winnberg, A., Matthews, H.E., & Baud, B. 1983, A&A, 128, 230

Lindqvist, M., Habing, H.J., Winnberg, A., & Matthews H.E. 1992a, A&AS, 92, 43

Lindqvist, M., Habing, H.J., & Winnberg, A. 1992b, A&A, 259, 118

McGinn, M.T., Sellgren, K., Becklin, E.E., & Hall, D.N.B. 1989, ApJ, 338, 824

Sevenster, M., Habing, H.J., Lindqvist, M., & Killeen, F. 1997, A&AS, 122, 79

Sjouwerman, L., van Langevelde, H.J., Winnberg, A., & Lindqvist, M. 1998, A&AS, 128, 35

te Lintel Hekkert, P., Caswell, J.L., Habing, H.J., Haynes, R.F., & Norris, R.P. 1991, A&AS, 90, 327

SiO MASER SURVEY TOWARDS THE STELLAR CLUSTER AT THE GALACTIC CENTER

Hiroshi Imai
Joint Institute for VLBI in Europe, Postbus 2, 7990 AA Dwingeloo, the Netherlands
imai@jive.nl

Shuji Deguchi, Jun-ichi Nakashima, Atsushi Miyazaki
Nobeyama Radio Observatory, Minamimaki, Minamisaku, Nagano 384-1305, Japan
deguchi@nro.nao.ac.jp, junichi@nro.nao.ac.jp, amiya@nro.nao.ac.jp

Osamu Kameya, Takahiro Fujii
VERA Project Office, National Astronomical Observatory
Mitaka, Tokyo 181-8588, Japan
kameya@miz.nao.ac.jp, fujii@cc.nao.ac.jp

Yoshifusa Ita, Yoshikazu Nakada
Institute of Astronomy, the University of Tokyo, Mitaka, Tokyo 181-8588, Japan
yita@mtk.ioa.s.u-tokyo.ac.jp, nakada@kiso.ioa.s.u-tokyo.ac.jp

Hideki Izumiura
Okayama Astrophysical Observatory, Kamogata, Asakuchi, Okayama 719-0232, Japan
Izumiura@oao.nao.ac.jp

Ian S. Glass
South African Astronomical Observatory, PO Box 9, Observatory 7935, South Africa
isg@saao.ac.za

Abstract We report preliminary results of a long-term survey of SiO v=1 and v=2 J=1–0 masers towards 418 large-amplitude variables within 15′ of the Galactic center with the Nobeyama 45-m telescope. We have detected 194 masers out of 338 surveyed fields during 2001–2002. The

Y. Nakada et al. (eds.), Mass-Losing Pulsating Stars and their Circumstellar Matter, 373–376.

averaged detection rate of 57% is comparable with that obtained towards IRAS point sources in the Galactic bulge. The detection rate increases clearly with the stellar pulsation periods. The Galactic longitude–radial velocity diagram of the SiO sources revealed is quite similar to that of OH masers. A hole in the diagram indicates the presence of the non-circular streaming motions of stars in the nuclear disk of the Galaxy.

Keywords: Masers, long period variables, the dynamics of the Galaxy

1. Introduction

The dynamical behavior of the central region of the Galaxy has attracted much attention. Since the bar-like structure of the Galactic bulge was recognized, non-circular motions have be taken into account also at the inner part of the Galaxy. Most of radial velocity information of stars in this part comes from radio observations of SiO and OH masers, which are not affected by interstellar extinction. The masers arise in the envelopes of mass-losing stars on the Asymptotic Giant Branch (AGB), which can potentially be identified at infrared wavelengths. Previously blind surveys of OH and SiO maser around the Galactic center have been made (e.g., Sjouwerman et al. 1998; Deguchi et al. 2002), Recently, new samples of survey candidate stars are available on basis of infrared catalogs. Glass et al. (2001) have identified 418 large-amplitude variables (Miras or semiregulars; abbreviated as LAV hereafter) towards the nuclear disk of the Galaxy by near-infrared photometry.

2. Observations and results

Here we report on results of SiO maser surveys towards LAVs catalogued by Glass et al. (2001), which are located within $15'$ of the Galactic center. Simultaneous observations in the SiO v =1 and v =2 transitions at 42.122 and 42.821GHz, respectively, have been made as a long-term project of Nobeyama 45-m telescope during 2001–2003. Here the results of the observations during 2001–2002 are presented (see also Imai et al. 2002). The sources with long periods ($>$ 300 d) were selected preferably. The positions of the stars measured by Glass et al. (2001) are sufficiently accurate (a few $''$) for a telescope beam size (HPBW) of about $40''$. We observed totally 338 fields locating at positions of the LAV stars and detected SiO maser emission towards 194. No correlation was recognized between detection/non-detection of SiO emission and phase of the infrared light curve.

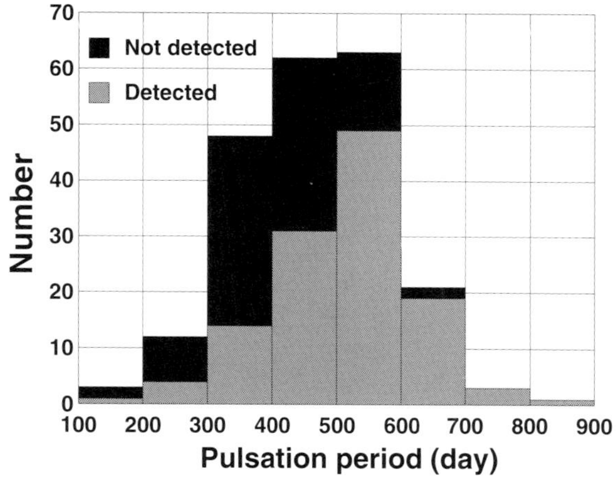

Figure 1 Histogram of maser detection/non-detection and stellar pulsation period. The SiO maser data include a sample observed during 2001 February–2002 March.

3. Discussion

3.1 Period–SiO Detection Rate

Figure 1 shows the period distribution of the observed sources. This figure clearly shows that the SiO detection rate increases with the period. Because of almost equal distances to the LAV stars, the correlation with period is much more striking. On the other hand, the SiO detection is twice as high as the detection rate of OH masers (Sjouwerman et al. 1998), indicating that an SiO maser survey can double the number of stars with radial velocity data near the Galactic center. The difference in the detection rates is likely because of a special physical condition of circumstellar envelopes in the Galactic center, for example, in which giant molecular clouds interact with the outer circumstellar envelopes and quench OH maser emission (Sjouwerman 2002 in private communication).

3.2 Radial Velocity Distribution

Figure 2 shows the longitude-velocity diagram for the detected sources. Most of the radial velocities fall within $|V_{\rm LSR}| < 150$ km s^{-1}. Two extremely high-velocity sources ($V_{\rm LSR} < -300$ km s^{-1}) occupy outlying positions. These sources might be outer bulge objects in highly eccentric orbits, seen by chance in the line of sight towards the Galactic center.

The best linear-fit slope of radial velocities against the galactic longitude is 2.1 ± 0.4 km s^{-1}pc^{-1} and much larger than that of the inner bulge out to 3° from the Galactic center, 0.14 km s^{-1}pc^{-1} (Deguchi et al. 2000), indicating that the nuclear disk is rotating more rapidly than the inner Bulge. The zero intercept of the slope at $\Delta l = 0$ is negative, which is consistent with previous SiO maser observations (Izumiura et al. 1995; Deguchi et al. 2000). This can be interpreted as the motion

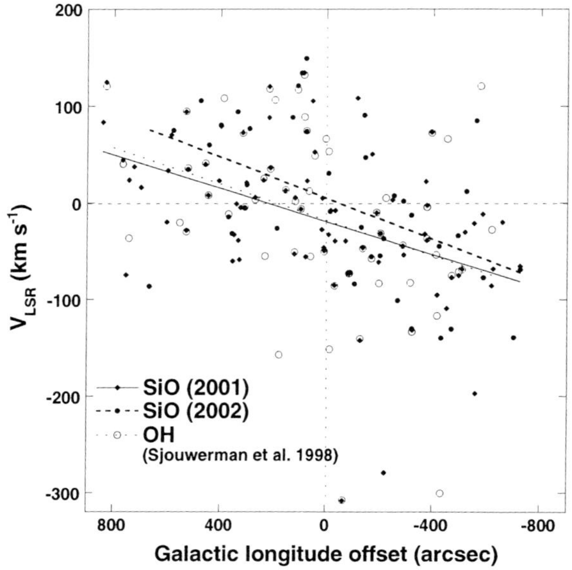

Figure 2 Radial velocity – Galactic longitude diagram of SiO and OH maser sources. The SiO maser data include a sample observed during 2001 February-2002 March. The best linear fits are shown by lines.

of the local standard of rest toward the dynamical center of the Galaxy (Izumiura et al. 1995).

The overall structure of the SiO l–v diagram shown in figure 2 is quite similar to the OH l–v diagram in the same region (Sjouwerman et al. 1998). We can recognize a hole in the SiO and OH diagrams at ($\Delta l = 150''$, $V_{\rm LSR} = 80$ km s^{-1}). Non-circular streaming motions of stars may be present in the nuclear disk of the Galaxy and the hole in the diagram can be a signature of bar structures. An alternative interpretation is that the stars within $100''$ from the Galactic center have a large circular motion (16.8 ± 0.7 km s^{-1}pc^{-1}; Deguchi et al. 2002). The magnitude of the circular motion is comparable with that of the circumnuclear ring revealed by the molecular line observations (e.g., Wright et al. 2001). It is possible that the rapidly rotating AGB stars near the Galactic center were born in the circumnuclear ring (Levine et al. 1995).

References

Imai, H., Deguchi, S., Fujii, T. et al. 2002, PASJ, 54, L19
Deguchi, S., Fujii, T., Miyoshi, M., & Nakashima, J. 2002, PASJ, 54, 61
Deguchi, S., Fujii, T., Izumiura, H. et al. 2000, ApJS, 128, 571
Glass, I. S., Matsumoto, S., Carter, B. S., & Sekiguchi, K. 2001, MNRAS, 321, 77
Izumiura, H., Deguchi, S., Hashimoto, O. et al. 1995, ApJ, 453, 837
Levine, D., Figer, D. F., Morris, M., & McLean, I. S. 1995, ApJ, 447, L101
Sjouwerman, L. O., van Langevelde, H. J., Winnberg, A., & Habing, H. J. 1998, A&AS, 128, 35
Wright, M. C. H., Coil, A. L., McGary, R. S., Ho, P. T. P., & Harris, A. I. 2001, ApJ, 551, 254

43 GHz SiO MASERS IN LATE-TYPE STARS WITH 86 GHz SiO MASERS IN THE INNER GALAXY

Loránt Sjouwerman
National Radio Astronomy Observatory (NRAO), PO.Box 0, Socorro NM 87801, U.S.A.
lsjouwerman@aoc.nrao.edu

Maria Messineo, Harm Habing
Sterrewacht, PO.Box 9513, 2300 RA Leiden, The Netherlands
messineo@strw.leidenuniv.nl, habing@strw.leidenuniv.nl

Abstract We present Very Large Array (VLA) observations of 43 GHz (J=1-0, v=0 and v=1) SiO maser emission in a sub-sample of late-type stars with relatively strong (> 0.5 Jy) 86 GHz SiO masers in the Inner Galaxy.

Keywords: Masers, Galactic center, AGB stars, Circumstellar matter, Astrometry

Messineo et al. (2002) report on a survey for 86 GHz masers in an IR-selected sample of late-type stars in the Inner Galaxy. Because the velocities of the circumstellar 86 GHz SiO masers indicate that they are located in the Inner Galaxy, their main objective is to use the detections both to study the intrinsic maser properties of the sample as well as to use the kinematic data – combined with that of other samples of stars, e.g. OH/IR stars – to study the dynamical structure of the Inner Galaxy.

Here, we selected a sample of relatively strong (> 0.5 Jy) 86 GHz (J=2-1, v=1) SiO masers within 2 degrees of Sgr A* in order to investigate the relative occurrence of the 86 and 43 GHz (J=1-0, v=0 and v=1) SiO maser lines. We observed 39, and detected 38 sources, mostly in both of the 43 GHz lines. In addition we found some non-targeted 43 GHz SiO masers. That the detection rate is so high, probably has to do both with the selection of the stronger 86 GHz SiO masers, and the time interval of roughly a year (about the expected stellar pulsation period of 300-400 days) between the 86 and 43 GHz observations.

The 43 GHz SiO masers can be used to measure stellar positions with VLBI, and thus proper motions over a time scale of years, e.g. with respect to the black hole Sgr A* (Sjouwerman et al. 1998). In particular, the Japanese VERA array is built for this purpose. However, the

Y. Nakada et al. (eds.), Mass-Losing Pulsating Stars and their Circumstellar Matter, 377–378.

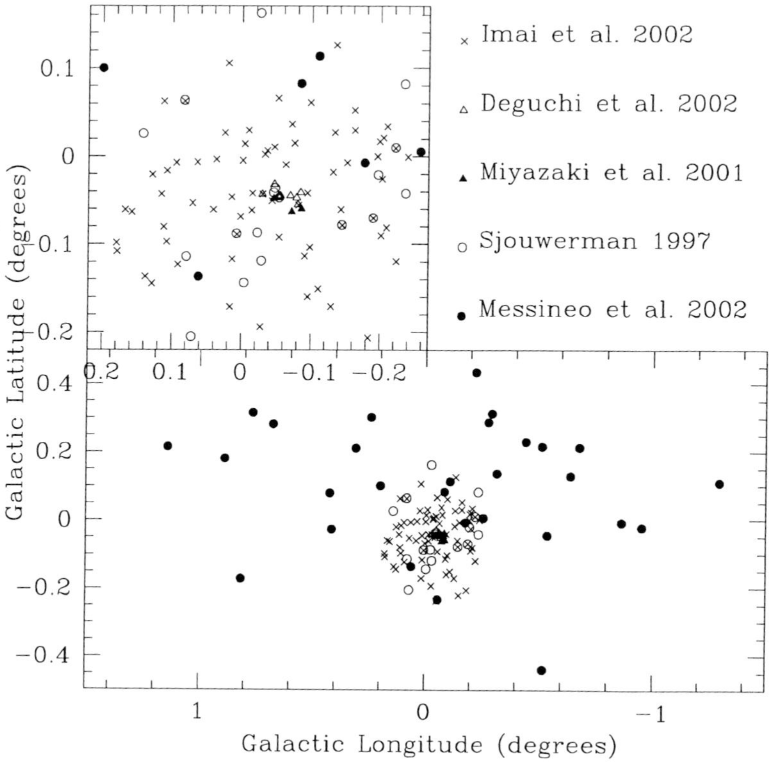

Figure 1. Current status of 43 GHz SiO masers in the Inner Galaxy.

dual-beam VERA system is limited to a minimum separation between calibrator and target source of half a degree. The most interesting science are the stellar proper motions very close to the black hole Sgr A*. The 43 GHz detections presented here may serve as calibrators for such a program, now that numerous candidate 43 GHz masing stars have been found close to Sgr A* to measure the proper motions of (Figure 1).

References

Deguchi, S., et al. 2002, PASJ, 54, 61
Imai, H., et al. 2002, PASJ, 54, L19
Messineo, M., et al. 2002, A&A, in print
Messineo, M., et al. 2002, in these proc.
Miyazaki, A. et al. 2001, PASJ, 53, 501
Sjouwerman, L. 1997, PhD-th. Onsala, Sweden
Sjouwerman, M., et al. 1998, A&A, 339, 897
Sjouwerman, L., et al. 2002, A&A, in print

VI

VERA AND OTHER FUTURE PROJECTS

ASTROMETRY OF CIRCUMSTELLAR MASERS

H.J. van Langevelde
Joint Institute for VLBI in Europe
Postbus 2, 7990 AA Dwingeloo, the Netherlands
langevelde@jive.nl

W.H.T. Vlemmings
Sterrewacht Leiden
Niels Bohr weg 2, 2300 RA Leiden, the Netherlands
vlemming@strw.leidenuniv.nl

Abstract The circumstellar masers around evolved stars offer an interesting possibility to measure stellar parameters through VLBI astrometry. In this paper the application of this technique is discussed, including the accuracy and the uncertainties of the method. The different maser species (OH, H_2O, SiO) have slightly different characteristics and applications. This paper does not concern astrometry of maser spots to study the kinematics of the envelope, but concentrates on attempting to measure the motion of the underlying star.

Keywords: masers – stars: circumstellar matter – techniques: interferometric – techniques: astrometry

Introduction

There are a number of scientific justifications for astrometry of stars with circumstellar masers. Historically, the registration of radio single dish and infrared data was so uncertain that proper identification of maser bearing stars was a problem. Most of these problems have been overcome; the accuracy of connected element interferometers is sufficient to make these identifications unambiguously. However, there is still a lot of uncertainty when comparing VLBI data from different experiments. Notably, when maser distributions from different transitions are overlayed, a common origin is usually assumed. Astrometry can be done to do such comparisons properly.

Y. Nakada et al. (eds.), Mass-Losing Pulsating Stars and their Circumstellar Matter, 381–388.

However, the most tantalizing application of astrometry of circumstellar masers is the accurate determination of the motion of the star on the sky. Proper motions of a star anywhere in the Galaxy (10 km/s at 8 kpc corresponds to $\mu = 0.25$ mas/year) can in principle be obtained with VLBI, as can the parallax for such distances ($\pi = 0.125$ mas). A direct distance estimate is important for studying the physics of the mass-losing stars. For many Mira variables Hipparcos measured the distance, but for the most enshrouded stars VLBI offers the best approach. Both the proper motion and distance are inputs for studying the stellar population and Galactic kinematics (Habing, these proceedings).

In addition, detailed astrometric monitoring of evolved stars could reveal information about the existence of binary components, or even planets in these systems. Moreover, by monitoring its position new facts can be learned about maser physics. Are spots persistent, what (turbulent) motions can be measured, is the brightness a function of stellar cycle? Finally it is worth mentioning that the maser astrometry could tie the observational reference frames, in particular in the infrared where these stars are very bright.

Carrying out astrometric monitoring projects with spectral line VLBI is extensive work that requires some special techniques and assumptions. There are a number of technical advancements and initiatives that will make improvements to this field which will make it much more accessible.

1. Methods & Techniques

The most straightforward way of doing VLBI astrometry is to use "phase referencing" to a bright and close calibrator in order to overcome the spatial and temporal fluctuations of the ionosphere and/or troposphere. The calibrator is usually an extragalactic source with a fixed and known position, which is sufficiently bright to calibrate the atmospheric and instrumental effects within a fraction of the coherence time. Because the modern VLBI correlators have incorporated the best geodetic models, phase referencing results obtained in this way can be accurate to a few mas. This is already sufficient for a number of applications. In other cases, higher accuracy is required and some adjustments are necessary to improve the model. In such cases one probably needs to return to the "totals" in order to use astrometric/geodetic software. To get accuracies below 1 mas is not trivial; better models of the atmospheric behavior and structure of the reference source are required.

For maser astrometry an additional complication can be that one needs to observe in mixed bandwidth mode, applying a narrow band for the maser detection and simultaneous wide bands for the reference source. Obviously the maser dictates the observing frequency, which rules out some calibration schemes used in continuum and geodetic VLBI.

Ideally the masers should be bright and persistent. The brightness relates to the size of individual maser features, which sets an upper limit to the resolution one can obtain. A fundamental issue remains to link the motion of the maser spot to the stellar properties. One way forward is to assume that the masers are on a linear path with respect to the star (e.g. a constant outflow). In this case the parallax of individual spots equals the parallax of the star, and the average motion may be assumed to be the stellar motion. In other cases one may be able to deduce the position of the star from the distribution of the masers, for example when they form a ring. If we can determine the position of the star, one can in principle measure the motion and the parallax. Note that only when one can consistently determine the stellar position with (much) better than 1 AU accuracy, a useful parallax can be determined. In some cases there is evidence that special maser spots exists that correspond to the stellar continuum amplified by the maser shell. Such spots are then tied very accurately to the stellar position. In other cases one has to worry about the systematic and turbulent motions in the maser shell. In any case, there may be additional motions involved, for example when masers occur in binary stars.

2. OH Masers

The OH masers usually originate in large shells at 1000 AU from the mass-losing stars. As the outflow is constant at that point, the maser radiates most effectively radially, from the front and back cones of the shell. Three 18 cm transitions can give reasonably bright maser emission. However, the VLBI spots are usually resolved at 20 mas, sometimes degraded further by angular broadening in the interstellar medium, notably in the direction of the Galactic centre (Van Langevelde et al. 1992).

Although the shells are large, the "amplified stellar image" paradigm makes the OH maser interesting for astrometry. Radio continuum, amplified by the maser results in a spot that is fixed on the stellar position, bright and persistent. The observation that the most blue-shifted spots in two OH transitions coincides is a very good confirmation of this idea (Sivagnanam 1990).

Recently, comparison of a proper motion study of U Her with Hipparcos agreed to 15 mas (van Langevelde et al. 2000), thus confirming that the most blue-shifted and most persistent spot indeed follows the stellar

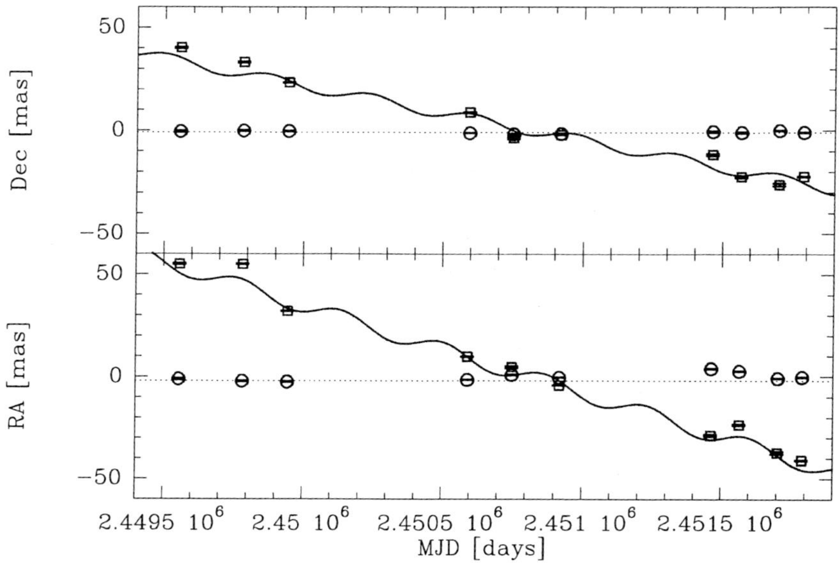

Figure 1. Proper motion of the most blue-shifted spot in the 1667 MHz maser of U Her. A motion of $\mu_{\rm vlbi} = -15.57 \pm 0.56, -9.66 \pm 0.61$ mas/yr and a parallax of $\pi_{\rm vlbi} = 3.85 \pm 1.14$ mas are fitted to the data. For comparison the position of a secondary reference source is also displayed.

trajectory in this case. Therefore the maser motion can be used to obtain the parallax and proper motion of the star. U Her was observed 10 times in 6 years on the NRAO VLBA which resulted in the first VLBI parallax of a mass-losing star (Fig 1; Van Langevelde et al. 2000). The motion agrees within the error with the Hipparcos data, from which no reliable parallax could be obtained.

If the amplified stellar image is a general feature in circumstellar OH masers, it will allow astrometric measurements of a large number of stars. However, the modest brightness of OH masers limits the application of this technique to stars closer than 1 kpc.

3. H_2O masers

The 22 GHz masers are much brighter than the OH masers in circumstellar shells. Therefore positional accuracies in the order of 50 μas are in principle achievable. Of course at this high frequency the coherence time is much shorter and phase referencing VLBI at 22 GHz is quite a bit more challenging.

Additional complications arise from the fact that the circumstellar H_2O masers are rather messy. The structures are often ring-like on scales of 100 AU, as expected from tangential amplification. However, large departures have been observed, as well as considerable variability. If the structures are persistent enough one can attempt to track individual maser spots over a year and measure the parallax of the star from these (e.g. Kurayama et al., these proceedings). If this approach works the water maser observations offer the interesting possibility to measure parallaxes throughout a large fraction of the Galaxy.

There have been claims that in water masers the amplified stellar image can also occur (Marvel 1996, Colomer et al. 2000). To check this we have observed U Her at 22 GHz with MERLIN, which yields a comparable resolution as our previous OH VLBI results. Using the same position calibrators we find, somewhat surprisingly, that the brightest H_2O maser coincides with the optical position and the OH image. Although the MERLIN observations do not resolve individual maser spots observable with VLBI, we take this as a clue that amplification of background radiation may play a role in water masers too (Fig 2; Vlemmings et al. 2002).

4. The use of SiO masers

The SiO masers around evolved stars can be even brighter. At 43 GHz the spots sizes allow 10μas accurate positions. At the same time the technical difficulties are considerable. The coherence time is only of the order of a minute and in addition there are fewer suitable extragalactic calibrators.

The masers themselves are located close to the star, within 10 AU. They are bright and abundant. Monitoring of their structure indicates that the masers are showing motions on the time-scale of the variability of the star (Diamond & Kemball 1999). Often these motions are not linear and it will be a challenge to measure a parallax from the combined motions of SiO masers. As the masers are in a ring close to the star, it will be possible to estimate the stellar position from the distribution of SiO masers. The proper motion of the star can be measured this way, even if the individual spots cannot be resolved. Proper motions of individual stars can probably be measured throughout the Galaxy using the 43 GHz transition.

A very suitable target for stellar astrometry by SiO masers are the Galactic center stars. In this region there is an abundance of maser stars

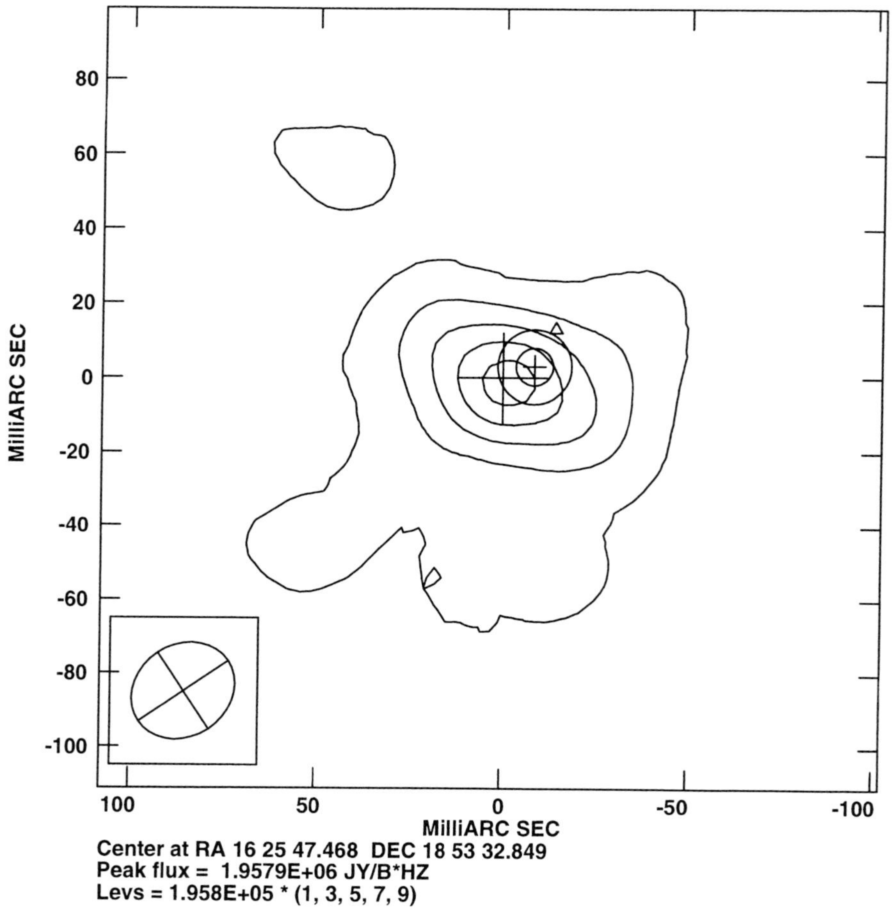

Figure 2. The location of the U Her H_2O masers with respect to the stellar positions determined from the Hipparcos observations, as determined from phase referencing to an extragalactic calibrator. The star is denoted by two circles indicating the star itself and the stellar radio photo-sphere. The error bars on the maser feature are the errors due to the positional fitting. The errors on the stellar position are due to the link to the radio-reference frame and due to the errors in the proper motion used to transpose the optical position. The triangle denotes the stellar position when using the phase referencing results on the secondary calibrator.

close together (Sjouwerman, these proceedings). In addition, SgrA* is a suitable calibrator; it is bright at short wavelengths and, as it is is the location of a massive black hole, it is stationary. In fact, an important results from circumstellar maser astrometry is the tie between the IR and radio positions in the Galactic centre. By locking the positions of the IR sources to the radio maser counterparts, it can be shown that the motion of stars in the Galactic centre is consistent with motion around a massive black hole located at the position of SgrA* (Menten et al. 1997; Eckart et al. 2002).

Several groups have explored the possibility to start monitoring SiO masers in the Galactic centre region with VLBI (Sjouwerman et al. 1998; Deguchi, these proceedings; Imai, these proceedings; Reid, private communication). The basic goal is to add two more components to the measurement of kinematics in this dynamically interesting region (Winnberg, these proceedings).

5. Future

There are many planned improvements of ordinary VLBI that will facilitate more astrometry of circumstellar masers. Several modern, big telescopes are under construction that will observe at high frequencies, necessary for H_2O and SiO masers. Disk-based recording systems will improve the reliability and productivity of VLBI, and will be especially advantageous for time consuming astrometric monitoring. In addition there are several indications that the atmospheric modeling can be improved considerably. Other improvements of the geodetic models are important too, for example accurate telescope positions.

However, the future for astrometry of masers clearly lies in instrumental setups that overcome the temporal variability of the atmosphere at high frequencies. Some of this can be done with "cluster–cluster" VLBI, but the dedicated VERA project provides a far more ambitious approach that will soon produce interesting results (Sasao, these proceedings).

A number of studies are required to establish new techniques for astrometry of circumstellar masers. Careful analysis of errors and consistency checks will be necessary to determine how to use the masers optimally to determine the underlying stellar motions. But in potential this method enables the measurement of proper motions and distances of mass-losing stars throughout the Galaxy. This will progress the study of stellar evolution as well as our knowledge of the structure of the Galaxy.

6. Acknowledgements

The National Radio Astronomy Observatory is a facility of the National Science Foundation operated under cooperative agreement by Associated Universities, Inc. MERLIN is a National Facility operated by the University of Manchester at Jodrell Bank Observatory on behalf of PPARC. We acknowledge the continuous support by Phil Diamond, Harm Habing and Richard Schilizzi.

References

Colomer, F., Reid, M.J., Menten, K.M., & Bujarrabal, V. 2000, A&A, 355, 979

Diamond, P.J., & Kemball, A.J. 1999, in IAU Symposium #191, Asymptotic Giant Branch Stars, ed. Le Bertre, Lebre & Waelkens, p. 195

Eckart, A., Genzel, R., Ott, T., & Schödel, R. 2002, MNRAS, 331, 917

Marvel, K.B. 1996, Ph.D. thesis, New Mexico State Univ.

Menten, K.M., Reid, M.J., Eckart, A., & Genzel, R. 1997, ApJ, 475, L111

Sivagnanam, P., Diamond, P.J., Le Squeren, A.M., & Biraud, F. 1990, A&A, 229, 171

Sjouwerman, L.O., van Langevelde, H.J., & Diamond, H.J. 1998, A&A, 339, 897

Van Langevelde, H.J., Frail, D.A., Cordes, J.M., & Diamond, P.J. 1992, ApJ, 396, 686

Van Langevelde, H.J., Vlemmings, W., Diamond, P.J., Baudry, A., & Beasley, A.J. 2000, A&A, 357, 945

Vlemmings, W.H.T., Van Langevelde, H.J., & Diamond, P.J., 2002, A&A, in press

ASTROMETRY OF MASS–LOSING PULSATING STARS WITH VERA

Current status and observational prospects of VERA Project

Tetsuo Sasao and VERA Project Team
VERA Project Office, National Astronomical Observatory, Japan
2-21-1 Osawa, Mitaka, Tokyo, 181–8588 JAPAN
sasao@miz.nao.ac.jp

Keywords: phase-referencing VLBI, astrometry, masers, period–luminosity relation

1. A Dedicated Phase–Referencing VLBI Array

The major observational goal of Project VERA(VLBI Exploration of Radio Astrometry) is to measure annual trigonometric parallaxes and proper motions of astronomical masers around evolved as well as young stars with an accuracy of 10 micro arcsecond (μas) level. An advanced phase–referencing VLBI technology allowing simultaneous observation of adjacent target and reference sources separated up to 2.2 degree is the key for achieving the high accuracy.

At 22 GHz (1.4 cm), the minimum fringe spacing of VERA with the longest baseline of 2300 km (Mizusawa–Ishigakijima) is roughly 1.4 cm / 2300 km $\cong$ 1 milliarcsecond (mas). Therefore, if we can calibrate the differential fringe phase between the target and reference sources at a level of 3.6 degree, then we readily obtain the 1 mas $\times$ (3.6 deg / 360 deg) $=$ 10 μas accuracy. Model simulations and experiments have shown that the atmospheric phase fluctuation, which has been the largest error source in the VLBI astrometry, can be reduced to the level of a few degrees in the differential fringe phase observable through the phase–referenced integration for several minutes to an hour, as far as the separation angle of the target and reference sources is around two degree or smaller (Sasao et al., 1993; Asaki et al., 1996; Sasao, 1997; Hachisuka, 2001).

Y. Nakada et al. (eds.), Mass-Losing Pulsating Stars and their Circumstellar Matter, 389–392.

2. Radio Sources to be Observed

Figure 1 shows sky distributions of water maser sources listed in Arcetri Catalog (Brand et al., 1994, and references therein) and SiO maser sources brought from various survey results. We expect that roughly 1000 masers brighter than a few Jy are detectable with VERA. Figure 2 shows sky distribution of reference source candidates, i.e. compact extragalactic continuum sources (Peck and Beasley, 1998; Honma et al., 2000). At present, more than 75 % of water maser sources have reference source candidates within 2.2 degree. This percentage will increase with more thorough surveys of compact continuum sources in the Galactic plane region and in the southern hemisphere.

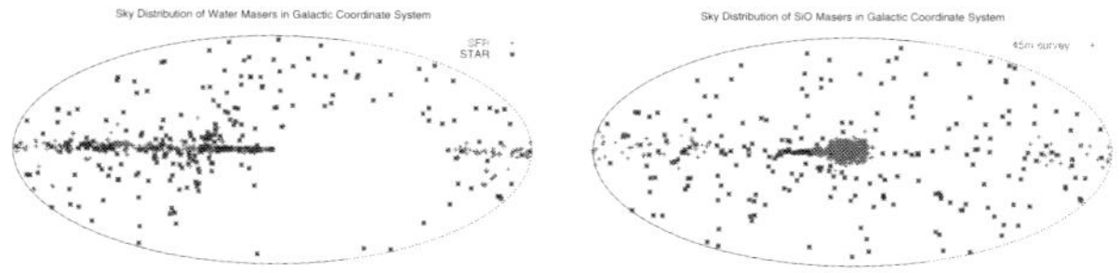

Figure 1. Sky distribution of water(left) and SiO(right) maser sources

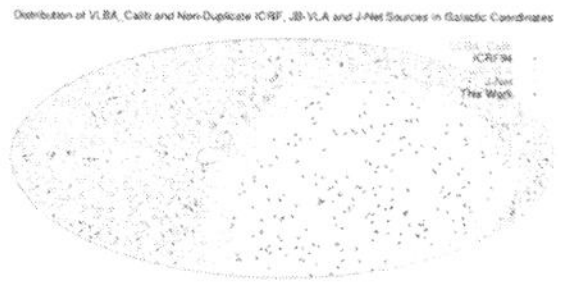

Figure 2. Sky distribution of reference source candidates

3. Masers around AGB Stars

Figure 3 (left) shows sky distribution of 69 Miras surrounded by water masers and having adjacent reference source candidates within 2.2 degree. Among them, 35 have roughly estimated distance values. Their distribution on the Galactic plane is shown in the same figure (right). Since they are within a few kpc from the Sun, their parallaxes must be well measurable with VERA.

As an example, relative proper motions of water maser features around a semiregular variable RT Vir revealed by a series of VLBA observations (Imai et al., 2002) are shown in figure 4 together with expected parallax ellipses for two extreme distance estimates for the star (120 and 460 kpc, respectively; preferred value by Imai et al., 2002, is 220 kpc). It

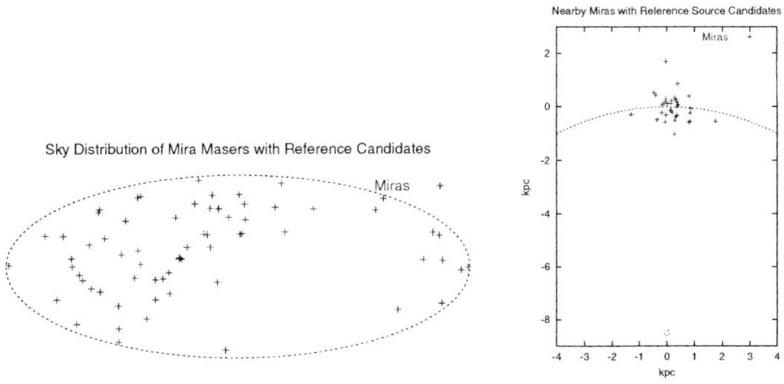

Figure 3. Sky and Galactic plane distributions of Miras with water masers

is evident that the parallax is large enough to be detected by the VLBI accuracy if we are given an extragalactic source as a reference point.

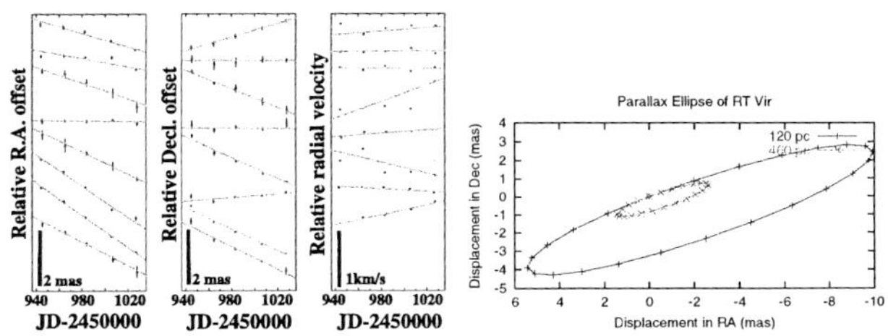

Figure 4. Relative proper motions of water masers around RT Vir (left two plots) and expected parallax ellipses (right) for two extreme distance estimates of the star

10% accuracy in distance determination corresponds to 0.2 mag accuracy in the absolute magnitude. With such an accuracy, the period–luminosity relation will be firmly established for field Miras.

4. Implications of VERA for AGB Studies

Roles of VERA could be summarized as follows.

VLBI monitoring of hundreds of water and SiO masers

Multi–epoch and multi–line maps with 'absolute' positional reference

Maser light curves, asymmetries, shock propagation,...

Trigonometric parallax distances

Kinematics in the Galaxy

⇓

Physics of the mass-loss flow

Region and mechanism of acceleration

Collimation mechanism

Period-luminosity relation of field AGB stars

Zero–point : Directly calibrated cosmic distance scale

Multiple PL sequences and pulsational modes

Metallicity dependence?

Comparative study of mass-loss history

Unfortunately, the number of target AGB stars with strong masers are rather limited for VERA. So, let the forthcoming space astrometric missions like GAIA and JASMINE do a lot more.

References

Asaki, Y., Saito, M., Kawabe, R., Morita, K., & Sasao, T. 1996, Radio Science, 31, 1615

Brand, J., Cesaroni, R., Caselli, P., Catarzi, M., Codella, C., Comoretto, G., Curioni, G.P., Curioni, P., Di Franco, S., Felli, M., Giovanardi, C., Olmi, L., Palagi, F., Palla, F., Panella, D., Pareschi, G., Rossi, E., Speroni, N., & Tofani, G. 1994, A&AS, 103, 541

Hachisuka, K. 2001, PhD Thesis, The Graduate University for Advanced Studies

Honma, M., Oyama, T., Hachisuka, K., Sawada-Satoh, S., et al. 2000, PASJ, 52, 631

Imai, H., Shibata, K-M., Marvel, K.B., Diamond, P.J., Sasao, T., Miyoshi, M., Inoue, M., Migenes, V., & Murata, Y. 2002, The 3-D Kinematics of Water Masers around the Semiregular Variable RT Vir, to be submitted to ApJ

Peck, A.B., & Beasley, A.J. 1998, Proceedings of IAU Colloquium 164, Radio Emission from Galactic and Extragalactic Compact Sources, ed. J.A. Zenzus, G.B. Taylor, & J.M. Wrobel (ASP Conference Series), 144, 155

Sasao, T., Kawano, N., Hara, T., Kuji, S., Sato, K-H., Tamura, Y., Kameya, O., Iwadate, K., Tsuruta, S., Asari, K., & Yasuda, S. 1993, in Proceedings of the International Workshop for Reference Frame Establishment and Technical Development in Space Geodesy, 91

Sasao, T. 1997, in Proceedings of Technical Workshop for APT and APSG 1996, 70

Question(P. Bastien) : What fraction of the time is a maser 'on' (or detectable)? This fraction could be used to estimate the completeness of surveys.

Answer : The time variability of masers is still poorly known, so we are not in position to estimate completeness of existing survey results. This means that we will have to regularly conduct single–dish flux check observations of masers before astrometric VLBI sessions in order to avoid unsuccessful runs.

UNVEILING THE MOLECULAR BIPOLAR OUTFLOW OF THE PECULIAR RED SUPERGIANT VY CANIS MAJORIS

Hiroko Shinnaga
Harvard-Smithsonian Center for Astrophysics, Submillimeter Array,
P.O. Box 824 Hilo, HI 96721, USA
hshinnaga@cfa.harvard.edu

Mark J. Claussen
National Radio Astronomy Observatory Array Operations Center,
P.O. Box O, Socorro, NM 87801, USA

Jeremy Lim, Dinh-Van-Trung
Institute of Astronomy and Astrophysics, Academia Sinica,
P.O.Box 23-141, Taipei 106, Taiwan

Masato Tsuboi
Institute of Astrophysics and Planetary Science,
Ibaraki University, Bunkyo, Mito, Ibaraki 310-8512, Japan

Abstract We carried out polarimetric spectral-line imaging of the molecular outflow of the peculiar red supergiant VY Canis Majoris in SiO $J = 1 - 0$ line in the ground vibrational state, which contains highly linearly-polarized velocity components, using the Very Large Array. We succeeded in unveiling the highly linearly polarized bipolar outflow for the first time at subarcsecond spatial resolution. The results clearly show that the direction of linear polarization of the brightest maser components is parallel to the outflow axis. The results strongly suggest that the linear polarization of the SiO maser is closely related to the outflow phenomena of the star. Furthermore, the results indicate that the linear polarization observed in the optical and infrared also occur due to the outflow phenomena.

Keywords: circumstellar matter — masers — polarization — stars : individual (VY Canis Majoris) — techniques: polarimetric

Y. Nakada et al. (eds.), Mass-Losing Pulsating Stars and their Circumstellar Matter, 393–396.

1. Introduction

VY Canis Majoris (VY CMa) is known as one of the brightest red supergiants at a distance of 1.5 kpc ((Herbig 1972; Lada & Reid 1978)). It is located at the ridge of the HII region S310 ((Lada & Reid 1978)). The star has a luminosity of $\sim 10^5\ L_\odot$, although its effective temperature is only $\sim$ 3000 K ((Le Sidaner & Le Bertre 1996; Smith et al.2001)). Those makes it the spectral class of M5Ib and an expected mass of ~ 25 $M_\odot$ based on theoretical models ((Chieffi et al.1998)). The extraordinary mass-loss rate of $\approx 2 \times 10^{-4}\ M_\odot\ \mathrm{yr}^{-1}$ ((Danchi et al.1994)) may imply that the star is close to experiencing a supernova.

SiO lower J lines in the v=1 and 2 states are known to be bright masers especially at the lower transitions, associated with cool, evolved, long-period variables. Using the bright SiO $v = 1$ and 2, $J = 1 - 0$ masers, VLBI imaging have been carried out to investigate the detailed structure of the inner circumstellar regions of the star ((Miyoshi et al.1994)), and revealed the presence of a ring-like structure, with a scale of $\sim 60 - 100$ milliarcsecond(mas) ((Miyoshi et al.1994; Miyoshi 2002)). On the other hand, the rotational lines in the $v = 0$ state is not usually maser. The possibility of maser phenomena in the $v = 0$, $J = 2 - 1$ line towards the star was pointed out by (Buhl et al.(1975)) because of the peaky profile of the spectrum, but not investigated further for more than two decades. In fact, the peaky components in the $v = 0$ state are highly linearly polarized, which strongly suggests that they are non-thermal emission, i.e., maser ((Shinnaga et al.1999)).

The nature of both the driving mechanism and the spatial-kinematic structure of the outflow from red supergiants is poorly understood. Indeed, no image of the thermal molecular-gas outflow from a red supergiant has hitherto been reported. Here we present the results of the polarimetric imaging observations of the outflow traced by SiO $J = 1-0$ in the $v = 0$ state at high angular resolution. In this paper, we concentrate on the linear polarization of the observations. A detail presentation of our results will be published elsewhere (H. Shinnaga et al. 2002, in preparation)

2. Observations

First, we carried out single-dish polarimetric observations using the millimeter polarimeter system ((Shinnaga et al.1999)) at Nobeyama. Subsequently, we carried out the imaging observation of VY CMa with the SiO $v = 0$, $J = 1-0$ transition at 43.423858 GHz in the B configuration of Very Large Array(VLA), which is operated by the National Radio Astronomy Observatory (NRAO), with two different settings. One has a velocity resolution of 2.7 km s^{-1} providing a total bandwidth of 12.5 MHz in 32 channels over a velocity range of 87 km s^{-1}. The other has a velocity resolution of 5.4 km s^{-1} providing the same total bandwidth and in 16 channels with full polarimetric information. The observations had been done on March in 2001 by using all the 27 antennas available of the array in total. The resulting size of the synthesized beam is to subarcsecond, $0.''42 \times 0.''17$. The attained rms noise levels is typically $\sim$ 2 mJy/beam with a velocity resolution of 5.4 kms^{-1} for the full polarization data set. The calibration of the polarization has been done by observing 0521+166(3C138).

3. Results and Discussion

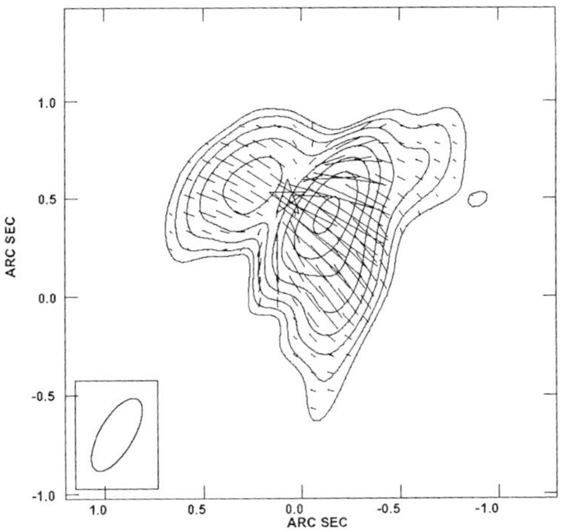

Figure 1. Polarization vectors of the SiO $J = 1 - 0$ transition in the $v = 0$ state superposed on the total intensity map at the peak velocity($V_{\rm LSR} = 19$ kms^{-1}) channel. The contours are drawn in a logarithmic order. The lowest contour is 5 σ level. The star marks the expected location where the star is, determined from this work.

Figure 1 shows the polarization imaging map of the molecular outflow in the velocity channel where the overall line intensity is strongest. The map clearly show the outflow is bipolar (like a "butterfly" shape). The south-western blue lobe is brighter than the north-eastern red lobe. The outflow terminal velocity is about 40 kms^{-1}. Both of brightest components are highly linearly polarized, $\sim 0.2''$ (i.e., 300 AU) apart from the center of the outflow. The direction of the linear polarization of the bright maser components is parallel to the outflow axis. The position angles of linear polarization derived from our single-dish polarimetry ((Shinnaga et al.1999)) are consistent with this imaging results. The polarization vectors of all the components well trace the flow directions of the outflow. These observational facts strongly suggest that the linear polarization is originated from the outflow phenomena.

This star also shows highly linearly polarized light in the optical and near infrared (NIR) ((Hashimoto et al.1970; Dyck et al.1971; Forbes 1971)). The observed directions of optical/NIR linear polarization are within the range of the direction of the SiO molecular outflow. This strongly suggests that the polarization in the optical and NIR also comes from the same origin, namely, outflow.

Optical and infrared images with high angular resolution ((Monnier et al.1999; Smith et al.2001)) clearly show the asymmetric nebular with many complicated structures surrounding the star. Several coincidences between the SiO image and optical/NIR images are identified. High resolution image in NIR, figure 2 of a paper by Monnier et al.(1999), show extended feature (~100 mas) towards east-south direction from the brightest spot. This may be a part of the root of the southern-western outflow, due to the coincidence in their directions. The SiO component towards this direction is also highly linearly polarized.

To date, several molecules are known to show maser lines at submillimeter wavelengths ((Humphreys et al.1997; Menten & Young 1995)). Using these masers as a probe, we will be able to investigate detailed structures by interferometric observations. The Sub-Millimeter Array (SMA), which is currently under construction at the summit of Mauna Kea, would greatly contribute towards further understanding of the nature of the star.

References

Buhl, D., Snyder, L.E., Lovas, F.J., et al. 1975. ApJL, 201, L29
Chieffi, A., Limongi, M., & Straniero, O. 1998, ApJ, 502, 737
Danchi, W.C., Bester, M., Degiacomi, C.G., et al. 1994, AJ, 107, 1469
Dyck, H.M., Forbes, F.F., & Shawl, S.J. 1971, ApJ, 76, 901
Forbes, F.F. 1971, ApJL, 165, L83
Hashimoto, J., Maihara, T., Okuda, H. et al. 1970, PASJ, 22, 335
Herbig, G. H. 1972, ApJ, 172, 375
Humphreys, E.M.L., Gray, M.D., Yates, J.A. et al. 1997, MNRAS, 287, 663
Lada, C., & Reid, M. 1978, ApJ, 219, 95
Le Sidaner, P. & Le Bertre, T. 1996, A&Ap, 314, 896
Menten, K.M., & Young, K. 1995, ApJL, 450, L67
Miyoshi, M., Matsumoto, K., Kameno, S., et al. 1994, Nature, 371, 29
Miyoshi, M. 2002, in this volume
Monnier, J.D., Tuthill, P.G., Lopez, B., et al. 1999, ApJ, 512, 351
Shinnaga, H., Tsuboi, M., and Kasuga, T. 1999, PASJ, 51, 175
Smith, N., Humphreys, R.M., Davidson, K., et al. 2001, AJ, 121, 1111

IR SPACE ASTROMETRY

JASMINE-Project

Naoteru Gouda*, Takuji Tsujimoto, Yukiyasu Kobayashi, Tadashi Nakajima, Naoki Yasuda, Yukitoshi Kan-ya
National Astronomical Observatory of Japan

Hideo Matsuhara
Institute of Space and Astronautical Science

Taihei Yano
The Institute of Physical and Chemical Research

Yoshiyuki Yamada
Graduate School of Science, Kyoto University

Munetaka Ueno
Graduate School of Arts and Science, University of Tokyo

Abstract We introduce a Japanese future plan of the IR space astrometry(JASMINE-project). JASMINE is an infrared(K-band) scanning astrometric satellite. JASMINE(I and/or II-project) is planned to be launched between 2013 and 2015 and will measure parallaxes, positions and proper motions with the precision of 10 microarcsec at K=12~14mag. JASMINE can observe about a few hundred million stars belonging to the disk and the bulge components of our Galaxy, which are hidden by the interstellar dust extinction in optical bands. Furthermore JASMINE will also measure the photometries of stars in K, J and H-bands. The main objective of JASMINE is to study the fundamental structure and evolution of the disk and the bulge components of the Milky Way Galaxy. Furthermore its important objective is to investigate stellar physics.

*e-mail:naoteru.gouda@nao.ac.jp

Y. Nakada et al. (eds.), Mass-Losing Pulsating Stars and their Circumstellar Matter, 397–404.

Keywords: astrometry, infrared, the Galaxy, stellar physics

1. Introduction

Astrometry is the part of astronomy which provides positions, trigonometric parallaxes(distances) and proper motions of objects on the celestial sphere. If we have furthermore information on radial velocities of the objects, we can see the 6-dimensional phase space densities of the objects. Needless to say, this information is most fundamental in many fields of astronomy and astrophysics. For example, the 6-dimensional phase space densities of stars in our Galaxy will give us dynamical structures of star clusters and the Milky Way. Moreover an accurate distance of a star brings us the accurate luminosity of the star. The information of the accurate luminosity is useful for distance indicators and necessary for analyzing stellar physics. So this information is very important for resolving the problem on pulsating stars which is one of the main subjects in this meeting. Furthermore motions of each star will bring us information on physical characters of binary systems and the information of the motions is also useful to detect planetary systems outside the solar system.

Hipparcos satellite has offered more accurate information on the astrometric parameters rather than that measured on the ground. The parallax accuracy is about 1 milli-arcsecond(mas). This accuracy corresponds to 10% distance error at 100pc. The proper motion accuracy of 1 mas/year given by Hipparcos is the velocity error of $\sim$ 5 km/s at 1kpc. The astrometric measurement by Hipparcos has certainly brought us a revolution in astronomy and astrophysics. But it is still a small revolution. We need a parallax accuracy of less than $\sim$ 10% error because errors larger than $\sim$ 10% cause bias effects when we convert the parallax to the distance. Then in Hipparcos we can measure distances of stars accurately only within 100pc. In this case, we have only a few interesting stellar objects in this area. Figure 1 shows "the Milky Way". The small circle(radius of 100pc) around the Sun is the horizon for distances accurate to 10% by Hipparcos.

In future, we need big revolutions by high accurate astrometric measurements. We require a parallax accuracy of $\sim$ 10μas. This accuracy corresponds to 10% distance error at 10kpc(see figure 1). The tangential velocity error is 1km/s at 20kpc(see figure 1). These astrometric measurements will certainly cause breakthrough in many fields of astronomy and astrophysics.

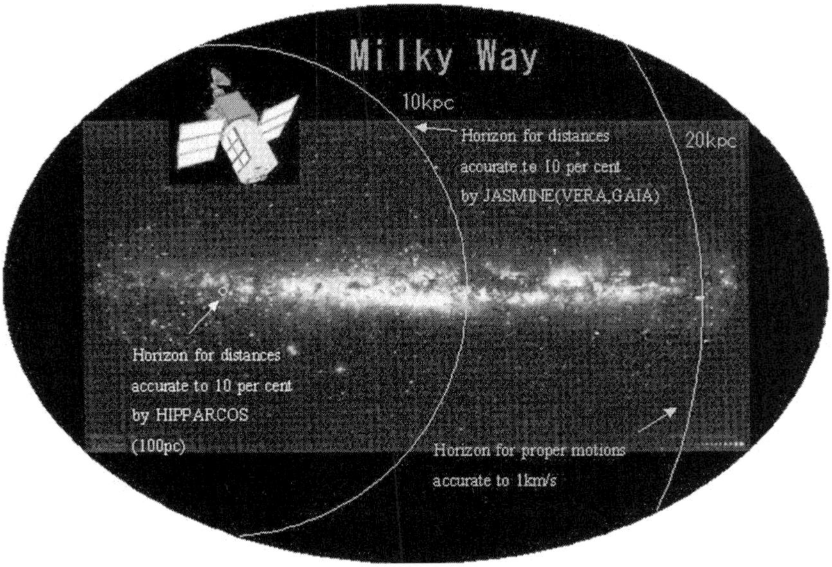

Figure 1. The horizons for distances accurate to 10% by Hipparcos(100pc) and JASMINE(VERA, GAIA)(10kpc) in the Milky Way. The horizon of proper motions accurate to 1km/s in JASMINE(VERA,GAIA)(20kpc) is also shown.

Table 1. Summary of Space Astrometry Projects

Mission	Agency	Launch	# of stars	Mag.limit	Accuracy
Hipparcos	ESA	1989	120000	12	1mas@V=10
DIVA	Germany	$\sim$ 2006	35 million	15	250μmas@V=10
FAME	USNO(U.S.A)	???	40 million	15	50μmas@V=9
SIM	NASA	$\sim$2009	10000	20	4μmas@V=20
GAIA	ESA	$\sim$ 2012	1 billion	20	10μmas@V=15

2. Future Space Astrometry Projects

Table1 summarizes the future space astrometry projects in the western countries. DIVA, FAME and GAIA are telescope-types and SIM is an interferometer-type. It should be noted that all space projects described here will observe in optical bands. On the other hand, in our JASMINE-project, we will operate the near infrared (K-band;2.2μm) because the near infrared measurement has some advantages which will be described in §3. Please refer to each web-page for detailed explanations of the other space astrometry projects (DIVA:http://www.ari.uni-heidelberg.de/diva/, FAME:http://www.usno.navy.mil/FAME/,

SIM:http://sim.jpl.nasa.gov/index.html, GAIA:http://astro.estec.esa.nl /SA-general/Projects/GAIA/).

Furthermore, VERA-project is the revolutionary important project of radio interferometer on the ground to measure the astrometric parameters with the accuracy of 10μas. It will start operating in 2003 or 2004.

3. JASMINE-project

JASMINE is a scanning satellite for infrared exploration(Gouda *et al.* 2002). JASMINE is planned to measure parallaxes, positions and proper motions of stellar objects in K-band with the precision of 10μas at $K = 12 \sim 14$mag. We can measure a few hundred million stars around the bulge and the disk of our Galaxy. Furthermore we will measure photometries in J, H and K-band. Infrared astrometry is important for the investigations of obscured objects when studying the bulge, the galactic plane and star formation regions: Interstellar dust hides the visible light of stellar objects located at low galactic latitudes. The interstellar extinction depends on the wavenumber and it is much smaller in K-band than that in optical bands. Then we can see more stars in K-band rather than those in optical bands.

We calculated the expected number of stars to be observed by JASMINE using our Galaxy model. This model is based on the "sky" model developed by Cohen and his collaborators(Wainscoat and Cohen 1992; Cohen 1994; Cohen, Sasseen and Bowyer 1994; Cohen 1995). We also took into account the dust map observed by DIRBE when we consider the dust extinction. Each panel of figure 2 shows the expected number of stars for $V \leq 15$, $K \leq 12$ and $K \leq 14$ at a distance r from the sun. Parallaxes of better than 10μas accuracy will be provided for brighter stars than $V = 15$, $K = 12$ and $K = 14$ for GAIA, JASMINE-I and JASMINE-II projects(see §4.1 and §4.2), respectively. We can see from figure 2 that the number of stars observed in K-band is much larger than that observed in V-band at the distance of more than a few kpc away from the sun on the galactic plane. Therefore we must operate in near infrared bands such as K-band if we would like to measure many objects in the galactic plane at a distance of more than a few kpc. This is one of the advantages for operating in K-band. Furthermore stars such as K-giants and M-giants which are considered to kinematically trace the galactic gravitational potential are bright in K-band. This is also an advantage for operating in K-band.

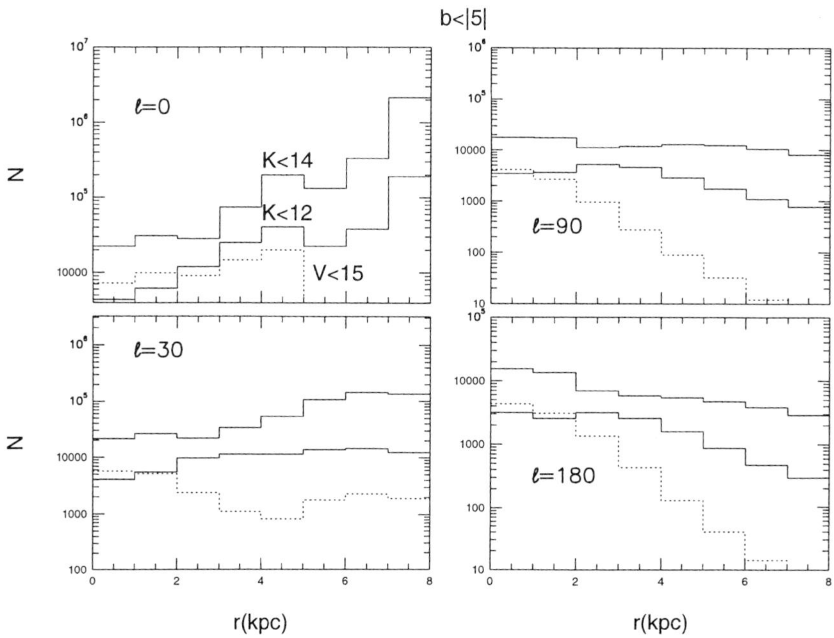

Figure 2. The expected number of stellar objects in the direction of the galactic longitude ℓ shown in each figure. The absolute value of the galactic latitude, b, of the field of view is restricted within 5°. The angular scale of the field of view along the galactic longitude is 1°. N is the expected number of stars for $\mathrm{V} \le 15$, $\mathrm{K} \le 12$ and $\mathrm{K} \le 14$, and r is the distance from the sun.

As for the scientific targets of JASMINE, we focus on the structure of the bulge at first. The morphology and kinematics of the bulge is very important information for solving the formation history of the bulge and our Galaxy itself. Furthermore it is very interesting to investigate the dynamical structure of the bulge from the point of view in physics(relaxation process, quasi-equilibrium structure *etc.*) of collisionless self-gravitating systems.

Second, the analysis of the stellar orbits on the disk plane is a big target. This analysis will clarify the origin of the spiral arms pattern and the nature of the stellar warp.

Third, JASMINE can pierce star formation regions where the visible light of stellar objects is buried by interstellar dusts.

The main objective of JASMINE is to study the fundamental structure and evolution of the disk and the bulge components of the Milky

Way Galaxy. Furthermore its important objective is to investigate stellar physics. Observations of extra-galaxies near our Galaxy are also interesting targets of our JASMINE project.

4. Preliminary Design of JASMINE

The accuracy of the astrometric parameters(position, parallax, proper motion per year) is roughly estimated as $\sigma \sim \lambda/D\sqrt{N}$. Here σ is an error dispersion. λ and D are wavelength of photons and a diameter of mirror, respectively. N is the number of detected photons of a target star during the observation period. Since observed number of photons obeys the Poisson distribution, the accuracy is proportional to $N^{-1/2}$. Then large N is required in order to get high accuracy. So we need a large mirror and furthermore many good detectors for achievement of large F.O.V.(field of view) on focal planes. In the optical bands, we can use CCD detectors in which TDI(time delayed integration)mode, that is, drift scan mode, can be operated. In TDI mode, the rate of charge transfer rate is synchronized with the scan velocity of the spinning satellite. By using TDI mode, we can gather large number of photons without distortion of stellar images. However we do not have a good IR detector in which TDI mode can be used. This is one of the technical challenges in our project(see §5).

We plan the following two options in our project, JASMINE-I and JASMINE-II. The size of the instruments is smaller in JASMINE-I than in JASMINE-II and then the cost is less expensive. So JASMINE-I has high possibility for its early realization although the K-band magnitude at which 10μas accuracy can be achieved is reduced and the survey area is narrower than in JASMINE-II.

4.1 JASMINE-I project

JASMINE-I is planed to be launched in around 2013 by a rocket like M-V rockets of ISAS(Institute of Space and Astronautical Science) in Japan. JASMINE will be designed for operation in Sun-Earth L2 point.

As for the payload, the diameter of the primary mirror for the astrometric filed is 1.8m. The diameter of the F.O.V. is 0.4°. The focal length is 66.6m. On the astro-focal plane, we put about 80 detectors, such as HgCdTe for the astrometric field. The detector has 2048 × 2048 pixels of size 18.5 μm.

JASMINE has other one or two small mirrors (the diameter is about 0.5m)beside the main mirror for the scientific targets. The extra mirror has the field of view separated by a large basic angle from the field of view of the main mirror. The objective of this extra small mirror is to measure the positions of bright stars in V-band (V$\leq$10mag) as "guide" stars in order to obtain absolute parallaxes by directly comparing very different parallax factors and improve the rigid system of the star positions and proper motions by linking widely separated regions of the sky.

The accuracy of parallaxes and proper motions is about 10μas at K=12mag: Many stellar objects can be observed in K-band toward the direction of the galactic plane. We found from our calculations based on our Galaxy model that the contamination of background faint stars makes it impossible to achieve the accuracy of 10μas around the bulge for the fainter stars above 13 or 14 mag. So we take 12 mag as the largest magnitude for obtaining the astrometric parameters with such high accuracy of 10μas. Furthermore, JASMINE will measure the photometries of stars in K, J and H-bands.

In this project, the primary objective is a survey of stellar objects around the galactic plane and the bulge. The area which can be surveyed within the mission life time $\sim$ 5 years is estimated to be about 3000($\sim 360^\circ \times 8^\circ$) square degree if we assume that TDI mode can be operated.

4.2 JASMINE-II project

Second one of our project is JASMINE-II project. JASMINE-II is planed to be launched around in 2015 by a H-IIA rocket which is larger than a M-V rocket. H-IIA rockets have been developed by NASDA(NAtional Space Development Agency of Japan). The first experimental launch of the H-IIA rocket succeeded on the 29th of August in 2001.

The main target in this project is the survey toward the whole galactic plane including the bulge. The accuracy of parallaxes and proper motions is about 10μas at K=14mag in the survey regions beside the galactic center region. The survey area is $360^\circ \times 10^\circ$. As for the payload, we will use a large primary mirror whose aperture is 3.6m$\times$1.8m for scientific targets. JASMINE has other small mirrors for measuring the positions of bright "guide" stars in V-band. On the astro-focal plane, we put about 130 detectors for the astrometric field.

5. Technical Challenges of JASMINE

We have many technical issues which have to be solved in order to realize our mission. The main issues are as follows.

We need a large mirror to get many photons for achievement of high accurate astrometry. Furthermore the mirror is expected to be very light for a launch. Then we have to develop ultra lightweight mirrors suitable for M-V and H-IIA rockets.

As mentioned above, TDI mode is very powerful for obtaining the high accuracy in measuring the centroids of stellar images. However, we do not have any infrared detectors with good quantum efficiency of detection in which TDI mode can be operated. We need develop a good infrared detector in which TDI can be operated. We have some ideas for making the detectors. If we will get such good infrared detectors in future, the detectors can be used for other survey projects in infrared astronomy and they must contribute to the enormous improvement in the fields of infrared astronomy.

DIVA, FAME and GAIA satellites will scan the whole sky in the way similar to Hipparcos. On the other hand, JASMINE will scan the restricted regions around the galactic center and the galactic disk plane. In this case, the way how to obtain the accurate parallaxes, positions and proper motions is a big problem which has never been considered. We are now investigating the design of instruments with small mirrors beside the main mirror and how to scan the survey area in order to effectively get the scientific objectives and furthermore obtain the accurate absolute parallaxes.

6. Summary

JASMINE is expected to measure parallaxes, proper motions and near-infrared photometries of about a few hundred million stars mainly within the bulge and disk components of our Galaxy. JASMINE aims at the high precision astrometry of 10μas level in K-band. The primary scientific targets of JASMINE are to clarify the structure and evolution of the bulge and the disk components of the Galaxy. Jasmine is a name of plants. The flower language of jasmine is elegance. Then we greatly hope that JASMINE will be *elegantly* successful in future.

References

Gouda, N., Tsujimoto, T., Kobayashi, Y., Nakajima, T., Yasuda, N. & Matsuhara, H. 2002, Ap. Space Sci., 280, 89

Cohen, M. 1994, AJ, 107, 582

Cohen, M. 1995, ApJ, 444, 874

Cohen, M., Sasseen, T. P. & Bowyer, S. 1994, ApJ, 427, 848

Wainscoat, R. J. & Cohen, M. 1992, ApJS, 83, 111

MIRA PROJECT AND THE OBSERVATIONS OF MIRA VARIABLES

Koichi Sato[1], Masanori Yoshizawa[1], Jun Nishikawa[1], Naoko Ohishi[1], and Takayuki Kotani[2]
[1]*MIRA Project, National Astronomical Observatory*
Mitaka, Tokyo 181-8588, JAPAN
[2]*Dept. of Astronomy, Univ. of Tokyo*
Bunkyo-ku, Tokyo 113-0033, JAPAN

Abstract We are now constructing the MIRA-I.2 interferometer, the second stage interferometer of MIRA (Mitaka optical and InfraRed Array Project) at Mitaka campus of National Astronomical Observatory of Japan. MIRA-I.2 system is an interferometer consisting of two 300mm siderostats which are placed on a 30m baseline. Fringe detection is made at 800nm with APD. Fringes of some bright stars were obtained with the test system of 6m baseline, on June, 2001. We are now constructing the regular system of 30m baseline. The fringes of Vega were detected with this system, on June, 2002. We discuss the possibility of observations of Mira variables with this interferometer.

Keywords: optical interferometer, Mira variable

Introduction

MIRA-I.1 interferometer, the first stage system of MIRA Project (Mitaka optical and InfraRed Array Project) at Mitaka campus of National Astronomical Observatory of Japan, was operated from October 1998 to March 1999 (Machida et al. 1998, Nishikawa et al. 2000). The basic design and the development of some elements of MIRA-I.2 have been carried out along with the development of MIRA-I.1 (Honma et al. 1998, Sato et al. 1998). We are now constructing the MIRA-I.2, the second stage interferometer of MIRA Project. Before setting up the regular system of 30m baseline of MIRA-I.2, observations with the test system of 6m baseline were carried out. The fringes of some bright stars were obtained at the end of June, 2001 with this test system (Sato et al. 2001). At the end of July, 2001, this test baseline system was closed. Af-

Y. Nakada et al. (eds.), Mass-Losing Pulsating Stars and their Circumstellar Matter, 405–408.

ter that, we have been making the regular system of 30m-baseline. The present status of the construction of the regular system is described.

1. Instrumentation

The outline of the MIRA-I.2 system is shown in Figure 1. The baseline lengths are 6m for the test system (baseline in north-south) and 30m for the regular system, respectively. The baseline of the regular system is inclined by 23 degrees from the north-south direction. The control is made with several personal computers connected to each instrument with direct wires and local area networks.

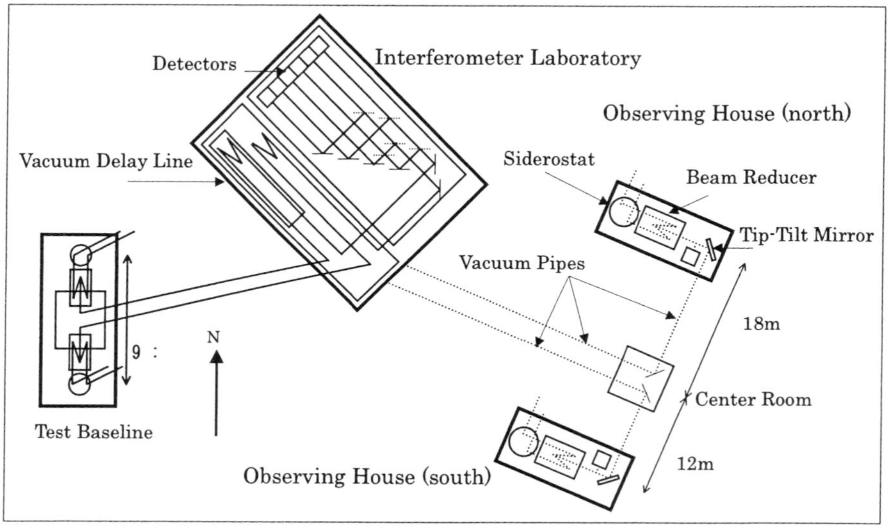

Figure 1. The outline of MIRA-I.2 system

The element telescope consists of a siderostat of 300mm diameter and a beam reducer. The beam reducer consists of a Newtonian type telescope of diameter 130mm and focal length 795mm. This beam reducer will be replaced by one of off-axis mirrors. A tip-tilt mirror is placed at each exit of the beam reducer. The tip-tilt system consists of a tip-tilt mirror driven by piezo-electric actuators in the observing house and a quadrant sensor in the interferometer laboratory. The quadrant sensor consists of a pyramid prism and four APD sensors.

The fine delay line in a vacuum chamber carries a retro-reflector of 200mm diameter. A fast delay line is used for delay modulation and fringe tracking. The fast delay line consists of a 100 μ m stroke PZT and its controller. A laser metrology system (HP5527B) is used for measuring the movements of the delay line and the modulation line.

After the lights are reflected from the delay line retro-reflectors, the beams are divided with the beam splitters. One beam is directed to the beam combiner followed by the fringe detector and the IICCD camera. Fringe detection is made with a photon counting APD at the wavelength of 800nm. The other is directed to the next beam splitters where the beams are divided toward the intensity sensors for the scintillation monitor and the quadrant sensors for the tip-tilt mirror control. Two artificial lights, the He-Ne laser and the white light (halogen lamp), are supplied from the back of one side of the beam combiner. The laser light is used to carry out the optical alignment. The white light is used for finding the position of the zero path delay and testing the visibility of the interferometer system. CCD camera is used to monitor the images of artificial lights. IICCD camera is used to monitor the star images.

2. Observations of Mira variables

Mira variables are observed with the optical and infrared interferometers such as Mark III, COAST, IOTA, and so on (e.g. Van Belle, 1996, Haniff, 1995). Here, we will discuss about the observations of Mira variables with the optical interferometer, especially the angular diameter observations. The fundamental parameters of stars, such as the mass, linear radius, effective temperature and luminosity, are considered. Determination of the mass of each star is difficult. Direct determinations, that is, the determinations by the orbital motion of binary stars, are not available. The mass is estimated regularly by spectroscopic observations and model calculations.

The angular diameter observations are important for the determination of linear radii of stars. The angular diameter determination of Mira variables, which have the extended and complicated atmosphere, has some problems. The limb-darkening law is not simple and dependent on the wavelength (e.g. Scholz, 1989). The definition of radius is also not clear. It will be necessary to observe the brightness distribution of stars precisely with the optical interferometer.

The effective temperature is determined by the angular diameter and bolometric flux. The linear radius is determined with the angular diameter and the distance. The Hipparcos parallaxes of Mira variables are not accurate enough, since Mira variables are rather bright, but distant. We can determine the distance of Mira variables by observing the variations of the angular diameter with the optical interferometer and the radial velocities. VERA Project (Kobayashi and Sasao, 2002) will determine the annual parallax of maser sources around the Mira variables accurately. The future space astrometry will give accurate parallaxes for Mira variables.

Angular diameters of bright Mira variables are large enough to detect with the MIRA-I.2 interferometer. At the shorter wavelengths, such as V-band, the Mira variables have very large magnitude variations and are faint even at the maximum. Only a few stars will be observed at the near maximum with the MIRA-I.2 interferometer. It is better to observe the Mira variables at near infrared bands, where those are bright and have the small amplitude of light variations. In order to determine the angular diameter more accurately, multi-wavelength and multi-baseline observations will be needed. We will introduce the near infrared detectors and the third element telescope in the near future for MIRA-I.2.

3. Conclusions

We obtained fringes of some bright stars with the test system of 6m baseline in June, 2001. The operation of the test baseline was stopped at the end of July, 2001. In August, we started the setup of the regular system of 30m baseline by making the modifications and improvements of each element. From the end of October,2001, the test observations of some bright stars have been made until now. The fringes of Vega were detected with the regular system, on June, just after this workshop. Hereafter, we will make the improvements of each component and complete the regular system which is fit for the regular observations of bright stars, such as, Mira variable, spectroscopic binaries and Cepheids, together with normal stars. We are planning to equip the near infrared detector and add the third element telescope in the near future.

References

Sasao, T., & Kobayashi, H. 2002, this issue

Haniff, C.A., Scholz, M., & Tuthill, P.G. 1995, MNRAS, 276, 640

Honma, Y., Nishikawa, J., & Kasuga, T. 1998, Proc. SPIE, 3350, 192

Machida, Y., Nishikawa, J., Sato, K., et al. 1998, Proc. SPIE 3350, 202

Nishikawa, J., Sato, K., Yoshizawa, M., et al. 2000, Proc. SPIE 4006, 68.1

Sato, K., Nishikawa, J., Yoshizawa, M., et al. 1998, Proc. SPIE 3350, 212

Sato, K., Yoshizawa, M., Nishikawa, J., et al. 2001, Proc. Liege Int. Astrophys. Coll., 36, 5

Scholz, M. & Takeda, Y. 1987, ApJ, 186, 200

van Belle, G.T., Dyck, H.M., Benson, J.A. & Lacasse, M.G. 1996, AJ, 112, 2147

MIZUSAWA 10m ANTENNA SYSTEM FOR STUDY OF PULSATING STARS

Osamu Kameya
National Astronomical Observatory of Japan
Mizusawa Astrogeodynamics Observatory
Hoshigaoka 2-12, Mizusawa, Iwate 023-0861, Japan
kameya@miz.nao.ac.jp

Abstract Mizusawa 10m antenna performs monitoring observation of light curve and polarization of the maser spots. The 10m antenna has 22GHz and 43GHz receivers, which are able to detect water vapor masers and SiO (J=1-0) masers around late-type stars. This system has good performance for linear polarization measurement and antenna temperature monitoring for several years.

Keywords: Monitoring, masers, polarization

1. Introduction

Late-type stars are often accompanied by molecular line masers (e.g. OH, water vapor, SiO, etc). Recent VLBI observations show that maser spots expand from the central stars, and change their distribution and velocity structure according to phase of the central stars (e.g. Imai et al. 1996, Ishitsuka et al. 2002). However, systematic monitoring observations of total intensity of the maser spots by using single dish radio telescopes are still important. Furthermore, polarization measurement of the maser spots may give us information on magnetic structure of circum stellar envelops. The Mizusawa 10m radio telescope may contribute to get this kind of information.

2. The observation system

The Mizusawa 10m radio telescope was constructed by National Astronomical Observatory of Japan in 1992, and has been used for many kind of researches: geodetic VLBI with S/X band, astronomical VLBI

Y. Nakada et al. (eds.), Mass-Losing Pulsating Stars and their Circumstellar Matter, 409–410.

with 22GHz band, experimental VLBI (for example, differential VLBI), and single dish observations (Kameya et al. 1996). This telescope has been used as one key station of Japanese VLBI Network (J-Net) from 1994 (Omodaka et al. 2001). Many astronomical results mainly dedicated to the observations of 22GHz water vapor maser towards many late-type stars and star forming regions. The 10m antenna is located at only 25m southwest of the VERA Mizusawa station. This close location aims at collaboration of VERA system.

We use 22GHz and 43GHz bands for water vapor masers and SiO J=1-0 masers, respectively. HPBW and aperture efficiency of these radio bands are 5.2' and 0.36 (K band), and 2.6' and 0.20 (Q band), respectively. The receivers at 22GHz and 43GHz bands use cooled HEMT amps on the 20K stage installed in one dewar. The receiver temperature and typical system noise temperature at zenith in winter season are 80K and 150K at 22GHz band, and 250K and 400K at 43GHz band, respectively. Either linear polarization, LHC, or RHC is selectable during the observation. For the backend, a 2048ch AOS with 40kHz resolution is available for observation near future.

Horiuchi and Kameya (2000) have used this system for measurement of the linear polarization of the highly polarized burst of a water vapor maser in Ori-KL from December 1998 to June 1999. The 10m system can measure only component along EL direction. Hence they measured variation of antenna temperature of the maser spots due to linear polarization assuming no variation in the intrinsic emission over one day.

The same method is available for the measurement of linear polarization towards late type stars. Even though physical condition of maser emitting region of circum stellar envelops is different from that of star forming regions, we may get some important information on magnetic field of circum stellar envelops. The performance check of the system towards the late type star VY CMa has started. The system will be used for research of time variation of polarization and total intensity of water maser and SiO masers towards some pulsating late-type stars together with those of some star forming regions.

References

Horiuchi, S., & Kameya, O. 2000, PASJ, 52, 545

Imai, H., et al. 1997, A&A, 317, L1

Ishitsuka, J.K., et al. 2002, PASJ, in press

Kameya, O., et al. 1996, in Proc. of Third East-Asian Meeting on Astronomy, Tokyo, "Ground-Based Astronomy in Asia", N. Kaifu, ed., 500

Omodaka, T., et al. 2001, Journal of the Communications Research Laboratory, Vol.48, No.1, pp23-29.

PHASE REFERENCE VLBI ASTROMETRY FOR MIRA-TYPE STARS

Kazuya Hachisuka
Departamento de Astronomía, Universitat de Valéncia, Spain
kazuya.hachisuka@uv.es

Makoto Miyoshi
National Astronomical Observatory, Japan
miyoshi@miz.nao.ac.jp

Abstract We carried out phase reference VLBI observations at 43 GHz between SiO masers in Mira type variables and adjacent extragalactic continuum source with VLBA. Main aim of this observation is determination of spatial distribution between SiO maser spots at $v = 1$ and $v = 2$, J=1→0 transition. Although the standard calibration was successful for each source, the phase reference calibration have not been successful. However, we will determine the absolute positional difference of each SiO maser spot within about 1 milli arcsecond due to the structure effect of reference source if the phase reference VLBI is successful.

Keywords: maser, phase-referencing VLBI

1. Introduction

Maser emission has been observed in Mira type variables with VLBI. Type of detected maser emission are OH masers (1665 and 1667 MHz), H_2O maser (22 GHz), SiO masers ($v = 1$ and $v = 2$, J=1→0 transition near 43 GHz). A traditional "onion shell" model for AGB stars contains SiO, H_2O and OH maser regions at increasing distances from the star. However, the precise region of them have been unknown with high resolution by means of VLBI. Especially, the relative spatial distribution of SiO maser spots between $v = 1$ and $v = 2$, J=1→0 transition is not revealed to be coincident or not, that is, it is not clear that the pumping mechanism is collisional or radiative. Although Miyoshi et al. (1994) reported that the emission of these two lines systematically arises from positionally coincident spots, Desmurs et al. (2000) reported that these spots are systematically not coincident. Both observations were able to determine only relative position of maser spots from reference maser spot in each transition by standard VLBI calibration. Therefore, the

Y. Nakada et al. (eds.), Mass-Losing Pulsating Stars and their Circumstellar Matter, 411–412.

superposition of images for both two transitions was performed by reference maser spot as a reference position. This procedure was assumed that the each reference spot were actually spatially coincident. In phase reference VLBI technique between SiO masers in Mira type variables and adjacent extragalactic continuum source, it is not necessary this assumption. We have to take account of only the difference of structure for extragalactic continuum source as a reference position.

2. Observation and Results

We carried out observations of the $v = 1$ and $v = 2$, J=1→0 lines of SiO maser with NRAO Very Long Baseline Array in September 2000. Observed sources were R Leo and J0947+111 as positional reference source. We mentioned that this observation was not simultaneous. Therefore, the UV coverage was a little difference in each frequency.

Although the standard calibration was successful for each source, the phase reference calibration have not been successful yet with AIPS. The images of J0947+111 shown in Figure 1. Both structure was elongated to east. The maximum difference of each structure was roughly 1 milli arcseconds. Therefore, we think that we will determine the absolute positional difference of each SiO maser spot within about 1 milli arcsecond due to the effect of structure of extragalactic continuum source by phase reference VLBI technique.

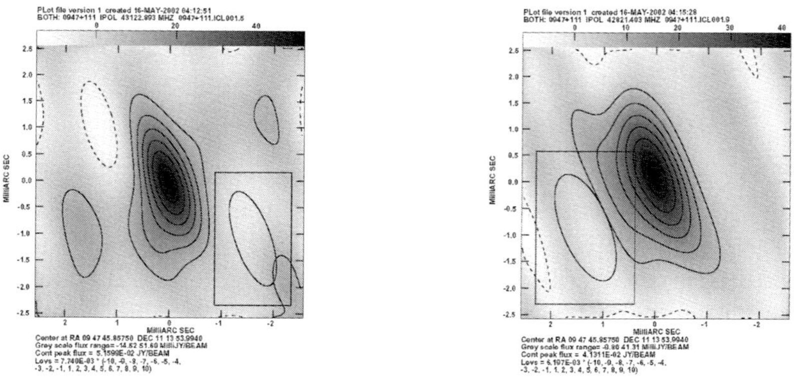

Figure 1. Image of J0947+111 at 43.122 GHz (left) and at 42.821 GHz (right). The beams are shown by the ellipses in the lower left-hand and right-hand corner of the images.

References

Miyoshi, M., et al. 1994, Nature, 371, 29
Desmurs, J. F., et al. 2000, A&A, 360, 189

VII

CONFERENCE SUMMARY

CONFERENCE SUMMARY

P.R. Wood
Research School of Astronomy and Astrophysics, Australian National University
wood@mso.anu.edu.au

1. Introduction

One of the pleasures of going to conferences is having long lunches with fellow astronomers. But when there is a busy conference program and you must get back to listen to the post-lunch speaker because you are to give the conference summary, the long lunch has to become a short one! So, in search of the short lunch, I discovered that some of the best bread and pastries ("fast food" in the current context!) in the world can be found in the basements of the large department stores in Sendai. I was expecting great seafood in Japan, but the excellent bread and pastries were a surprise... Anyway, I think I did make it to nearly all the conference talks and I found time to browse through most of the posters. Below are my impressions of the conference. My comments start at the stellar centre and work outwards. The list of names given below refer to speakers whose articles appear in these proceedings.

2. The deep stellar interior - convection

Undoubtedly the main deficiency in modern stellar evolution calculations is the lack of an accurate treatment of convection. This means that we don't have a good measure of the amount of mixing that occurs in stellar interiors and hence, by the time stars have reached the AGB, we are very unsure of the surface composition. Hopefully, ambitious projects such as the Djehuty project at LLNL will soon begin to make progress in the area of stellar mixing. Meanwhile, approximations can be used to explore the consequences of mixing during stellar evolution. Fujimoto showed us the consequences of the mixing of hydrogen into helium-burning regions in some of his calculations. He was able to come up with an origin for the R stars (carbon stars without binary companions) and he was able to explain the high [C/Fe] ratios in low-metallicity stars.

Y. Nakada et al. (eds.), Mass-Losing Pulsating Stars and their Circumstellar Matter, 415–419.

These days, AGB evolution seems to be done more by "synthetic" evolution calculations than physically-based calculations. There is a clear need to get back to some real physics again and attempt more of these difficult calculations.

3. The stellar envelope - pulsation

The pulsational stability of stellar envelopes is determined by energy transport processes. In AGB stars, convection carries most of the energy through the envelope and thus stability depends on convection. As with the evolution of AGB stars, reliable models for the pulsational stability of AGB stars awaits a good treatment of convection. In the mean time, observations can provide us with constraints. In particular, the chaotic dynamics used to explain the behaviour of RV Tauri stars (Buchler, Ambika, Takeuti) has the exciting potential, when applied to the semi-regular AGB stars, to give us a direct indication of the stability of the various pulsation modes.

Pulsational periods are not as dependent on convection theory as stability. Three of the 4 recently found period-luminosity (PL) relations in AGB stars (Wood, Noda, Cioni) can be explained in terms of pulsation theory. Explaining the fourth sequence (D) is a challenge. Are thick circumbinary disks involved (Jura)? Or perhaps binary companions (Izumiura) or episodic dust ejections (Winters)? Or even strange modes (Wood)? More observations of these stars are required, especially in the mid-infrared.

There was a suggestion (Blommaert) that the usual Mira PL relation does not apply at high mass loss rates. Detailed pulsation and evolution models are required to understand this result. Another set of Mira variables which need an explanation are those of short period but red colour (Whitelock, Feast). Are these stars pulsating in an overtone? Good distances could answer this question. Perhaps the very accurate distances obtainable using the motion of maser spots (van Langevelde) could be used for this purpose.

The Luminous Blue Variables sit on the opposite side of the HR-diagram to the AGB stars but they also show pulsations and extreme mass loss. Dorfi showed that the Fe opacity bump could drive pulsations in these stars but there was no significant mass loss in the models, in striking contrast to observations. Clearly, more work is required here.

4. The stellar atmosphere - Mira sizes

One of the enduring problems with AGB stars has been the pulsation mode of the Mira variables. The period ratios in multimode semiregular variables and the large pulsation velocity amplitudes of the Mira variables seem to have definitively shown that the Miras are fundamental mode pulsators (Wood). But if the Miras are fundamental mode pulsators, directly-observed Mira radii have generally been more than 50% larger than expected. Now, at this conference, observations from many sources seem to have come together and explained these large apparent diameters as due to a "layer" of H_2O at $\sim$2 R_* (Perrin, Takami, Yamamura, Cami, Reid, Ryde). Why is this layer there? Why have the models used to derive observational diameters in the past (Scholz) not predicted these layers? The observers and theorists seem on the verge of solving this long-standing problem after more than 25 years!

5. Stellar winds - circumstellar matter

One of the prime requirements for understanding AGB star evolution and their enrichment of the interstellar medium is an accurate knowledge of the mass loss rate as a function of stellar parameters. In recent years, theoretical models for mass loss rates from carbon stars have been produced (Winters) and similar models for oxygen-rich stars are under development (Jeong). However, even for the C stars, a wide set of models covering the appropriate ranges of M, L, [Fe/H], C/O ratio, P and pulsation amplitude still needs to be made. Observationally, attempts are being made to empirically derive the dependence of mass loss rate on these parameters (Groenewegen, Ojha, Glass, Blommaert, Nakada, Elitzur, Bujarrabal). Much work still needs to be done in this area.

The circumstellar region provides an environment for the production of a number of masers. The pumping mechanism for the SiO maser is still not fully understood - since the maser is close to the stellar atmosphere it is influenced by the pulsation and dust-induced shocks in this region. Humphreys showed dynamical models that seemed to reproduce the observed velocity and general radial position (Yi, Miyoshi) of the maser spots. These sorts of models have the potential to provide valuable constraints on the dynamical motions and densities in the near-circumstellar environment critical for defining the mass loss rates in theoretical mass loss models.

One of the most fascinating features of AGB (and post-AGB) stars are the multiple mass-loss rings seen around them (de Laverny, Speck). A variety of explanations for these rings arose during the conference: the dust drift instability (Simis); changes in pulsation period and amplitude (Zijlstra); dust-induced shocks (Winters); and flares from the central AGB stars (Rudnitskij - Hα and H_2O flares; Deguchi - SiO maser flares). Can more than one of these explanations be correct?

6. Asymmetries

It is always convenient to make the simplest assumptions when attempting to solve a problem. In particular, spherical symmetry is generally assumed in theoretical modelling. But clumpiness is observed in dusty circumstellar shells, especially around C stars (Monnier, Niccolini), polarization is common in AGB stars (Bastien), magnetic fields have been measured in AGB stars and supergiants (Vlemmings, Shinnaga), and the ISM around PN has been found to be asymmetric, perhaps indicating an asymmetric AGB wind (Weinberger). We need to know if asymmetries are second order effects or if they are significant prime factors affecting AGB mass loss. The VLT Interferometer should provide very interesting new data here (Richichi).

7. Surveys

Massive surveys, recent and ongoing, have provided enormous advances in our knowledge of mass-losing, pulsating stars and the galactic environments they inhabit. Cioni described results from the EROS, MACHO and DENIS surveys and derived an LMC distance modulus and tilt, as well as the C/M star ratio across the LMC. Similar studies will undoubtedly be done in the SMC and, with future surveys, other local-group galaxies (Tanabe, Ita, Menzies, Yanagisawa). Similarly, various surveys of the Galactic centre region, bar and bulge are being done (Habing, Sjouwerman, Werman, Deguchi, Winnberg, Messineo, Blommaert, Imai, Ojha, Glass). These surveys include SiO and OH maser observations so that velocity and stellar population information, as well as distance, can be obtained. Results so far seem far from clear: Is the bulge rotating as a solid body or as an isothermal sphere (Winnberg)? Is there evidence for a recent star burst at the Galactic Centre?

Finally, we heard of the incredibly precise astrometry we can expect from future developments such as VERA in the radio (Kobayashi, Sasao), GAIA in the optical (Kucinskas) and JASMINE in the near-IR (Gouda). Results from these projects should completely define the dynamical state of the Galaxy. But it will be a while before results from these instruments are available - there is much to keep us occupied in the mean time.

Acknowledgments

I would like to thank the LOC for their smooth handling of the conference arrangements, the fine morning and afternoon teas, and the interesting excursion. I would particularly like to thank Prof. S. Deguchi for acting as host for the Japan Society for the Promotion of Science Fellowship which I received during May 2002.

List of Participants

Aikawa, Toshiki (Japan) Tohoku Gakuin University
Ambika, G (India) Maharajas College
Bastien, Pierre (Canada) University of Montreal
Blommaert, Joris (Belgium) Institute of Astronomy - Leuven
Buchler, J. Robert (USA) University of Florida
Bujarrabal, Valentin (Spain) National Astronomical Observatory
Cami, Jan (The Netherlands) University of Amsterdam
Cioni, Maria-Rosa (Germany) European Southern Observatory
de frutos Granda Marta (Spain)
de Laverny, Patrick (France) Observatoire de la Cote d'Azur
Deguchi, Shuji (Japan) NAOJ
Dorfi, Ernst (Austria) Wien University
Elitzur, Moshe (USA) University of Kentucky
Etoka, Sandra (UK) Jodrell Bank Observatory
Feast, Michael (South Africa) University of Cape Town
Frankowski, Adam (Poland) N. Copernicus Astronomical Center
Fujii, Takahiro (Japan) NAOJ
Fujimoto, Masayuki (Japan) Hokkaido University
Glass, Ian (South Africa) South African Astronomical Observatory
Goto, Miwa (USA) NAOJ
Gouda, Naoteru (Japan) NAOJ
Groenewegen, Martin (Belgium) Institute of Astronomy
Habing, Harm (The Netherlands) Leiden Observatory
Hachisuka, Kazuya (Japan) NAOJ
Hashimoto, Osamu (Japan) Gunma Astronomical Observatory
Honma, Mareki (Japan) NAOJ
Humphreys, Elizabeth (Sweden) Onsala Space Observatory
Imai, Hiroshi (The Netherlands) JIVE
Ishida, Toshihito (Japan) Nishi-Harima Astronomical Observatory
Ishitsuka, Jose (Japan) University of Tokyo
Isogai, Mizuki (Japan) Tohoku University
Ita, Yoshifusa (Japan) University of Tokyo
Izumiura, Hideyuki (Japan) NAOJ
Jeong, Kyung Sook (France) Observatoire de Paris

Jiang, Biwei (P. R. China)
National Astronomical Observatories, Chinese Academy of Sciences
Jimenez-Esteban, Francisco (Spain) European University
Jura, Martha (USA)
Jura, Michael (USA) University of California, Los Angels
Jurgenson, Colby (USA) University of Denver
Justtanont, Kay (Sweden) Stockholm Observatory
Kameya, Osamu (Japan) NAOJ
Kan-ya, Yukitoshi (Japan) NAOJ
Kiyota, Seiichiro (Japan) VSOLJ
Kobayashi, Hideyuki (Japan) NAOJ
Kucinskas, Arunas (Lithuania)
Institute of Theoretical Physics and Astronomy
Kurayama, Tomoharu (Japan) NAOJ
Leeuw, Lerothodi (South Africa)
South African Astronomical Observatory
Liseau, Rene (Sweden)
Manabe, Seiji (Japan) NAOJ
Matsumura, Masafumi (Japan) Kagawa University
Matsuura, Mikako (UK)
University of Manchester Institute of Science and Technology
Matsuyama, Gyusuke (Japan) University of Tokyo
Mennesson, Bertrand (USA)
Jet Propulsion Laboratory - California Institute of Technology
Menzies, John (South Africa) South African Astronomical Observatory
Messineo, Maria (The Netherlands) Leiden Observatory
Mito, Hiroyuki (Japan) University of Tokyo
Miyata, Takashi (Japan) University of Tokyo
Miyoshi, Makoto (Japan) NAOJ
Monnier, John (USA) Harvard-Smithsonian Center for Astrophysics
Motiduki, Nanako (Japan) NAOJ
Murakawa, Koji (USA) NAOJ
Muraki, Yasushi (Japan) Nagoya University
Nakada, Yoshikazu (Japan) University of Tokyo
Nakamura, Yasuhisa (Japan) Fukushima University
Niccolini, Gilles (France) Observatoire de la Cote d'Azur
Noda, Sachiyo (Japan) NAOJ
Ojha, Devendra (India) Tata Institute of Fundamental Research
Omodaka, Toshihiro (Japan) Kagoshima University
Onaka, Takashi (Japan) University of Tokyo
Perrin, Guy (France) Observatoire de Paris
Reid, Mark (USA) Harvard-Smithsonian Center for Astrophysics

Richichi, Andrea (Germany) European Southern Observatory
Rudnitskij, Georgij (Russia) Sternberg Astronomical Institute
Ryde, Nils (USA) The University of Texas at Austin
Saha, L.M. (India) University of Delhi
Saijo, Keiichi (Japan) National Science Museum
Saio, Hideyuki (Japan) Tohoku University
Saito, Yasumichi (Japan) Iwate University
Sasao, Testuo (Japan) NAOJ
Sato, Koichi (Japan) NAOJ
Scholz, Michael (Germany)
University of Heidelberg and University of Sydney
Scholz-Asmus, Jutta (Germany)
Seki, Munezo (Japan) Tohoku University
Shibata, Katsunori (Japan) NAOJ
Shinnaga, Hiroko (USA) Harvard-Smithsonian Center for Astrophysics
Simis, Yvonne (Germany) Astrophysical Institute Potsdam
Sivagnanam, Philippe (France) Observatoire de Paris
Sjouwerman, Lorant (USA) NRAO
Speck, Angela (USA) University of Illinois at Urbana-Champaign
Sudou, Hiroshi (Japan) Tohoku University
Sung, Hyun-il (Korea South) Korea Astronomy Observatory
Takami, Hideki (Japan) NAOJ
Takeuti, Mine (Japan) Tohoku University
Takeuti, Sadako (Japan)
Tamura, Shin'ichi (Japan) Tohoku University
Tanabe, Toshihiko (Japan) University of Tokyo
Tanaka, Yasuo (Japan) Ibaraki University
Ujihara, Hideki (Japan) NAOJ
van Langevelde, Huib (The Netherlands) JIVE
Vlemmings, Wouter (The Netherlands) Leiden Observatory
Weinberger, Ronald (Austria) University of Innsbruck
Whitelock, Patricia (South Africa)
South African Astronomical Observatory
Winnberg, Anders (Sweden) Onsala Space Observatory
Winters, Jan Martin (Germany)
Max-Planck-Institute for Radioastronomy
Wood, Margaret (Australia)
Wood, Peter (Australia) Mount Stromlo Observatory
Yadomaru, Yasushi (Japan) Misato Observatory
Yamamura, Issei (Japan)
The Institute of Space and Astronautical Science
Yanagisawa, Kenshi (Japan) NAOJ

Yi, Jiyune (Sweden) Onsala Space Observatory
Yoshioka, Kazuo (Japan) The University of the Air

Topic Index

Author Index

Previously published in Astrophysics and Space Science Library book series:

- **Volume 282**: **Radio Recombination Lines**
  Authors: M.A. Gordon, R.L. Sorochenko
  Hardbound, ISBN 1-4020-1016-8, November 2002
- **Volume 280**: **Organizations and Strategies in Astronomy III**
  Editor: André Heck
  Hardbound, ISBN 1-4020-0812-0, September 2002
- **Volume 279**: **Plasma Astrophysics , Second Edition**
  Author: Arnold O. Benz
  Hardbound, ISBN 1-4020-0695-0, July 2002
- **Volume 278**: **Exploring the Secrets of the Aurora**
  Author: Syun-Ichi Akasofu
  Hardbound, ISBN 1-4020-0685-3, August 2002
- **Volume 277**: **The Sun and Space Weather**
  Author: Arnold Hanslmeier
  Hardbound, ISBN 1-4020-0684-5, July 2002
- **Volume 276**: **Modern Theoretical and Observational Cosmology**
  Editors: Manolis Plionis, Spiros Cotsakis
  Hardbound, ISBN 1-4020-0808-2, September 2002
- **Volume 275**: **History of Oriental Astronomy**
  Editor: S.M. Razaullah Ansari
  Hardbound, ISBN 1-4020-0657-8, December 2002
- **Volume 274**:**New Quests in Stellar Astrophysics: The Link Between Stars and Cosmology**
  Editors: Miguel Chávez, Alessandro Bressan, Alberto Buzzoni,Divakara Mayya
  Hardbound, ISBN 1-4020-0644-6, June 2002
- **Volume 273**: **Lunar Gravimetry**
  Author: Rune Floberghagen
  Hardbound, ISBN 1-4020-0544-X, May 2002
- **Volume 272**:**Merging Processes in Galaxy Clusters**
  Editors: L. Feretti, I.M. Gioia, G. Giovannini
  Hardbound, ISBN 1-4020-0531-8, May 2002
- **Volume 271**: **Astronomy-inspired Atomic and Molecular Physics**
  Author: A.R.P. Rau
  Hardbound, ISBN 1-4020-0467-2, March 2002
- **Volume 270**: **Dayside and Polar Cap Aurora**
  Authors: Per Even Sandholt, Herbert C. Carlson, Alv Egeland
  Hardbound, ISBN 1-4020-0447-8, July 2002
- **Volume 269**:**Mechanics of Turbulence of Multicomponent Gases**
  Authors: Mikhail Ya. Marov, Aleksander V. Kolesnichenko
  Hardbound, ISBN 1-4020-0103-7, December 2001
- **Volume 268**:**Multielement System Design in Astronomy and Radio Science**
  Authors: Lazarus E. Kopilovich, Leonid G. Sodin
  Hardbound, ISBN 1-4020-0069-3, November 2001
- **Volume 267: The Nature of Unidentified Galactic High-Energy Gamma-Ray Sources**
  Editors: Alberto Carramiñana, Olaf Reimer, David J. Thompson
  Hardbound, ISBN 1-4020-0010-3, October 2001

- **Volume 266: Organizations and Strategies in Astronomy II**
  Editor: André Heck
  Hardbound, ISBN 0-7923-7172-0, October 2001
- **Volume 265: Post-AGB Objects as a Phase of Stellar Evolution**
  Editors: R. Szczerba, S.K. Górny
  Hardbound, ISBN 0-7923-7145-3, July 2001
- **Volume 264: The Influence of Binaries on Stellar Population Studies**
  Editor: Dany Vanbeveren
  Hardbound, ISBN 0-7923-7104-6, July 2001
- **Volume 262: Whistler Phenomena**
  Short Impulse Propagation
  Authors: Csaba Ferencz, Orsolya E. Ferencz, Dániel Hamar, János Lichtenberger
  Hardbound, ISBN 0-7923-6995-5, June 2001
- **Volume 261: Collisional Processes in the Solar System**
  Editors: Mikhail Ya. Marov, Hans Rickman
  Hardbound, ISBN 0-7923-6946-7, May 2001
- **Volume 260: Solar Cosmic Rays**
  Author: Leonty I. Miroshnichenko
  Hardbound, ISBN 0-7923-6928-9, May 2001
- **Volume 259: The Dynamic Sun**
  Editors: Arnold Hanslmeier, Mauro Messerotti, Astrid Veronig
  Hardbound, ISBN 0-7923-6915-7, May 2001
- **Volume 258: Electrohydrodynamics in Dusty and Dirty Plasmas**
  Gravito-Electrodynamics and EHD
  Author: Hiroshi Kikuchi
  Hardbound, ISBN 0-7923-6822-3, June 2001
- **Volume 257: Stellar Pulsation - Nonlinear Studies**
  Editors: Mine Takeuti, Dimitar D. Sasselov
  Hardbound, ISBN 0-7923-6818-5, March 2001
- **Volume 256: Organizations and Strategies in Astronomy**
  Editor: André Heck
  Hardbound, ISBN 0-7923-6671-9, November 2000
- **Volume 255: The Evolution of the Milky Way**
  Stars versus Clusters
  Editors: Francesca Matteucci, Franco Giovannelli
  Hardbound, ISBN 0-7923-6679-4, January 2001
- **Volume 254: Stellar Astrophysics**
  Editors: K.S. Cheng, Hoi Fung Chau, Kwing Lam Chan, Kam Ching Leung
  Hardbound, ISBN 0-7923-6659-X, November 2000

Missing volume numbers have not yet been published.
For further information about this book series we refer you to the following web site: http://www.wkap.nl/prod/s/ASSL

To contact the Publishing Editor for new book proposals:
Dr. Harry (J.J.) Blom: harry.blom@wkap.nl